T0211082

Using R for Modelling and Quantitative Methods in Fisheries

Chapman & Hall/CRC
The R Series

Series Editors
John M. Chambers, Department of Statistics, Stanford University, California, USA
Torsten Hothorn, Division of Biostatistics, University of Zurich, Switzerland
Duncan Temple Lang, Department of Statistics, University of California, Davis, USA
Hadley Wickham, RStudio, Boston, Massachusetts, USA

Recently Published Titles

Statistical Computing with R, Second Edition
Maria L. Rizzo

Geocomputation with R
Robin Lovelace, Jakub Nowosad, Jannes Muenchow

Advanced R, Second Edition
Hadley Wickham

Dose Response Analysis Using R
Christian Ritz, Signe Marie Jensen, Daniel Gerhard, Jens Carl Streibig

Distributions for Modelling Location, Scale, and Shape
Using GAMLSS in R
Robert A. Rigby, Mikis D. Stasinopoulos, Gillian Z. Heller, Fernanda De Bastiani

Hands-On Machine Learning with R
Bradley Boehmke and Brandon Greenwell

Statistical Inference via Data Science
A ModernDive into R and the Tidyverse
Chester Ismay and Albert Y. Kim

Reproducible Research with R and RStudio, Third Edition
Christopher Gandrud

Interactive Web-Based Data Visualization with R, plotly, and shiny
Carson Sievert

Learn R
Pedro J. Aphalo

Using R for Modelling and Quantitative Methods in Fisheries
Malcolm Haddon

For more information about this series, please visit: https://www.crcpress.com/Chapman--HallCRC-The-R-Series/book-series/CRCTHERSER

Using R for Modelling and Quantitative Methods in Fisheries

Malcolm Haddon

CSIRO, Oceans and Atmosphere, Hobart
IMAS, University of Tasmania

CRC Press
Taylor & Francis Group
Boca Raton London New York

CRC Press is an imprint of the
Taylor & Francis Group, an **informa** business

A CHAPMAN & HALL BOOK

CRC Press
Taylor & Francis Group
6000 Broken Sound Parkway NW, Suite 300
Boca Raton, FL 33487-2742

International Standard Book Number-13: 978-0-367-46989-4 (Hardback)
978-0-367-46988-7 (Paperback)

Library of Congress Cataloging-in-Publication Data

Names: Haddon, Malcolm, author.
Title: Using R for modelling and quantitative methods in fisheries / Malcolm Haddon.
Description: Boca Raton : CRC Press, [2021] | Summary: "The book has evolved and adapted from an earlier book by the same author and provides a detailed introduction to analytical methods commonly used by fishery scientists, ecologists, and advanced students using the open source software R as a programming tool"-- Provided by publisher.
Identifiers: LCCN 2020014277 (print) | LCCN 2020014278 (ebook) | ISBN 9780367469894 (hardback) | ISBN 9780367469887 (paperback) | ISBN 9781003032601 (ebook)
Subjects: LCSH: Fisheries--Mathematical models. | R (Computer program language)
Classification: LCC SH331.5.M48 H343 2021 (print) | LCC SH331.5.M48 (ebook) | DDC 333.95/611015118--dc23
LC record available at https://lccn.loc.gov/2020014277
LC ebook record available at https://lccn.loc.gov/2020014278

Visit the Taylor & Francis Web site at
http://www.taylorandfrancis.com

and the CRC Press Web site at
http://www.crcpress.com

Contents

Preface

This new book, *Using R for Modelling and Quantitative Methods in Fisheries*, has evolved and been adapted from my earlier book, *Modelling and Quantitative Methods in Fisheries* (Haddon, 2001, 2011). The new book aims to introduce an array of analytical methods useful when conducting fisheries stock assessments, but at the same time, many of those methodologies are more generally applicable in ecology. By concentrating on general analytical methods, while retaining some of the earlier focus on fisheries modelling, the idea was to make the new volume applicable, and hopefully useful, to a wider ecological audience, rather than just fisheries scientists. Like the original book, there is still a mixture of text and worked examples. However, time has passed and software tools that are easily and freely available for such analyses, such as R (R Core Team, 2019), have moved along in their development, as has the background knowledge of many students of the natural sciences. When that was considered, and having received numerous email requests for R versions of the earlier example boxes, it occurred to me that writing a new book focussed upon many of the analytical techniques, and using R for those analyses, might be a useful contribution.

The earlier book was supposed to be an introductory text; however, as stated in the *Preface* to the second edition (Haddon, 2011) "... keeping it an introductory text that is accessible to the majority is something of a challenge when covering some of the more important though intrinsically complex methods." The challenge at that time came from trying to implement such advanced methods using Microsoft Excel, which required such things as worksheet maps, macros, and related strategies (most would hopefully agree that implementing a Markov Chain Monte Carlo analysis in Excel is not really a sensible undertaking). It is far more sensible to attempt to implement such things in a programming language such as R. This remains an introductory book and the intent is to illustrate how such methods work and to provide a framework with which one can experiment and learn. The hope is that this will provide a foundation from where the reader can move to more production-based methods.

Some examples from the earlier book have been re-used, but I have taken the opportunity to expand and generate new ones for implementation in R. I have attempted to include more practical details and example code needed to conduct the work. To complement and streamline the code sections in this book I have also developed an R package, **MQMF**, available at both https://www.github.com/haddonm/MQMF, and published to CRAN, https://cran.r-project.org/. This is required for the examples in the book to work. The example R-code chunks are all provided in the help sections of six functions called chapter2, chapter3, ..., chapter7. Each example is in the sequence it is found in each chapter along with section headings to simplify finding the particular example of interest. Within RStudio, one just has to select the section of code wanted and press {ctrl}{enter}, as usual. Otherwise, select the code and paste it into the R console or into an editor to adjust each script according to your own wishes.

I have no pretensions about being a computer programmer but am rather a scientific programmer. This means that the code I write is likely not as efficient or as well structured as it could be. Hopefully, however, its intent is clear and relatively simple to understand. One very nice aspect of R code is that it is usually open source, so you must feel free to modify or improve the code found here in any way you wish.

As with all introductory books, the question arises of what to include and where to stop. The survey of methods included remains relatively broad and the depth with which each is treated varies. My primary hope is that the reader finds something in here that they consider to be useful to them in their work and their understanding of the analyses.

Malcolm Haddon
Hobart, January 2020

Acknowledgements

I would especially like to thank Oceans and Atmosphere, CSIRO, Hobart, for giving me the opportunity of being a CSIRO Honorary Fellow. Continued access to the library and the office has facilitated the writing of this book immensely. Besides the practicalities, it is also valuable to remain within such an active working environment with so many fine people with whom I have mixed and collaborated while working in fisheries science. In addition, I would like to thank the University of Tasmania for appointing me an Adjunct Professor with the Institute of Marine and Antarctic Studies. Even after I left the University of Tasmania, late in 2008, I continued to work with and collaborate with the many excellent staff there, particular Dr Craig Mundy, who I also thank for the permissions to use various data from the abalone research program in a number of data-sets included in the **MQMF** R package.

I would like also to thank the many people who have contacted me through email and who have asked interesting questions. They help me identify the areas that I have not made as clear as I could.

Finally, I would like to thank Rob Calver, my editor at Chapman & Hall/CRC, who arranged to let me publish an online GitBook version of this book for free, either on GitHub pages or at www.bookdown.org. It seems to me that one of the great advantages of Bookdown is that it can facilitate such a diversity of publishing options. The Open Source community, of which R, CRAN, and parts of RStudio constitute some of the best examples, is a remarkable achievement and one for which we should all celebrate and applaud the people involved.

Production Details

This book was written using Bookdown (Xie, 2017) within RStudio (RStudio, 2019), and I have used the Chapman & Hall/CRC LaTeX class (kranz.cls) as modified by Yihui Xie and with some very minor trial and error edits to the page fractions by myself. The code chunks are written using the `Consolas` font and will appear in slightly grey boxes. Package names will be bolded (**MQMF**) and function names, `function()`, will end in round brackets when they are merely being referred to in the text. The brackets are to distinguish them from other `code` text, which will generally refer to variables, parameters and function arguments. The code chunks relating to figures will generally refer to colours, but the book is printed in black and white, so different patterns or specific orders are also used to distinguish different lines. Elsewhere, one would generally use colour or pattern, but not both. I have kept the references to colour in the code blocks and in the figure legends so that the code remains general, but hopefully not confusing.

Console output from a code chunk will generally be pre-fixed with #, while whole line comments within code chunks will always be indented by a single space. No matter what the code chunks try to impose (*font=7* implies a serif font), the LaTeX class used to define how the pdfengine renders the figures is set up to use sans serif fonts. If the code chunks are used separately, they will therefore render the plots differently.

1

On Modelling

1.1 Characteristics of Mathematical Models

1.1.1 General

Formal fishery stock assessments are generally based upon mathematical models of population production processes and of the population dynamics of the stock being fished. The positive production processes include the recruitment dynamics (which add numbers) and individual growth (which adds biomass), while the negative production processes include natural mortality and fishing mortality (that includes selectivity), which both reduce numbers and hence biomass. The population dynamics includes details such as the time-steps used in the modelled dynamics, whether biomass or numbers are modelled (at age or at size, or both), details of spatial structuring, and other specifics to the case in hand. Such a plethora of potential details means there is a vast potential array of diverse mathematical models of biological populations and processes. Nevertheless, it is still possible to make some general statements regarding such models.

All models constitute an abstraction or simulation by the modeller of what is currently known about the process or phenomenon being modelled. Mathematical models are only a subset of the class of all models, and models may take many forms, ranging from a physical representation of whatever is being modelled (think of ball and stick models of DNA, as produced by Watson and Crick, 1953), diagrammatic models (such as a geographical map), and the more abstract mathematical representations being discussed here. We can impose some sort of conceptual order on this diversity by focussing on different properties of the models and on some of the constraints imposed by decisions made by the modellers.

1.1.2 Model Design or Selection

As abstractions, models are never perfect copies of what is known about the modelled subject, so there must be some degree of selection of what the modeller considers to be a system's essential properties or components. This notion of "essential properties or components" is making the assumption that all parts of a system are not all equally important. For example, in a model of the human blood circulation system a superficial vein somewhere in the skin would not be as important as the renal artery. If that assumption is accepted then a fundamental idea behind modelling is to select the properties to be included in order that the behaviour of the model may be expected to exhibit a close approximation to the observable behaviour of the modelled system. This selection of what are considered to be the important properties of a system permits, or even forces, the modeller to emphasize particular aspects of the system being modelled. A road map shows roads greatly magnified in true geographical scale because that is the point of the map. A topological map emphasizes different things, so the purpose for which a model is to be used is also important when determining what structure to use.

The selection of what aspects of a system to include in a model is what determines whether a model will be generally applicable to a class of systems, or is so specialized that it is attempting to simulate the detailed behaviour of a particular system (for system one might read a fished stock or population). However, by selecting particular parts of a natural system the model is also being constrained in what it can describe. The assumption is that despite not being complete it will provide an adequate description of the process of interest and that those aspects not included will not unexpectedly distort the representation of the whole (Haddon, 1980).

Of course, in order to make an abstraction one first needs to understand the whole, but unfortunately, in the real world, there is a great deal that remains unknown or misunderstood. Hence it is quite possible that a model becomes what is known as "misspecified". This is where the model's dynamics or behaviour, fails to capture the full dynamics of the system being studied. In some of the examples illustrated later in this book we will see the average predicted biomass trajectory for a stock fail to account for what appear to be oscillations in stock size that exhibit an approximate 10-year cycle (see, for example, the surplus production model fitted to the *dataspm* data-set in the *Bootstrap Confidence Intervals* section of the chapter on *Surplus Production Models*). In such a case there is an influence (or influences) acting with an unknown mechanism on the stock size in what appears to be a patterned or repeatable manner. Making the assumption that the pattern is meaningful, because the mechanism behind it is not included in the model structure then, obviously, the model cannot account for its influence. This is classical misspecification, although not all misspecifications are so clear or have such clear patterns.

Model design, or model selection, is complex because the decisions made when putting a model together will depend on what is already known and the use to which the model is to be put.

1.1.3 Constraints Due to the Model Type

A model can be physical, verbal, graphical, or mathematical, however, the particular form chosen for a model imposes limits on what it can describe. For example, a verbal description of a dynamic population process would be a challenge for anyone as invariably there is a limit to how well one can capture or express the dynamic properties of a population using words. Words appear to be better suited to the description of static objects. This limitation is not necessarily due to any lack of language skills on the part of the speaker. Rather, it is because spoken languages (at least those of which I am aware) do not seem well designed for describing dynamic processes, especially where more than one variable or aspect of a system is changing through time or relative to other variables. Happily, we can consider mathematics to be an alternative language that provides excellent ways of describing dynamic systems. But even with mathematics as the basis of our descriptions there are many decisions that need to be made.

1.1.4 Mathematical Models

There are many types of mathematical models. They can be characterized as descriptive, explanatory, realistic, idealistic, general, or particular; they can also be deterministic, stochastic, continuous, and discrete. Sometimes they can be combinations of some or all of these things. With all these possibilities, there is a great potential for confusion over exactly what role mathematical models can play in scientific investigations. To gain a better understanding of the potential limitations of particular models, we will attempt to explain the meaning of some of these terms.

Mathematical population models are termed dynamic because they can represent the present state of a population/fishery in terms of its past state or states, with the potential to describe future states. For example, the Schaefer model (Schaefer, 1957) of stock biomass dynamics (of which we will be hearing more) can be partly represented as:

$$B_{t+1} = B_t + rB_t \left(1 - \frac{B_t}{K}\right) - C_t \qquad (1.1)$$

where the variable C_t is the catch taken during time t, and B_t is the stock biomass at the start of time t (B_t is also an output of the model). The model parameters are r, representing the population growth rate of biomass (or numbers, depending on what interpretation is given to B_t, perhaps $= N_t$), and K, the maximum biomass (or numbers) that the system can attain (these parameters come from the logistic model from early mathematical ecology; see the *Simple Population Models* chapter). By examining this relatively simple model one can see that expected biomass levels at one time ($t+1$) are directly related to catches and the earlier biomass (time $= t$; the values are serially correlated).

The influence of the earlier biomass on population growth is controlled by the combination of the two parameters r and K. By accounting for the serial correlations between variables from time period to time period, such dynamic state models differ markedly from traditional statistical analyses. Serial correlation means that if we were to sample a population each year then, strictly, the samples would not be independent, which is a requirement of more classical statistical analyses. For example, in a closed population the number of two-year old fish in one year cannot be greater than the number of one-year fish the year before; they are not independent.

1.1.5 Parameters and Variables

At the most primitive level, mathematical models are made up of variables and parameters. A model's variables must represent something definable or measurable in nature (at least in principle). Parameters modify the impact or contribution of a variable to the model's outputs, or are concerned with the relationships between the variables within the model. Parameters are the things that dictate quantitatively how the variables interact. They differ from a model's variables because the parameters are the things estimated when a model is fitted to observed data. In **Equ**(1.1), B_t and C_t are the variables and r and K are the parameters. There can be overlap as, for example, one might estimate the very first value in the B_t series, perhaps B_{init} and hence the series would be made up of one parameter with the rest a direct function of the B_{init}, r and K parameters and the C_t variable.

In any model, such as **Equ**(1.1), we must either estimate or provide constant values for the parameters. With the variables, either one provides observed values for them (e.g., a time-series of catches, C_t) or they are an output from the model (with the exception of B_{init} as described above). Thus, in **Equ**(1.1), given a time-series of observed catches plus estimates of parameter values for B_{init}, r, and K, then a time-series of biomass values, B_t, is implied by the model as an output. As long as one is aware of the possibilities for confusion that can arise over the terms observe, estimate, variable, parameter, and model output, one can be more clear about exactly what one is doing while modelling a particular phenomenon. The relation between theory and model structure is not necessarily simple. Background knowledge and theory may be the drive behind the selection of a model's structure. The relationships proposed between a set of variables may constitute a novel hypothesis or theory about the organization of nature, or simply a summary of what is currently known, ready to be modified as more is learnt.

A different facet of model misspecification derives from the fact that the parameters controlling the dynamics of a population are often assumed to be constant through time, which should generally be acknowledged to be an approximation. If the population growth rate r or carrying capacity K varied randomly through time, and yet were assumed to be constant, this would be an example of what is known as process error. Such process error would add to the observable variation in samples from the population, even if they could be collected without error (without measurement error). If the parameters varied in response to some factor external to the population (an environmental factor or biological factor such as a predator or competitor) then such non-random responses have the potential to lead to an improved understanding of the natural world. So, an important aspect of the decisions made when constructing a model is to be explicit about the assumptions being made about the chosen structure.

1.2 Mathematical Model Properties

1.2.1 Deterministic vs Stochastic

We can define a model parameter as *a quantitative property (of the system being modelled) that is assumed either to remain constant over the period for which data are available, or to be modulated by environmental variation.* Roughly speaking, models in which the parameters remain constant on the timescale of the model's application are referred to as deterministic. With a given set of inputs, because of its constant parameters, a deterministic model will always give the same outputs for the same inputs. Because the relationships between the model variables are fixed (constant parameters), the output from a given input is "determined" by the structure of the model. One should not be confused by situations where parameters in a deterministic model are altered sequentially by taking one of an indexed set of predetermined values (e.g., a recruitment index or catchability index may alter and be changed on a yearly basis). In such a case, although the estimated parameters are expected to change through time, they are doing so in a repeatable, deterministic fashion (constant over a longer timescale), and the major property that a given input will always give the same output still holds.

Deterministic models contrast with stochastic models in which at least one of the parameters varies in a random or unpredictable fashion over the time period covered by the model. Thus, given a set of input values, the associated output values will be uncertain. The parameters that vary will take on a random value from a predetermined probability distribution (either from one of the classical probability density functions (PDFs) or from a custom distribution). Thus, for example, when simulating a fish stock, each year the recruitment level may attain a mean value plus or minus a random amount determined by the nature of a random variate, **Equ**(1.2):

$$R_y = \bar{R}e^{N(0,\sigma_R^2)-\sigma_R^2/2} \tag{1.2}$$

where R_y is the recruitment in year y, $\bar{R}$ is the average recruitment across years (which may itself be a function of stock size), $N\left(0, \sigma_R^2\right)$ is the notation used for a random variable whose values are described in this example by a Normal distribution with mean = zero (i.e., has both positive and negative values) and variance σ_R^2. By including the Normal distribution in an exponential term, this designates Log-Normal variation, and $-\sigma_R^2/2$ is a bias correction term for Log-Normal errors within recruitment time-series (Haltuch *et al*, 2008).

A simulation model differs from a model with estimated parameters. The objectives of these two types of model are also different as the former might be used to explore the implications of different management scenarios while the latter might be used to estimate the current state of depletion of a stock.

Given a set of input data (assumed to be complete and accurate; watch out for those assumptions), a deterministic model expresses all of its possible responses. However, stochastic models form the basis of so-called Monte Carlo simulations where the model is run repeatedly with the same input data, but for each run new random values are produced for the stochastic parameters, as with **Equ**(1.2). For each run a different output is produced, and these are tabulated or graphed to see what range of outcomes could be expected from such a system. Even if the variation intrinsic to a model is normally distributed, it does not imply that a particular output can be expected to be normally distributed about some mean value. If there are nonlinear aspects in the model, skew and other changes may arise.

Future population projections, risk assessments, and determining the impact of uncertainty in one's data all require the use of Monte Carlo modelling. Simulation testing of model structures is a very powerful tool. Details of running such projections are given in the chapters *On Uncertainty* and *Surplus Production Models*.

1.2.2   Continuous vs Discrete Models

Early fishery modellers used continuous differential equations to design their models, so the time steps in the models were all infinitesimal (Beverton and Holt, 1957). At that time computers were still very much in their infancy and analytical solutions were the culture of the day. Early fishery models were thus formed using differential calculus, and parts of their structures were determined more by what could be solved analytically than because they reflected nature in a particular accurate manner. At the same time, the application of these models reflected or assumed equilibrium conditions. Fortunately, we can now simulate a population using easily available computers and software, and we can use more realistic, or more detailed, formulations. While it may not be possible to solve such models analytically (i.e., if the model formulation has that structure it follows that its solution must be this), they can usually be solved numerically (informed and improving trial and error). Although both approaches are still used, one big change in fisheries science has been a move away from continuous differential equations toward difference equations, which attempt to model a system as it changes through discrete intervals (ranging from infinitesimal to yearly time-steps).

There are other aspects of model building that can limit what behaviours can be captured or described by a model. The actual structure or form of a model imposes limits. For example, if a mathematical modeller uses difference equations to describe a system, the resolution of events cannot be finer than the time intervals with which the model is structured. This obvious effect occurs in many places. For example, in models that include a seasonal component the resolution is quite clearly limited depending on whether the available data are for weeks, months, or some other interval. For example, in the *Static Models* chapter we fit a seasonal growth curve using data collected at mostly weekly intervals, obviously, if the data had been collected yearly then describing seasonal growth would be impossible.

1.2.3 Descriptive vs Explanatory

Whether a model is discrete or continuous, and deterministic or stochastic, is a matter of model structure and clearly influences what can be modelled. The purpose for which a model is to be used is also important. For a model to be descriptive it only needs to mimic the empirical behaviour of the observed data. A fine fit to individual growth data, for example, may usually be obtained by using polynomial equations:

$$y = a + bx + cx^2 + dx^3 ... + mx^n \tag{1.3}$$

in which no attempt is made to interpret the various parameters used (usually one would never use a polynomial greater than order six, with order two or three being much more common). Such descriptive models can be regarded as black boxes, which provide a deterministic output for a given input. It is not necessary to know the workings of such models; one could even use a simple look-up table that produced a particular output value from a given input value by literally looking up the output from a cross-tabulation of values. Such black box models would be descriptive and nothing else. Even though empirical descriptive models may make assumptions, if a particular case fails to meet those assumptions this does not mean the model need be rejected completely, merely that one must be constrained concerning which systems to apply it to. Such purely descriptive models need not have elements of realism about them except for the variables being described although it is common that their parameters can be give interpretations (such as the maximum size achievable). But again, what matters is how well such models describe the available data not whether the values attributed to their parameters make biological sense. In the *Model Parameter Estimation* chapter we will examine an array of three growth curves, including the famous von Bertalanffy curve. That section will enable a deeper discussion of the use of such descriptive models.

Explanatory models also provide a description of the empirical observations of interest, but in addition they attempt to provide some justification or explanation, a mechanism, for why the particular observations noted occurred instead of a different set. With explanatory models it is necessary to take into account the assumptions and parameters, as well as the variables that make up the model. By attempting to make the parameters and variables, and how the variables interact, reflect nature, explanatory models attempt to simulate real events in nature. A model is explanatory if it contains theoretical constructs (assumptions, variables, or parameters), which purport to relate to the processes of nature and not only to how nature behaves.

1.2.4   Testing Explanatory Models

Explanatory models are, at least partly, hypotheses or theories about the mechanisms and structure of nature and how it operates. They should thus be testable against observations from nature. But how do we test explanatory models? Can fitting a model to data provide a test of the model? If the expected values for the observed data, predicted by a model, account for a large proportion of the variability within the observed data, then our confidence that the model adequately describes the observations can be great. But the initial model fitting does not constitute a direct test of the structure of the model. A good fit to a model does not test whether the model explains observed data; it only tests how well the model describes and is consistent with the data (Haddon, 1980). The distinction between explanation and description is very important. A purely descriptive or empirical model could provide just as good a fit to the data, which hopefully makes it clear that we need further independent observations against which to really test the model's structure. What requires testing is not only whether a model can fit a set of observed data (i.e., not only the quality of fit), but also whether the model assumptions are valid and whether the interactions between model variables, as encoded in one's model, closely reflect nature.

Comparing the now fitted model with new observations does constitute a test of sorts. Ideally, given particular inputs, the model would provide a predicted observation along with confidence intervals around the expected result. An observation would be said to be inconsistent with the model if the model predicted that its value was highly unlikely given the inputs. But with this test, if there is a refutation, there is no indication of what aspect of the model was at fault. This is because it is not a test of the model's structure but merely a test of whether the particular parameter values are adequate (given the model structure) to predict future outcomes! We do not know whether the fitting procedure was limited because the data available did not express the full potential for variation inherent in the population under study. Was it the assumptions or the particular manner in which the modeller has made the variables interact that was at fault? Was the model too simple, meaning were important interactions or variables left out of the structure? We cannot tell without independent tests of the assumptions or of the importance of particular variables.

If novel observations are in accord with the model, then one has gained little. In practice, it is likely that the new data would then be included with the original and the parameters re-estimated. But the same could be said about a purely empirical model. What are needed are independent tests that the structure chosen does not leave out important sources of variation; to test this requires more than a simple comparison of expected outputs with real observations.

While we can be content with the quality of fit between our observed data and those predicted from a model, we can never be sure that the model we settle on is the best possible. It is certainly the case that some models can appear less acceptable because alternative models may fit the data more effectively.

However, any discussion over which curve or model best represents a set of data depends not only upon the quality of fit, but also upon other information concerning the form of the relationship between the variables. An empirical model with a parameter for every data point could fit a data-set exactly but would not provide any useful information. Clearly, in such cases, criteria other than just quality of numerical fit must be used to determine which model should be preferred. In the *Static Models* chapter, we consider methods for *Objective Model Selection*, which attempt to assess whether increasing the number of parameters in a model is statistically justifiable. Any explanatory model must be biologically plausible. It might be possible to ascribe meaning even to the parameters of a completely arbitrary model structure. However, such interpretations would be ad hoc and only superficially plausible. There would be no expectation that the model would do more than describe a particular set of data. An explanatory model should be applicable to a new data-set, although perhaps with a new set of particular parameters to suit the new circumstances.

Precision may not be possible even in a realistic model because of intrinsic uncertainty either in our estimates of the fitted variables (observation error) or in the system's responses, perhaps to environmental variation (process error in the model's parameters). In other words, it may not be possible to go beyond certain limits with the precision of our predicted system outcomes (the quality of fit may have intrinsic limits).

1.2.5 Realism vs Generality

Related to the problem of whether or not we should work with explanatory models is the problem of realism within models. Purely descriptive models need have nothing realistic about them. But it is an assumption that if one is developing an explanatory model, then at least parts of it have to be realistic. For example, in populations where ages or sizes can be distinguished, age- or size-structured models would be considered more realistic than a model that lumped all age or size categories into one. But a model can be a combination of real and empirical.

For a model to be general, it would have a very broad domain of applicability, that is, it could be applied validly in many circumstances. There have been many instances in the development of fisheries science where a number of models describing a particular process (e.g., individual growth) have been subsumed into a more general mathematical model of which they are special cases (see *Static Models* chapter). Usually this involves increasing the number of parameters involved, but nevertheless, these new models are clearly more mathematically general. It is difficult to draw conclusions over whether such more general equations/models are less realistic. That would be a matter of whether the extra parameters can be realistically interpreted or whether they are simply ad hoc solutions to combining disparate equations into one that is more mathematically general. With more complex phenomena, such as age-structured models, general models do not normally give as accurate predictions as more specialized models tuned to a particular situation. It is because of this that modellers often consider mathematically general models to be less realistic when dealing with particular circumstances (Maynard-Smith, 1974).

1.2.6   When Is a Model a Theory?

All models may be considered to have theoretical components, even supposedly empirical models. It becomes a matter of perception more than model structure. With simple models, for example, the underlying assumptions can begin to take on the weight of hypothetical assertions. Thus, if one were using the logistic equation to describe the growth of a population, it imports the assumption that density-dependent compensation of the population growth rate is linearly related to population density. In other words, the negative impact on population growth of increases in population size is linearly related to population size (see the *Simple Population Models* chapter). This can be regarded either as a domain assumption (that is, the model can only apply validly to situations where density-dependent effects are linearly related to population density) or as a theory (nonlinear density-dependent effects are unimportant in the system being modelled). It is clearly a matter of perception or modelling objective as to which of these two possibilities is the case. This is a good reason one should be explicit concerning the interpretation of one's model's assumptions.

If one were to restrict oneself purely to empirical relationships, the only way in which one's models could improve would be to increase the amount of variance in the observations accounted for by the model. There would be no valid expectation that an empirical model would provide insights into the future behaviour of a system. An advantage of explanatory/theoretical models is that it should be possible to test the assumptions, the relationships between variables, and the error structures, independently from the quality of fit to observed outcomes.

It should, therefore, be possible to present evidence in support of a model, which goes beyond the quality of fit. Those models where the proposed structure is not supported in this way may as well be empirical.

1.3 Concluding Remarks

Writing and talking about models, their use and construction is sometimes valuable as providing a reminder of the framework within which we work. A theoretical understanding of the strengths and weaknesses of mathematical models will always have value if you are to become a modeller. However, often the best way to understand models and their properties is to actually use them and explore their properties by manipulating their parameters and examining how they operate in practice. Hopefully you will find that using R as a programming language makes such explorations relatively simple to implement.

The material to follow includes very general methods and others that are more specific. An objective of the book is to encourage you, and perhaps provide you with a beginning, to develop your own analytical functions, perhaps by modifying some from this book. You might do that so that your own analyses become quicker and easier, and to some extent automated, leaving you more time to think about and interpret your findings. The less time you need to spend mechanically conducting analyses the more time there is for thinking about your scientific problems and exploring further than you might if you were using other less programmatic analyses. A major advantage of using R to implement your modelling is that any work you do should become much more easily repeatable and thus, presumably, more defensible. Of course, the range of subjects covered here only brushes the surface of what is available but tries to explore some of the fundamental methods, such as maximum likelihood estimation. Remember there are an enormous number of R packages available and these may assist you in implementing your own models be they statistical or dynamic.

2

A Non-Introduction to R

2.1 Introduction

This book is a branched and adapted version of some of the material described in *Modelling and Quantitative Methods in Fisheries* (Haddon, 2011). The principle changes relate to using R (R Core Team, 2019) to expand upon and implement the selected analyses from fisheries and ecology. This book is mostly a different text with more emphasis on how to conduct the analyses, and the examples used will be, in many places, different in detail to those used in the earlier book.

The use of R to conduct the analyses is a very different proposition to using something such as Microsoft Excel (as in Haddon, 2001, 2011). It means the mechanics of the analyses and the outcomes are not immediately in front of the analyst. Although, even when using Excel if the analyses involved the use of a macro in Visual Basic for Applications then that too involved a degree of abstraction. Using R as a programming language is simply taking that abstraction to completion across all the examples. Generally, I try to adhere to the principle of "if it is not broken do not fix it", so you can assume there are good reasons to make the move to R.

2.2 Programming in R

The open source software, R, is obviously useful for conducting statistical analyses, but it also provides an extremely flexible and powerful programming environment. For a scientific programmer R has many advantages as it is a very high-level language with the capacity to perform a great deal of work with remarkably few commands. These commands can be expanded by writing your own scripts or functions to conduct specialized analyses or other actions like data manipulations or customized plots. Being so very flexible has many advantages but it also means that it can be a challenge learning R's syntax. The very many available, mostly open source, R packages (see https://cran.r-project.org/) enhance its flexibility but also have the potential to increase the complexity of its syntax.

Fortunately, there are many fine books already published that can show people the skills they may need to use R and become an R programmer (e.g., Crawley, 2007; Matloff, 2011; Murrell, 2011; Venables and Ripley, 2002; Wickham, 2019; etc.). Wickham's (2019) book is titled *Advanced R*, but do not let that "Advanced" put you off, it is full of excellent detail, and, if first you want to get a taste, you can see a web version at https://adv-r.hadley.nz/. Other learning resources are listed in the appendix to this chapter.

This present book is, thus, not designed to teach anyone how to program using R. Indeed, the R code included here is rarely elegant nor is it likely to be foolproof, but it is designed to illustrate as simply as possible different ways of fitting biological models to data using R. It also attempts to describe some of the complexities, such as dealing with uncertainty, that you will come across along the way and how to deal with them. The focus of this book is on developing and fitting models to data, getting them running, and presenting their outputs. Nevertheless, there are a few constructs within R that will be used many times in what follows, and this chapter is aimed at introducing them so that the user is not overly slowed down if they are relatively new to R, though it needs to be emphasized that familiarity with R is assumed. We will briefly cover how to get started, how to examine the code inside functions, printing, plotting, the use of factors, and the use and writing of functions. Importantly we will also consider the use of R's non-linear solvers or optimizers. Many examples are considered, and example data-sets are included to speed these along. However, if the reader has their own data-sets then these should be used wherever possible, as few things encourage learning and understanding so well as analyzing one's own data.

2.2.1 Getting Started with R

Remember this book does not aim to teach anyone how to use R, rather it will try to show ways that R can be used for analyses within ecology, especially population dynamics and fisheries. To get started using R with this book a few things are needed:

- Download the latest version of R from https://cran.r-project.org/ and install it on your computer.

- Download RStudio, which can be obtained for free from the downloads tab on the RStudio web-site https://rstudio.com/. RStudio is now a mature and well-structured development environment for R and its use is recommended. Alternatively, use some other software of your own choosing for editing text files and storing the code files. Find the way of operating, your own work flow, that feels best for you. Ideally you should be able to concentrate on what you are trying to do rather than on how you are doing it.

- Download the book's package **MQMF** from CRAN or github. From CRAN this is very simple to do using the 'Packages' tab in RStudio. It is also available on github at https://github.com/haddonm/MQMF from where you can clone the R project files for the package or install it directly using the **devtools** package as in `devtools::install_github("https://github.com/haddonm/MQMF")`.

- The possession of at least some working knowledge of R and how to use it. If you are new to R then it would be sensible to use one or more of the many introductory texts available and a web search on 'introductory R' will also generate a substantial list. The documentation for RStudio will also get you started. See also this chapter's appendix.

To save on space only some of the details of the functions used in the analyses will be described in this book's text. Examining the details of the code from the **MQMF** package will be an important part of understanding each of the analyses and how they can be implemented. Read each function's help file (type `?functionname` in the console) and try the examples provided. In RStudio one can just select the example code in the help page and, as usual, use {ctrl}{enter} to run the selected lines. Alternatively, select and copy them into the console to run the examples, varying the inputs to explore how they work. The package functions may well contain extra details and error capture, which are improvements, but which would take space and time in the exposition within the book.

2.2.2 R Packages

The existence of numerous packages ('000s) for all manner of analyses (see https://cran.r-project.org/) is an incredible strength of R, but it would be fair to say that such diversity has also led to some complexity in learning the required R syntax for all the available functions. Base R, that which gets installed when R is installed, automatically includes a group of seven R packages (try `sessionInfo()` in the console to see a list). So using packages is natural in R. Normally, not using other people's packages would be a very silly idea as it means there would be a good deal of reinventing of perfectly usable wheels, but as the intent here is to illustrate the details of the analyses rather than focus on the R language, primarily the default or 'base' installation will be used, though in a few places suggestions for packages to consider and explore by yourself will be made. For example, we will be using the package **mvtnorm** when we need to use a multivariate Normal distribution. Such packages are easily installed within RStudio using the [Packages] tab. Alternatively one can type `install.packages("mvtnorm")` into the console and that should also do the job. Other packages will be introduced as they are required. Wherever I write "install library(xxxxxxx)", this should always be taken as saying to install the identified library and all its dependencies. Many packages require and call other packages so one needs to have their dependencies as well as the packages themselves. For example, the **MQMF** package requires the installation of the packages **mvtnorm** and **MASS**. If you type `packageDescription("MQMF")` into the console this will list the contents of the package's DESCRIPTION file (all packages have one). In there you will see that it imports both **MASS** and **mvtnorm**, hence they need to be installed as well as **MQMF**.

This book focusses upon using Base R packages, and this was a conscious decision. Currently there are important developments occurring within the R community, in particular there is the advent of the so-called 'tidyverse', which is a collection of libraries that aim to improve, or at least alter, the way in which R is used. You can explore this new approach to using R at https://www.tidyverse.org/, whose home page currently states "The tidyverse is an opinionated collection of R packages designed for data science. All packages share an underlying design philosophy, grammar, and data structures". Many of these innovations are indeed efficient and very worthwhile learning. However, an advantage of Base R is that it is very stable with changes only being introduced after extensive testing for compatibility with previous versions. In addition, developing a foundation in the use of Base R enables a more informed selection of which parts of the tidyverse to adopt should you wish to. Whatever the case, here we will be using as few extra libraries as possible beyond the 'base' installation and the book's own package.

2.2.3   Getting Started with MQMF

The **MQMF** package has been tested and is known to behave as intended in this book on both PC, Linux (Fedora), and Apple computers. On Apple computers it will be necessary to install the .tar.gz version of the package, on PCs, either the .zip or the .tag.gz can be installed, RStudio can do this for you. One advantage of the **tar.gz** version is this is known as a source file and the library's **tar.gz** archive contains all the original code for deeper examination should you wish. This code can also be examined in the **MQMF** github repository.

After installing the **MQMF** package it can be included in an R session by the command `library(MQMF)`. By using `?"MQMF"` or `?"MQMF-Package"` some documentation is immediately produced. At the bottom of that listing is the package name, the version number, and a hyperlink to an 'index'. If you click that 'index' link a listing of all the documentation relating to the exported functions and data-sets within **MQMF** is produced; this should work with any package that you examine. Each exported function and data-set included in the package is listed, and more details of each can be obtained by either pressing the required link or using the ? operator; try typing, for example, `?search` or `?mean` into the console. The resulting material in RStudio's 'Help' tab describes what the function does and usually provides at least one example of how the function may be used. In this book, R code that can be entered into the console in RStudio or saved as a script and sent to R is presented in slightly grey boxes in a `Consolas` font (a mono-spaced font in which all characters use the same space). Packages names will be bolded and function names will `function()` end in round brackets when they are merely being referred to in the text. The brackets are to distinguish them from other `code` text, which will generally refer to variables, parameters and function arguments.

Of course, another option for getting started is to do all the installations required and then start working through the next chapter, though it may be sensible to complete the rest of this chapter first as it introduces such things as plotting graphs and examining the workings of different functions.

2.2.4 Examining Code within Functions

A useful way of improving one's understanding of any programming language is to study how other people have implemented different problems. One excellent aspect of R is that one can study other people's code simply by typing a function name into the console omitting any brackets. If **MQMF** is loaded using `library(MQMF)`, then try typing, for example, `plot1`. If RStudio assists by adding in a set of brackets then delete them before pressing return. This will print the code of `plot1()` to the console. Try typing `lm` all by itself, which lists the contents of the linear model function from the **stats** package. There are many functions which when examined, for example, `mean`, only give reference to something like `UseMethod("mean")`. This indicates that the function is an S3 generic function which points to a different method for each class of object to which it can be applied (it is assumed the reader knows about different classes within R, such as vector, matrix, list, etc; if not search for 'classes in R', which should enlighten you). There are, naturally, many classes of objects within R, but in addition, one can define one's own classes. We will talk a little more about such things when we discuss writing functions of our own. In the meantime, when finding a `UseMethod("mean")` then the specific methods available for such functions can be listed by using `methods("mean")`; try, for example, `methods("print")`. Each such generic function has default behaviour for when it gets pointed at an object class for which it does not have a specific definition. If you type `getS3method("print","default")` then you get to see the code associated with the function and could compare it with the code for `getS3method("print","table")`. In some cases, for example, `print.table` will work directly, but the `getS3method` should always work. One can use S3 classes oneself if you want to develop customized printing, plotting, or further processing of specialized R objects (of a defined class) that you may develop (Venables and Ripley, 2002; Wickham, 2019).

Once a package has been loaded, for example `library(MQMF)`, then a list of exported objects within it can be obtained using the `ls` function, as in ls("package:MQMF") or ls.str("package:MQMF"); look up the two ls functions using ?ls.str. Alternatively, without loading a package, it is possible to use, for example, mvtnorm::dmvnorm, MQMF::plot1, or MQMF::'%ni%', where the :: allows one to look at functions exported into the 'Namespace' of packages, assuming that they have, at least, been installed. If the :: reports no such function exists then try three colons, :::, as functions that are only used inside a package are not usually exported, meaning it is not easily accessible without the ::: option (even if not exported its name may still be visible inside an exported function). Finally, without loading a library first it is possible to see the names of the exported functions using getNamespaceExports("MQMF"), whose output can be simplified by using a sort(getNamespaceExports("MQMF")) command.

2.2.5   Using Functions

R is a language that makes use of functions, which lends itself to interactive use as well as using scripts (Chambers, 2008, 2016; Wickham, 2019). This means it is best to develop a good understanding of functions, their structure, and how they are used. You will find that once you begin to write chunks of code it is natural to encapsulate them into functions so that repeatedly calling the code becomes simpler. An example function from **MQMF**, countones, can be used to illustrate their structure.

```
#make a function called countones2, don't overwrite original
countones2 <- function(x) return(length(which(x == 1)))  # or
countones3 <- function(x) return(length(x[x == 1]))
vect <- c(1,2,3,1,2,3,1,2,3)  # there are three ones
countones2(vect)  # should both give the answer: 3
countones3(vect)
set.seed(7100809) # if repeatability is desirable.
matdat <- matrix(trunc(runif(40)*10),nrow=5,ncol=8)
matdat #a five by eight matrix of random numbers between 0 - 9
apply(matdat,2,countones3)  # apply countones3 to 8 columns
apply(matdat,1,countones3)  # apply countones3 to 5 rows
```

```
# [1] 3
# [1] 3
#      [,1] [,2] [,3] [,4] [,5] [,6] [,7] [,8]
# [1,]    5    9    6    0    5    3    1    7
# [2,]    8    8    5    2    1    5    8    2
# [3,]    7    9    5    1    7    1    0    8
# [4,]    5    1    7    9    2    6    2    9
# [5,]    5    5    4    1    6    0    2    4
# [1] 0 1 0 2 1 1 1 0
# [1] 1 1 2 1 1
```

The countones2() function takes a vector, x, and counts the number of ones in it and returns that count; it can be used alone or, as intended, inside another function such as apply() to apply the countones() function to the columns or rows of a matrix or data.frame. The apply() function, and its relatives such as lapply() and sapply(), which are used with lists (see the help file for lapply()) are all very useful for applying a given function to the columns or rows of matrices and data.frames. Data.frames are specialized lists allowing for matrices of mixed classes. One can produce very short functions that can be used within the apply() family and **MQMF** includes countgtone, countgtzero, countones, countNAs, and countzeros as example utility functions; examples are given in each of their help files.

A function may have input arguments, which may also have default values and it should return an object (perhaps a single value, a vector, a matrix, a list, etc.) or perform an action. Ideally it should not alter the working or global environment (Chambers, 2016; Wickham, 2019) although functions that lead to the printing or plotting of objects are an exception to this:

```
#A more complex function prepares to plot a single base graphic
#It has the syntax for opening a window outside of Rstudio and
#defining a base graphic
plotprep2 <- function(plots=c(1,1),width=6, height=3.75,usefont=7,
                      newdev=TRUE) {
  if ((names(dev.cur()) %in% c("null device","RStudioGD")) &
      (newdev)) {
    dev.new(width=width,height=height,noRStudioGD = TRUE)
  }
  par(mfrow=plots,mai=c(0.45,0.45,0.1,0.05),oma=c(0,0,0,0))
  par(cex=0.75,mgp=c(1.35,0.35,0),font.axis=usefont,font=usefont,
      font.lab=usefont)
} #  see ?plotprep; see also parsyn() and parset()
```

Hopefully, this plotprep2() example reinforces the fact that if you see a function that someone has written, which you could customize to suit your own needs more closely, then there is no reason you should not do so. It would, of course, be good manners to make reference to the original, where that is known, as part of the documentation but a major advantage of writing your own functions is exactly this ability to customize your work environment (see the later section on *Writing Functions*).

2.2.6 Random Number Generation

R offers a large array of probability density functions (e.g., the Normal distribution, the Beta distribution, and very many more). Of great benefit to modellers is the fact that R also provides standard ways for generating random samples from each distribution. Of course, these samples are actually pseudo-random numbers because truly random processes are remarkably tricky things to simulate. Nevertheless, there are some very effective pseudo-random number generators available in R. Try typing RNGkind() into the console to see what the current one in use is on your own computer. If you type RNGkind, without brackets, in the code you will see the list of generators available within the *kinds* vector.

Pseudo-random number generators produce a long string of numbers, each dependent on the number before it in a long loop. Very early versions of such generators had notoriously short loops, but the default generator in R, known as the Mersenne-Twister, is reported to have a period of $2^{19937} - 1$, which is rather a large number. However, even with a large loop it has to start somewhere and that somewhere, is termed the seed. If you use the same seed each time you would obtain the same series of *pseudo random* values each time you tried to use them. Such repeatability can be very useful if a particular simulation that uses random numbers behaves in a particular manner it helps to know the seed used so it can be repeated. In the help file for ?RNGkind it notes that "Initially, there is no seed; a new one is created from the current time and the process ID when one is required. Hence different sessions will give different simulation results, by default". But the help also states: "Most of the supplied uniform generators return 32-bit integer values that are converted to doubles, so they take at most 2^{32} distinct values and extremely long runs will return duplicated values". $2^{32} \sim 4.295$ billion (thousand million) so be warned. It also warns that if a previously saved workspace is restored then the previous seed will come back with it. Such tucked-away details are worth remembering so one is fully aware of what an analysis is actually doing.

To make our own simulations repeatable we would want to save the seed used along with the results from a particular trial. One sets the seed using the function set.seed(), as in set.seed(12345). I deliberately used the 12345 to illustrate that it is not necessarily a good idea to just make up a seed in your head. We all may have good imaginations (or not), but when it comes to making up a series of random seeds it is best to use a different approach. Despite people's best intentions just making up a seed often leads to the repeat use of particular numbers, which risks obtaining biased outcomes to your simulations. One method of generating such seeds is to use the **MQMF** function getseed(). It uses the system time (Sys.time()) to obtain an integer whose digits are then reordered pseudo-randomly. Check out its code and see how it works. It seems reasonable to only use set.seed() when one really needs to be able to repeat a particular analysis that uses random numbers exactly.

```
#Examine the use of random seeds.
seed <- getseed()  # you will very likely get different naswers
set.seed(seed)
round(rnorm(5),5)
```

```
# [1]   0.27147 -1.01258 -0.74700 -0.30161 -1.13909
set.seed(123456)
round(rnorm(5),5)
```

```
# [1]   0.83373 -0.27605 -0.35500  0.08749  2.25226
set.seed(seed)
round(rnorm(5),5)
```

```
# [1]   0.27147 -1.01258 -0.74700 -0.30161 -1.13909
```

2.2.7 Printing in R

Most printing implies that output from an analysis is written to the console although the option of writing the contents of an object to a file is also present (see ?save and ?load, see also ?write.table and ?read.table). If you would like to keep a record of all output sent to the console then one can also do that using the sink() function, which diverts the output from R to a named text file. By using sink(file="filename", split=TRUE) the *split=TRUE* argument means it is possible to have the output on the screen as well as sent to a file. Output to a file can be turned off by repeating the sink() call, only with nothing in the brackets. If the path to the 'filename' is not explicitly included then the file is sent to the working directory (see ?getwd). The sink() function can be useful for recording a session but if a session crashes before a sink has been closed off it will remain operational. Rather than try to remember this problem it is possible to include early on in a script something like on.exit(suppressWarnings(sink())). The on.exit() function can be useful for cleaning up after running an analysis that might fail to complete.

2.2.8 Plotting in R

Using R simplifies the production of high-quality graphics and there are many useful resources both on the *www* and in books that describe the graphic possibilities in R (e.g., Murrell, 2011). Here we will focus only on using the base graphics that are part of the standard installation of R. It is possible to gain the impression that the base graphics are less sophisticated or capable than other major graphics packages implemented in R (for example, you should explore the options of using either the **lattice**, **ggplot2**, **graph**, **vcd**, **scatterplot3d**, and other graphics packages available; you may find one or more of these may suit your programming style better than the base graphics). However, with base or traditional graphics it is possible to cover most requirements for plotting that we will need. How one generates graphs in R provides a fine illustration of how there are usually many ways of doing the same thing in any given programming language. For particular tasks some ways are obviously more efficient than others, but whatever the case it is a good idea to find which approaches, methods, and packages you feel comfortable using and stick with those, because the intent here is to get the work completed quickly and correctly, not to learn the full range of options within the R language (which, with all available packages is now very extensive). Having said that, it is also the case that the more extensive your knowledge of the R language and syntax the more options you will have for solving particular problems or tasks; there are trade-offs everywhere!

Among other things, **MQMF** contains an array of functions that perform plotting functions. A nice thing about R is that diagrams and plots can be easily generated and refined in real time by adjusting the various graphics parameters until one has the desired form. Once finalized it is also possible to generate publication quality graphic files ready to be included in reports or sent off with a manuscript. As a start examine the MQMF::plotprep code to see how it defines the plot area, the margins, and other details of how to format the plot. plotprep() can be used as a shortcut for setting up the format for base graphic plots (external to the RStudio plot window) and if variations are wanted then parsyn() writes the core syntax for the par() function to the console, which can be copied into your own code and used as a starter to a base graphics plot. If you are content with the plotting window in RStudio then all you would need is parsyn(). Once familiar with the syntax of the par() function you could then just use the MQMF::parset() function as a shortcut or wrapper function.

```
#library(MQMF)   # The development of a simple graph  see Fig. 2.1
data("LatA")  #LatA = Length at age data; try properties(LatA)
#The statements below open the RStudio graphics window, but opening
#a separate graphics window using plotprep is sometimes clearer.
#plotprep(width=6.0,height=5.0,newdev=FALSE)
setpalette("R4") #a more balanced, default palette see its help
par(mfrow=c(2,2),mai=c(0.45,0.45,0.1,0.05))  # see ?parsyn
par(cex=0.75, mgp=c(1.35,0.35,0), font.axis=7,font=7,font.lab=7)
hist(LatA$age) #examine effect of different input parameters
hist(LatA$age,breaks=20,col=3,main="") # 3=green #try ?hist
hist(LatA$age,breaks=30,main="",col=4) # 4=blue
hist(LatA$age, breaks=30,col=2, main="", xlim=c(0,43), #2=red
     xlab="Age (years)",ylab="Count")
```

2.2.9 Dealing with Factors

One aspect of R that can cause frustration and difficulties is the use of factors to represent categorical variables. These convert the categorical variable into levels (the different values) of a factor, even factors that already have numerical values. It is common, for example, to use at least some categorical variables when standardizing fisheries cpue, thus, if one has depth categories from 50 – 600 m in steps of 50, these need to be treated in the standardization as fixed

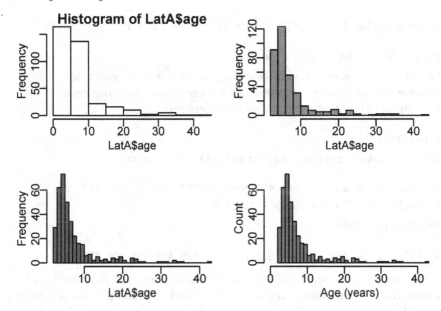

FIGURE 2.1 An example series of histograms using data from the **MQMF** data-set LatA, illustrating how one can iterate towards a final plot. See also the function uphist().

categorical factors rather than numeric values. But once converted to a set of factors they no longer behave as the original numerical variables, which is generally what is wanted when being plotted. However, if originally they were *numeric*, they can be recovered using the function facttonum(). In the example code that explores some issues relating to factors, note the use of the function try() (see also tryCatch()). This is used to prevent the R-code stopping due to the error of trying to multiple a factor by a scalar. This is useful when developing code and when writing books using bookdown and Rmarkdown.

```
 #Dealing with factors/categories can be tricky
DepCat <- as.factor(rep(seq(300,600,50),2)); DepCat
```

```
#  [1] 300 350 400 450 500 550 600 300 350 400 450 500 550 600
# Levels: 300 350 400 450 500 550 600
```

```
try(5 * DepCat[3], silent=FALSE) #only returns NA and a warning!
```

```
# [1] NA
```

```
as.numeric(DepCat) # returns the levels not the original values
```

```
#  [1] 1 2 3 4 5 6 7 1 2 3 4 5 6 7
```

```
as.numeric(levels(DepCat)) #converts 7 levels not the replicates
```

```
# [1] 300 350 400 450 500 550 600
```

```
DepCat <- as.numeric(levels(DepCat))[DepCat] # try ?facttonum
 #converts replicates in DepCat to numbers, not just the levels
5 * DepCat[3]   # now treat DepCat as numeric
```

```
# [1] 2000
```

```
DepCat <- as.factor(rep(seq(300,600,50),2)); DepCat
```

```
#  [1] 300 350 400 450 500 550 600 300 350 400 450 500 550 600
# Levels: 300 350 400 450 500 550 600
```

```
facttonum(DepCat)
```

```
#  [1] 300 350 400 450 500 550 600 300 350 400 450 500 550 600
```

It can often be simpler to take a copy of a data.frame and factorize variables within that copy so that if you want to do further manipulations or plotting using the original variables you do not need to convert them back and forth from numbers to factors.

When data is read into a data.frame within R the default behaviour is to treat strings or characters as factors. You can discover the default settings for very many R options by typing options() into the console. If you search for $stringsAsFactors (note the capitals), you will find it defaults to TRUE. Should you wish to turn this off it can be done easily using options(stringsAsFactors=FALSE). As is obvious from the code chunk on factors some care is needed when working with categorical variables.

2.2.10 Inputting Data

As with all programming languages (and statistical packages) a common requirement is to input one's data ready for analysis. R is very good at data manipulation and such things are well described in books like Crawley (2007). Of course, there are also many R packages that deal with data manipulation as well. A particularly useful one is **dplr**, which is part of the **tidyverse** group. but here we will merely be mentioning a few ways of getting data into R. A very common and low-level approach is simply to use text files as an input. I still often use comma-separated-value (.csv) files as being simple to produce and able to be opened and edited from very many different editors. One can use read.csv() or read.table() with very little trouble. Once data has been read into R it can then be saved into more efficiently stored formats by save(object,file="filename.RData"), after which it can be loaded back into R using load(file="filename.RData"). Check out ?save and ?load for details.

In the examples for this book, we will be using data-sets that are part of the **MQMF** package, but, by copying the formats used in those data-sets, you will be able to use your own data once you have it loaded into R.

2.3 Writing Functions

If you are going to use R to conduct any analyses you will typically type your R code using a text editor and at very least save those scripts for later use. There will be many situations where you will want to repeat a series of commands perhaps with somewhat different input values for the variables and or parameters used. Such situations, and many others, are best served by converting your code into a function, though it is not impossible to use source() as an intermediary step. You will see many examples of functions in the following text and in the **MQMF** package, and to make full use of the ideas expressed in the text you will find it helpful to learn how to write your own functions.

The structure of a function is always the same:

```
#Outline of a function's structure
functionname <- function(argument1, fun,...) {
  # body of the function
  #
  # the input arguments and body of a function can include other
  # functions, which may have their own arguments, which is what
  # the ... is for. One can include other inputs that are used but
  # not defined early on and may depend on what function is brought
  # into the main function. See for example negLL(), and others
  answer <- fun(argument1) + 2
  return(answer)
} # end of functionname
functionname(c(1,2,3,4,5),mean)  # = mean(1,2,3,4,5)= 3 + 2 = 5
```

```
# [1] 5
```

We will develop an example to illustrate putting together such functions.

2.3.1 Simple Functions

Very often one might have a simple formula such as a growth function or a function describing the maturity ogive of a fish species. The von Bertalanffy curve, for example, is used rather a lot in fisheries work.

$$\hat{L}_t = L_\infty\left(1 - e^{-K(t-t_0)}\right) \tag{2.1}$$

This equation can be rendered in R in multiple ways, the first as simple code within a code chunk which steps through the ages in a loop, or is vectorized, or more sensibly, uses the equation translated into a function that can be called multiple times rather than copying the explicit line of code (Figure 2.1).

```r
# Implement the von Bertalanffy curve in multiple ways
ages <- 1:20
nages <- length(ages)
Linf <- 50; K <- 0.2; t0 <- -0.75
 # first try a for loop to calculate length for each age
loopLt <- numeric(nages)
for (ag in ages) loopLt[ag] <- Linf * (1 - exp(-K * (ag - t0)))
 # the equations are automatically vectorized so more efficient
vecLt <- Linf * (1 - exp(-K * (ages - t0))) # or we can convert
 # the equation into a function and use it again and again
vB <- function(pars,inages) { # requires pars=c(Linf,K,t0)
  return(pars[1] * (1 - exp(-pars[2] * (inages - pars[3]))))
}
funLt <- vB(c(Linf,K,t0),ages)
ans <- cbind(ages,funLt,vecLt,loopLt)
```

TABLE 2.1: Three differnt ways of generating the same growth curve. *loopLt* uses a for loop, *vecLT* vectorizes the vB equation, and *funLT* is produced by the vB() function.

ages	funLt	vecLt	loopLt	ages	funLt	vecLt	loopLt
1	14.766	14.766	14.766	11	45.232	45.232	45.232
2	21.153	21.153	21.153	12	46.096	46.096	46.096
3	26.382	26.382	26.382	13	46.804	46.804	46.804
4	30.663	30.663	30.663	14	47.383	47.383	47.383
5	34.168	34.168	34.168	15	47.857	47.857	47.857
6	37.038	37.038	37.038	16	48.246	48.246	48.246
7	39.388	39.388	39.388	17	48.564	48.564	48.564
8	41.311	41.311	41.311	18	48.824	48.824	48.824
9	42.886	42.886	42.886	19	49.037	49.037	49.037
10	44.176	44.176	44.176	20	49.212	49.212	49.212

The use of vectorization makes for much more efficient code than when using a loop, and once one is used to the idea the code also becomes more readable. Of course one could just copy across the vectorized line of R that does the calculation, vecLt <- Linf * (1 - exp(-K * (ages - t0))), instead of using a function call. But one could also include error checking and other details into the function and using a function call should aid in avoiding the accidental introduction of errors. In addition, many functions contain a large number of lines of code so using the function call also makes the whole program more readable and easier to maintain.

```
#A vB function with some error checking
vB <- function(pars,inages) { # requires pars=c(Linf,K,t0)
  if (is.numeric(pars) & is.numeric(inages)) {
    Lt <- pars[1] * (1 - exp(-pars[2] * (inages - pars[3])))
  } else { stop(cat("Not all input values are numeric! \n")) }
  return(Lt)
}
param <- c(50, 0.2,"-0.75")
funLt <- vB(as.numeric(param),ages) #try without the as.numeric
halftable(cbind(ages,funLt))
```

```
#    ages    funLt  ages    funLt  ages    funLt
# 1     1 14.76560    8 41.31130    15 47.85739
# 2     2 21.15251    9 42.88630    16 48.24578
# 3     3 26.38167   10 44.17579    17 48.56377
# 4     4 30.66295   11 45.23154    18 48.82411
# 5     5 34.16816   12 46.09592    19 49.03726
# 6     6 37.03799   13 46.80361    20 49.21178
# 7     7 39.38760   14 47.38301    NA       NA
```

2.3.2 Function Input Values

Using functions has many advantages. Each function has its own analytical environment and a major advantage is that each function's environment is insulated from analyses that occur outside of the function. In addition, the global environment, in which much of the work gets done, is isolated from the workings within different functions. Thus, within a function you could declare a variable *popnum* and changes its values in many ways, but this would have no effect on a variable of the same name outside of the function. This implies that the scope of operations within a function is restricted to that function. There are ways available for allowing what occurs inside a function to directly affect variables outside but rather than teach people bad habits I will only give the hint to search out the symbol ->>. Forgetting the exceptional circumstances, the only way a function should interact with the environment within which it finds itself (one can have functions, within functions, within functions,...) is through its arguments or parameters and through what it returns when complete. In the vB() function we defined just above, the arguments are *pars* and *inages*, and it explicitly returns *Lt*. A function will return the outcome of the last active calculation but in this book we always call return() to ensure what is being returned is explicit and clear. Sometimes one can sacrifice increased brevity for increased clarity.

When using a function one can use its argument names explicitly, `vB(pars=param, inages=1:20)`, or implicitly `vB(param,1:20)`. If used implicitly then the order in which the arguments are entered matters. The function will happily use the first three values of 1:20 as the parameters and treat $param=c(50,0.02,-0.75)$ as the required ages if they are entered in the wrong order. For example, from $vB(ages, param)$ one dutifully obtains the following predicted lengths: 1.0000 -269.4264 -1807.0424. If the argument names are used then the order does not matter. It may feel like more effort to type out all those names but it does make programming easier in the end and acts like yet another source of self-documentation. Of course, one should develop a programming style that suits oneself, and, naturally, one learns from making mistakes.

Software is, by its nature, obedient. If we feed the `vB()` function its arguments back to front, as we saw, it will still do its best to return an answer or fail trying. With relatively simple software and functions it is not too hard to ensure that any such parameter and data inputs are presented to the functions in the right order and in recognizable forms. Once we start generating more inter-linked and complex software where the output of one process forms the input to others then the planning involved needs to extend to the formats in which arguments and data are used. Error checking then becomes far more important than when just using single functions.

2.3.3　R Objects

We are treating R as a programming language and when writing any software one should take its design seriously. Focussing solely on R there are two principles that assist with such designing (Chambers, 2016, p4).

1. Everything that exists in R is an object.
2. Everything that happens in R is a function call.

This implies that every argument or parameter and data-set that are input to any function is an R object, just as whatever is returned from a function (if anything) is an R object.

2.3.4　Scoping of Objects

You may have wondered why I used the name *inages* instead of *ages* in the `vB()` function arguments above. This is purely to avoid confusion over the scope of each variable. As mentioned above, in R all operations occur within environments, with the global environment being that which is entered when using the console or a simple script. It is important to realize that any function you write has its own environment which contains the scope of the internal variables. While it is true that even if an R object is not passed to a function as an argument that object can still be seen within a function. However, workings internal to a function are hidden from the global or calling environment. If you define a variable within a function but use the same name as one outside a function, then it will first search within its own environment when the name is used and use the internally defined version rather than the externally defined version. Nevertheless, it is better practice to mostly use names for objects within functions that differ from those used outside the function.

```
# demonstration that the globel environment is 'visible' inside a
# a function it calls, but the function's environment remains
# invisible to the global or calling environment
vBscope <- function(pars) { # requires pars=c(Linf,K,t0)
  rhside <- (1 - exp(-pars[2] * (ages - pars[3])))
  Lt <- pars[1] * rhside
  return(Lt)
}
ages <- 1:10; param <- c(50,0.2,-0.75)
vBscope(param)
```

```
#   [1] 14.76560 21.15251 26.38167 30.66295 34.16816 37.03799 39.38760
        41.31130
#   [9] 42.88630 44.17579
```

```
try(rhside)    # note the use of try() which can trap errors ?try
```

```
# Error in try(rhside) : object 'rhside' not found
```

2.3.5 Function Inputs and Outputs

Like most software, R functions invariably have inputs and outputs (though both may be *NULL*; a function's brackets may be empty, e.g., parsyn()). When inputting vectors of parameters or matrices of data it is sensible to adopt a standard format and by that I mean set a **standard** format. That way any function dealing with such data can make assumptions about its format. Ideally, it remains a good idea to make data checks before using any input data but that can be simplified by adhering to any standard format decided upon.

It is important to know and understand the structure of the data being fed into each function because a series of vectors needs to be referenced differently to a matrix, which behaves differently to a data.frame. It is usually best to reference a matrix explicitly by its column names as there are no guarantees that all such data-sets will be in the order found in *schaef* (Table 2.2). We can compare schaef[,2] with schaef[,"catch"], with schaef$catch.

```
#Bring the data-set schaef into the working of global environment
data(schaef)
```

TABLE 2.2: The fishery dependent data for yellowfin tuna fishery
used by Schaefer (1957) when describing surplus production models
as a means of stock assessment. From the *schaef* data-set.

year	catch	effort	cpue	year	catch	effort	cpue
1934	60913	5879	10.3611	1945	89194	9377	9.5120
1935	72294	6295	11.4844	1946	129701	13958	9.2922
1936	78353	6771	11.5719	1947	160134	20381	7.8570
1937	91522	8233	11.1165	1948	200340	23984	8.3531
1938	78288	6830	11.4624	1949	192458	23013	8.3630
1939	110417	10488	10.5279	1950	224810	31856	7.0571
1940	114590	10801	10.6092	1951	183685	18726	9.8091
1941	76841	9584	8.0176	1952	192234	31529	6.0971
1942	41965	5961	7.0399	1953	138918	36423	3.8140
1943	50058	5930	8.4415	1954	138623	24995	5.5460
1944	64094	6397	10.0194	1955	140581	17806	7.8951

```
#examine the properties of the data-set schaef
class(schaef)
a <- schaef[1:5,2]
b <- schaef[1:5,"catch"]
c <- schaef$catch[1:5]
cbind(a,b,c)
mschaef <- as.matrix(schaef)
mschaef[1:5,"catch"]   # ok
d <- try(mschaef$catch[1:5]) #invalid for matrices
d  # had we not used try()eveerything would have stopped.
```

```
# [1] "data.frame"
#           a      b      c
# [1,] 60913 60913 60913
# [2,] 72294 72294 72294
# [3,] 78353 78353 78353
# [4,] 91522 91522 91522
# [5,] 78288 78288 78288
#   1934  1935  1936  1937  1938
# 60913 72294 78353 91522 78288
# Error in mschaef$catch : $ operator is invalid for atomic vectors
# [1] "Error in mschaef$catch : $ operator is invalid for atomic vectors\n"
# attr(,"class")
# [1] "try-error"
```

```
# attr(,"condition")
# <simpleError in mschaef$catch: $ operator is invalid for atomic vectors>
```

By testing the *class* of the *schaef* object we could see it is a data.frame and
not a matrix. So handling such inputs may not be as simple as they might
appear at first sight. Yes, the order of the columns may change but the names
might be different too. So it is up to us to at least make some checks and work
to make our software at least partly foolproof. Often we are the only people
to use the software we write, but I have found this does not mean I do not
need to try to make it foolproof (it seems that sometimes I am foolish).

Notice with *schaef* that we are using all lowercase column headings, which is
not always common. As we have seen, column names are important because
it is often better to reference a data.frame as if it were a matrix. Typically,
when programming, there are usually more than one option we can use. We
could coerce an input data matrix into becoming a data.frame, but we can
also force the column names to be lowercase. I know the column names are
already lowercase, but here we are talking about a function receiving any given
data-set.

```
#Convert column names of a data.frame or matrix to lowercase
dolittle <- function(indat) {
   indat1 <- as.data.frame(indat)
   colnames(indat) <- tolower(colnames(indat))
   return(list(dfdata=indat1,indat=as.matrix(indat)))
} # return the original and the new version
colnames(schaef) <- toupper(colnames(schaef))
out <- dolittle(schaef)
str(out, width=63, strict.width="cut")
```

```
# List of 2
# $ dfdata:'data.frame':    22 obs. of  4 variables:
#  ..$ YEAR  : int [1:22] 1934 1935 1936 1937 1938 1939 1940 1..
#  ..$ CATCH : int [1:22] 60913 72294 78353 91522 78288 110417..
#  ..$ EFFORT: int [1:22] 5879 6295 6771 8233 6830 10488 10801..
#  ..$ CPUE  : num [1:22] 10.4 11.5 11.6 11.1 11.5 ...
# $ indat : num [1:22, 1:4] 1934 1935 1936 1937 1938 ...
#  ..- attr(*, "dimnames")=List of 2
#  .. ..$ : chr [1:22] "1934" "1935" "1936" "1937" ...
#  .. ..$ : chr [1:4] "year" "catch" "effort" "cpue"
```

So, in what follows we will try to only use lowercase column names for our
fishery dependent data; we will use standard names "year", "catch", "cpue",
etc., but we will also include a statement to convert *tolower* in our functions
where we expect to input data matrices. Such details become more important
when inputting more complex list objects containing data matrices of fishery
dependent data, age-composition data, vectors of biological constants related

to growth, maturity, recruitment, and selectivity, as well as other data besides. Deciding on a standard format and checking such inputs should always be included in the design of any somewhat more complex software.

In the case of *schaef*, this was data that was used with a surplus production model in an assessment. It is a reasonable idea to develop functions to pre-check a data-set for whether it contains all the required data, whether there are gaps in the data, and other details, before it is used in an assessment. One could even develop *S3* classes for specific analyses, which would allow one to add functions to generic functions like *print*, *plot*, and *summary*, as well as create new generic method functions of one's own.

Alternatively, it is simple to write functions to generate customizable plots for specific analyses to conduct particular tasks that are likely to be repeated many times. An example can be found with the function `plotspmdat()`, which will take typical fisheries data and plot up the time-series of catches and of cpue.

```
#Could have used an S3 plot method had we defined a class    Fig.2.2
plotspmdat(schaef) # examine the code as an e.g. of a custom plot
```

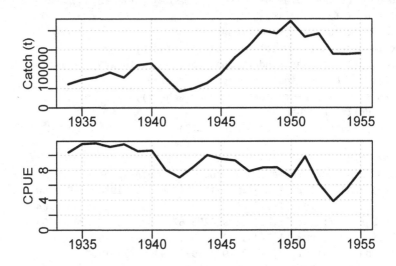

FIGURE 2.2 The catch and cpue data from the **MQMF** data-set *schaef* plotted automatically.

2.4   Appendix: Less-Traveled Functions

The following functions are listed merely to assist the reader in remembering some of the functions within R that tend to be easily forgotten because they are less often used, even though they are very useful when working on real problems.

`getNamespaceExports("MQMF")`

`getS3method("print","default")` often `print.default` works as well

`ls("package:MQMF")` lists the exported functions in the package

`ls.str("package:MQMF")` lists the functions and their syntax

`methods("print")` what S3 methods are available

`on.exit(suppressWarnings(sink()))` a tidy way to exit a run (even if it fails)

`packageDescription("MQMF")`

`RNGkind()` what random number generator is being used

`sessionInfo()`

`sink("filename",split=TRUE)` then after lots of screen output `sink()`,

`suppressWarnings(sink())` removes a sink quietly if a run crashes and leaves one open

2.5   Appendix: Additional Learning Resources

Other Books: these two have a **tidyverse** bent.

Wickham and Grolemund (2017) *R for Data Science* https://r4ds.had.co.nz/

Grolemund (2014) *Hands-On Programming with R* https://rstudio-education. github.io/hopr/

There are blogs devoted to R. I rather like https://www.r-bloggers.com/, but I get that delivered to my email. This blog lists many others.

There is even an open-source journal concerning R:

https://journal.r-project.org/

3

Simple Population Models

3.1 Introduction

Simple population models form the foundations of numerous fisheries models, indeed the separation of fisheries modelling from ecological population modelling is only an artificial distinction. Understanding how any population model operates should be useful for an improved understanding of fisheries models and so in this chapter we are going to use R to examine the dynamics of some of the fundamental ideas in population biology.

3.1.1 The Discrete Logistic Model

We will first examine the dynamics exhibited by what has become one of the classical population models in ecology, the discrete logistic model. This model crops up in very many places and in many different forms and is thus worth exploring in some detail. Here we will use the Schaefer (1954, 1957) version, which has four components or terms:

$$
\begin{aligned}
N_{t=0} &= N_0 \\
N_{t+1} &= N_t + rN_t\left(1 - \left(\frac{N_t}{K}\right)\right) - C_t
\end{aligned}
\tag{3.1}
$$

The term, N_t, is the numbers-at-time t (e.g., at the start of year t), and the last term is C_t, which is the catch in numbers taken in time t (e.g., during the year t). The first line reflects the initial conditions, defined as the numbers (or biomass) at time zero. Be wary of the sub-scripts as, with either numbers or biomass, they refer to particular times, usually the start of the identified time period t, whereas with catches they refer to the catches taken throughout the time period t. The equation is a summary of the dynamics across the time-step t; it is a difference equation not a continuous differential equation. The middle term of the second equation is a more complex component that defines what is termed the production curve in fisheries. This is made up of r, which is known as the intrinsic rate of population growth and K, which is the carrying capacity or maximum population size. If the dynamics are represented as being before exploitation began and at equilibrium then the N_0 would take the same value as K, but it is kept separate to allow for deviations from equilibrium at the beginning of any time-series. In the **MQMF** `discretelogistic()` function you will see another parameter, p, which will be examined in detail in the chapter on *Surplus Production Models*. It suffices here to state that with the default value of $p = 1.0$, the equation is equivalent to the Schaefer model in **Equ**(3.1). Here, the Schaefer model is written out using numbers-at-time although in fisheries it is commonly used directly with biomass-at-time (B_t). The fact that these terms can be altered in this way emphasizes that this model ignores biological facts and properties such as size- and age-structure within a population, as well as any differences between the sexes and other properties. The surprising thing is that such a simple model can still be as useful as it is in both theory and practice.

We can illustrate the production curve by plotting the central term from **Equ**(3.1) against a vector of N_t values. The sub-term rN_t represents unconstrained exponential population growth and this term is added to N_t in this discrete form of the equation. Thus, as long as $r > 0.0$ and catches are zero, we would expect un-ending positive population growth from the $N_t + rN_t$ part of the equation. However, the sub-term $(1 - (Nt/K))$ acts to constrain the exponential growth term and does so more and more as the population size increases. This acts as, and is known as, a density-dependent effect.

```
#Code to produce Figure 3.1. Note the two one-line functions
surprod <- function(Nt,r,K) return((r*Nt)*(1-(Nt/K)))
densdep <- function(Nt,K) return((1-(Nt/K)))
r <- 1.2; K <- 1000.0; Nt <- seq(10,1000,10)
par(mfrow=c(2,1),mai=c(0.4,0.4,0.05,0.05),oma=c(0.0,0,0.0,0.0))
par(cex=0.75, mgp=c(1.35,0.35,0), font.axis=7,font=7,font.lab=7)
plot1(Nt,surprod(Nt,r,K),xlab="Population Nt",defpar=FALSE,
      ylab="Production")
plot1(Nt,densdep(Nt,K),xlab="Population Nt",defpar=FALSE,
      ylab="Density-Dependence")
```

Density-dependence is an important part of population size regulation (May and McLean, 2007). As is apparent in **Figure** 3.1, the density-dependent term in the Schaefer model has a descending linear relationship with population size. This is literally what is meant by the phrase "linear density-dependence", which you will find in the population dynamics literature. This density-dependent term attempts to compensate for the exponential increases driven by the rN_t term and constrain the population size to no larger than the long-term average unfished population size K (see the next section for what happens when that compensation is insufficient).

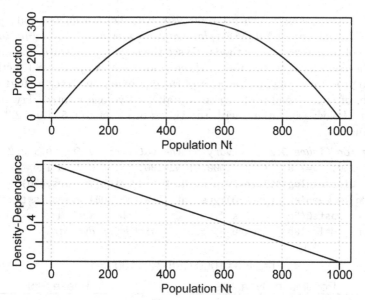

FIGURE 3.1 The full production curve for the discrete Schaefer (1954, 1957) surplus production model and the equivalent implied linear-density-dependent relationship with population size.

3.1.2 Dynamic Behaviour

The differential equation version of the Schaefer model was always known as being very stable with the maximum population size truly being constrained to the K value. It was with some surprise that the dynamics of the discrete version were found to be much more complex and interesting (May, 1973, 1976). We have implemented a version of the equation above in the function `discretelogistic()`, whose help file you can read via *?discretelogistic* and whose code can be seen by entering `discretelogistic` into the console (note no brackets). If you examine the code in this function you will see that it uses a *for* loop to step through the years to calculate the population size each year. Hopefully, as an R programmer, you will be wondering why I did not use a vectorized approach to avoid the use of the *for* loop. This highlights a fundamental aspect of population dynamics, the numbers at time t are invariably a function of the numbers at time $t - 1$ (initial conditions can differ from this, hence the N_0 parameter). The sequential nature of population dynamics means that we cannot use vectorization in any efficient manner. The sequential nature of population dynamics is such an obvious fact that the constraints and structure that it imposes on population models often go unnoticed.

The `discretelogistic()` function outputs (invisibly) a matrix of the time passed, *year* (as *year*), the numbers at time t, n_t (as *nt*), and the numbers at

time $t+1$, n_{t+1} (as *nt1*). This matrix is also given the class *dynpop* and a plot S3
method is included in **MQMF** (*plot.dynpop*; see *A non-introduction to R* for
the use of S3 classes and methods. To see the code type *MQMF:::plot.dynpop*
into the console, note the use of three colons not two). The S3 method is
used to produce a standardized plot of the outcome from discretelogistic(),
though, of course, one could write any alternative one wanted. Try running
the code below sequentially with the values of rv set to 0.5, 1.95, 2.2, 2.475,
2.56, and 2.8.

```
#Code for Figure 3.2. Try varying the value of rv from 0.5-2.8
yrs <- 100; rv=2.8;   Kv <- 1000.0; Nz=100; catch=0.0; p=1.0
ans <- discretelogistic(r=rv,K=Kv,N0=Nz,Ct=catch,Yrs=yrs,p=p)
avcatch <- mean(ans[(yrs-50):yrs,"nt"],na.rm=TRUE) #used in text
label <- paste0("r=",rv," K=",Kv," Ct=",catch, " N0=",Nz," p=",p)
plot(ans, main=label, cex=0.9, font=7) #Schaefer dynamics
```

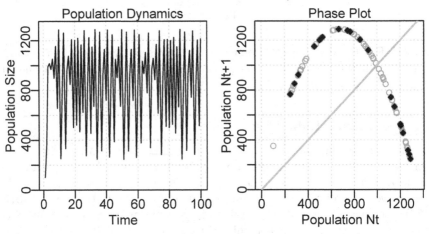

FIGURE 3.2 Schaefer model dynamics. The left plot is the numbers by year,
illustrating chaotic dynamics from an r-value of 2.8. The right plot is numbers-
at-time $t+1$ against time t, known as a phase plot. The final 20% of points
are in red to illustrate any equilibrium behaviour. The grey diagonal is the
1:1 line.

If you ran the code that gives rise to **Figure** 3.2, with the six values of rv
mentioned above, you should have seen a monotonically damped equilibrium
(one solid dot), then a damped oscillatory equilibrium (still one solid dot), then
something between a damped oscillatory equilibrium and a 2-cycle stable limit
cycle (somewhat smeared two solid dots). With r = 2.475 there is a 4-cycle
stable limit cycle (4 solid dots), and with r = 2.56 an 8-cycle stable limit cycle.
Finally, with r = 2.8 the dynamics end in chaos in which the outcome at each

step is not random but neither is it directly predictable, although all points fall on the parabola that can be seen in the right-hand plot. In Chaos theory this parabola is known as a strange attractor, although this is a very simple one. Such chaotic systems also depend on the starting conditions. If you change the initial numbers to 101 instead of 100 the time-line of numbers changes radically, although the parabola remains the same. Of course one can input any range of numbers desired, one could also alter the dynamics by using a slightly different model. While chaos is interesting and fun to play with (Gleick, 1988; Lauwerier, 1991) it seems to be rarely useful in real-world applications of modelling natural populations. Indeed, some have claimed that the apparent randomness in their observations was brought about by population responses to stochastic environmental effects rather than by the non-linearity introduced by under- and over-compensation in a density-dependent model (Higgins *et al*, 1997).

In practical terms, if you were to find a model was giving rise to apparently chaotic or cyclic behaviour, or even negative population numbers, it would be sensible to try to determine whether what was being observed was real biological population behaviour (obviously not the negative numbers) or, more simply, just a mathematical expression of unusual parameters. One could always force avoidance of a model dropping below 0 by including something like $max((B_t + (rB_t)(1 - (B_t/K)) - C_t), 0)$ in the calculation (see the discretelogistic() code. Analogously, values far above K could be avoided by $min((B_t + (rB_t)(1 - (B_t/K)) - C_t), 1.1K)$, although such additions to the code remain ad hoc (should, for example, any over-shoot above K be allowed, as in $1.1K$, or should it be $1.0K$?; some populations are naturally variable). In *Surplus Production Models* we will examine the use of penalty functions that can be used to prevent anomalous values arising when estimating such parameters while avoiding the use of blunt instruments such as max(...,0).

We can examine the actual population size each year by directing the invisible output from the discretelogistic() function into an object which can then be worked with, **Table 3.1**.

```
#run discrete logistic dynamics for 600 years
yrs=600
ans <- discretelogistic(r=2.55,K=1000.0,N0=100,Ct=0.0,Yrs=yrs)
```

When we set the r value to generate a 2-cycle stable limit (e.g., $r = 2.2$) this can be clearly seen in the repetition of the numbers in alternative years. Examining the actual numbers-at-time is clearly more accurate than trusting to a visual impression from a plot. This also means we can interrogate the outcome in as much detail as we wish. In the case above the mean of the last 50 years of catches was 859.675, which is obviously less than the K of 1000. Because of the phase relationship N_{t+1} vs N_t, the final line in the last year would not normally have an N_{t+1} value, so in the code the number of years input is incremented by one so that a final N_{t+1} is generated (see the code).

TABLE 3.1: The last 30 numbers-by-year from the *ans* object exhibiting an 8-cycle stable limit, if you try $r = 2.2$, within dis-cretelogistic(), it would lead to a 2-cycle stable limit bouncing between 746.24 and 1162.84. Try using plot(ans) with each change.

year	nt	nt1	year	nt	nt1	year	nt	nt1
571	493.9	1131.3	581	752.4	1227.4	591	515.6	1152.4
572	1131.3	752.4	582	1227.4	515.6	592	1152.4	704.5
573	752.4	1227.4	583	515.6	1152.4	593	704.5	1235.4
574	1227.4	515.6	584	1152.4	704.5	594	1235.4	493.9
575	515.6	1152.4	585	704.5	1235.4	595	493.9	1131.3
576	1152.4	704.5	586	1235.4	493.9	596	1131.3	752.4
577	704.5	1235.4	587	493.9	1131.3	597	752.4	1227.4
578	1235.4	493.9	588	1131.3	752.4	598	1227.4	515.6
579	493.9	1131.3	589	752.4	1227.4	599	515.6	1152.4
580	1131.3	752.4	590	1227.4	515.6	600	1152.4	704.5

3.1.3   Finding Boundaries between Behaviours

One can find when a stable limit cycle arises by averaging the rows of the *nt* and *nt1* columns within *ans* and examining the last 100 values (rounded to three decimal place). The names from the table() function identify the values of the cycle points, if there are any. If only one or two mean values occur in the table, these identify an asymptotic equilibrium or a 2-period stable limit cycle respectively. By examining the time-series of values in each *ans* object one could search for the first occurrence of the values identified and hence determine when the cyclic behaviour (to three decimal places) first arises. We round off the values and use 600 or more years because if we used all 15 decimal places any cycles beyond 8 might not be identified clearly (try changing the $r = 2.63$, and then change the round value to 5; try using plot(ans)).

```
#run discretelogistic and search for repeated values of Nt
yrs <- 600
ans <- discretelogistic(r=2.55,K=1000.0,N0=100,Ct=0.0,Yrs=yrs)
avt <- round(apply(ans[(yrs-100):(yrs-1),2:3],1,mean),2)
count <- table(avt)
count[count > 1] # with r=2.55 you should find an 8-cycle limit
```

```
# avt
# 812.64 833.99 864.65  871.5 928.45 941.88 969.92 989.93
#     12     13     12     13     12     13     12     13
```

We can set up a routine to search for the values of r that generate stable limit cycles of different periods, although the following code is only partially successful. By setting up a *for* loop that substitutes different values into the

r value input to discretelogistic(), we can search the final years of a long
time-series for unique values in the time-series of numbers. However, rounding
errors can lead to unexpected results especially at the boundaries between the
different types of dynamic behaviour. We can avoid some of those problems
by rounding off the numbers being examined to three decimal places but try
hashing that line out in the following code to see the issues get worse.

```
#searches for unique solutions given an r value   see Table 3.2
testseq <- seq(1.9,2.59,0.01)
nseq <- length(testseq)
result <- matrix(0,nrow=nseq,ncol=2,
                 dimnames=list(testseq,c("r","Unique")))
yrs <- 600
for (i in 1:nseq) {  # i = 31
   rval <- testseq[i]
   ans <- discretelogistic(r=rval,K=1000.0,N0=100,Ct=0.0,Yrs=yrs)
   ans <- ans[-yrs,] # remove last year, see str(ans) for why
   ans[,"nt1"] <- round(ans[,"nt1"],3) #try hashing this out
   result[i,] <- c(rval,length(unique(tail(ans[,"nt1"],100))))
}
```

In the example above there are a range of r values that give rise to the
maximum of 60 unique values in the last 100 observations. These represent
issues with rounding errors on computers. If the number of years run was
increased dramatically, then an equilibrium would eventually be expected to
arise. Whatever the case there is obviously a split or transition from single
equilibrium points, to 2-cycle, to 4-cycle, and finally to 8-cycle stable limits (Table 3.2). At each switch in behaviour there is some instability in the
outcome due to rounding errors.

TABLE 3.2: Schaefer model dynamics. The contents of result: the
number of unique population numbers (N) obtained over the last
60 years of 600 years of predicted dynamics from different values
of r. 100 implies non-equilibrium or even chaos.

r	Unique	r	Unique	r	Unique
1.90	1	2.14	2	2.38	2
1.91	1	2.15	2	2.39	2
1.92	1	2.16	2	2.40	2
1.93	1	2.17	2	2.41	2
1.94	1	2.18	2	2.42	2
1.95	1	2.19	2	2.43	2
1.96	1	2.20	2	2.44	4
1.97	1	2.21	2	2.45	100
1.98	7	2.22	2	2.46	4
1.99	100	2.23	2	2.47	4

r	Unique	r	Unique	r	Unique
2.00	100	2.24	2	2.48	4
2.01	4	2.25	2	2.49	4
2.02	2	2.26	2	2.50	4
2.03	2	2.27	2	2.51	4
2.04	2	2.28	2	2.52	4
2.05	2	2.29	2	2.53	4
2.06	2	2.30	2	2.54	4
2.07	2	2.31	2	2.55	8
2.08	2	2.32	2	2.56	8
2.09	2	2.33	2	2.57	100
2.10	2	2.34	2	2.58	100
2.11	2	2.35	2	2.59	100
2.12	2	2.36	2		
2.13	2	2.37	2		

3.1.4 Classical Bifurcation Diagram of Chaos

May (1976) produced a review of the dynamics of single species population models based on difference equations. In it he described the potential for chaotic dynamics from very simple models. In that article he plotted a diagram he termed a bifurcation diagram which shows how the equilibrium properties of the model change as a single parameter is changed. We can duplicate that diagram, **Figure** 3.3, to almost any degree of precision using the `discretelogistic()` function and the following code.

In amongst the chaos within the bifurcation plot there are occasional simplifications denoted by the white gaps in the inked space of the diagram. It is possible to examine some of those locations by replacing the numbers in the *testseq* vector of values with, for example, lower value = *2.82*; upper value = *2.87*; and inc = *0.0001*. These starting points provide more detail in amongst which there are previously obscured details. These too can be examined by changing the numbers to *testseq <- c(2.845,2.855,0.00001)*, whereupon bifurcations to 2-, 4- and higher cycles can be seen (Even more details can be seen by altering the *limy=0* argument to bound the y-axis in the region of interest inside *limy=c(600,750))* to bound the y-axis to the range desired. This diagram is plotted by using the final *taill <- 100* points at each sequential value of *r*. The number of years of simulation could be increased and the number of final points plotted increased to gain improved resolution in the chaotic regions. To see more detail expand the scale in the region in which details are wanted.

```
#the R code for the bifurcation function
bifurcation <- function(testseq,taill=100,yrs=1000,limy=0,incx=0.001){
  nseq <- length(testseq)
```

```
result <- matrix(0,nrow=nseq,ncol=2,
                  dimnames=list(testseq,c("r","Unique Values")))
result2 <- matrix(NA,nrow=nseq,ncol=taill)
for (i in 1:nseq) {
   rval <- testseq[i]
   ans <- discretelogistic(r=rval,K=1000.0,N0=100,Ct=0.0,Yrs=yrs)
   ans[,"nt1"] <- round(ans[,"nt1"],4)
   result[i,] <- c(rval,length(unique(tail(ans[,"nt1"],taill))))
   result2[i,] <- tail(ans[,"nt1"],taill)
}
if (limy[1] == 0) limy <- c(0,getmax(result2,mult=1.02))
parset() # plot taill values against taill of each r value
plot(rep(testseq[1],taill),result2[1,],type="p",pch=16,cex=0.1,
  ylim=limy,xlim=c(min(testseq)*(1-incx),max(testseq)*(1+incx)),
   xlab="r value",yaxs="i",xaxs="i",ylab="Equilibrium Numbers",
    panel.first=grid())
for (i in 2:nseq)
    points(rep(testseq[i],taill),result2[i,],pch=16,cex=0.1)
return(invisible(list(result=result,result2=result2)))
} # end of bifurcation
```

```
#Alternative r value arrangements for you to try; Fig 3.3
#testseq <- seq(2.847,2.855,0.00001) #hash/unhash as needed
#bifurcation(testseq,limy=c(600,740),incx=0.0001) # t
#testseq <- seq(2.6225,2.6375,0.00001) # then explore
#bifurcation(testseq,limy=c(660,730),incx=0.0001)
testseq <- seq(1.9,2.975,0.0005) # modify to explore
bifurcation(testseq,limy=0)
```

3.1.5 The Effect of Fishing on Dynamics

If we set up a default discretelogistic() run to have an 8-cycle stable limit when catches are zero, we can apply different levels of constant catch to see how this may influence the dynamics. Here we have applied catches of 0, 50, 200, and 325 individuals (although the N_t, K, and C_t could also be in tonnes). Of course the initial value N_0 must be greater than the catch but otherwise one is free to vary the values. In the four cases below we have assumed a p value of 1.0 (= Schaefer model); you could try $p = 1e - 08$ to explore the difference arising from using the equivalent to the Fox model.

```
#Effect of catches on stability properties of discretelogistic
yrs=50; Kval=1000.0
nocatch <- discretelogistic(r=2.56,K=Kval,N0=500,Ct=0,Yrs=yrs)
catch50 <- discretelogistic(r=2.56,K=Kval,N0=500,Ct=50,Yrs=yrs)
```

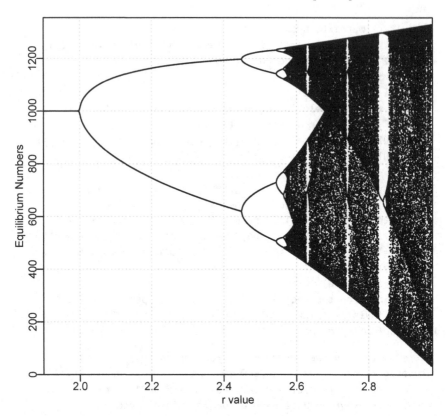

FIGURE 3.3 Schaefer model dynamics. A classical bifurcation diagram (May, 1976) plotting equilibrium dynamics against values of r, illustrating the transition from 2-, 4-, 8-cycle, and beyond, including chaotic behaviour.

```
catch200 <- discretelogistic(r=2.56,K=Kval,N0=500,Ct=200,Yrs=yrs)
catch300 <- discretelogistic(r=2.56,K=Kval,N0=500,Ct=300,Yrs=yrs)
```

We have put together a small function, `plottime()`, to simplify the repeated plotting of the final objects. We have also used the utility function `parsyn()` to aid with the syntax of the `par()` function for defining the bounds of a base graphics plot. Notice that producing such functions can simplify the subsequent code. I would recommend that individuals build up a collection of such useful functions that can be used with the function `source()` to introduce them to their own sessions (like a pre-package; do not forget to document your functions as you write them, RStudio even has an insert Roxygen skeleton command). Here, as catches increase one can see the dynamics begin to stabilize from 4- to 2-cycle, then asymptotic equilibrium. Notice that the mean stock size through time (the number at the bottom of each graph) declines as the constant catch increases.

```
#Effect of different catches on n-cyclic behaviour Fig. 3.4
plottime <- function(x,ylab) {
    yrs <- nrow(x)
    plot1(x[,"year"],x[,"nt"],ylab=ylab,defpar=FALSE)
    avB <- round(mean(x[(yrs-40):yrs,"nt"],na.rm=TRUE),3)
    mtext(avB,side=1,outer=F,line=-1.1,font=7,cex=1.0)
} # end of plottime
 #the oma argument is used to adjust the space around the graph
par(mfrow=c(2,2),mai=c(0.25,0.4,0.05,0.05),oma=c(1.0,0,0.25,0))
par(cex=0.75, mgp=c(1.35,0.35,0), font.axis=7,font=7,font.lab=7)
plottime(nocatch,"Catch = 0")
plottime(catch50,"Catch = 50")
plottime(catch200,"Catch = 200")
plottime(catch300,"Catch = 300")
mtext("years",side=1,outer=TRUE,line=-0.2,font=7,cex=1.0)
```

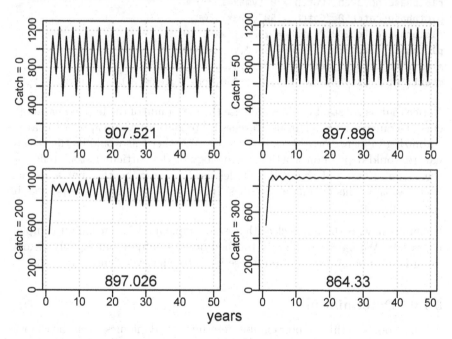

FIGURE 3.4 Schaefer model dynamics. The top-left plot depicts the expected unfished dynamics (8-cycle) while the next three plots, with decreasing average catches, illustrate the impact of the increased catches with no other factor changing.

We can do something very similar for the four phase plots. Here we have borrowed parts of the code from the MQMF:::plot.dynpop() function to define a small custom function to facilitate plotting four phase plots together.

```
#Phase plot for Schaefer model Fig. 3.5
plotphase <- function(x,label,ymax=0) { #x from discretelogistic
  yrs <- nrow(x)
  colnames(x) <- tolower(colnames(x))
  if (ymax[1] == 0) ymax <- getmax(x[,c(2:3)])
  plot(x[,"nt"],x[,"nt1"],type="p",pch=16,cex=1.0,ylim=c(0,ymax),
       yaxs="i",xlim=c(0,ymax),xaxs="i",ylab="nt1",xlab="",
       panel.first=grid(),col="darkgrey")
  begin <- trunc(yrs * 0.6) #last 40% of yrs = 20, when yrs=50
  points(x[begin:yrs,"nt"],x[begin:yrs,"nt1"],pch=18,col=1,cex=1.2)
  mtext(label,side=1,outer=F,line=-1.1,font=7,cex=1.2)
} # end of plotphase
par(mfrow=c(2,2),mai=c(0.25,0.25,0.05,0.05),oma=c(1.0,1.0,0,0))
par(cex=0.75, mgp=c(1.35,0.35,0), font.axis=7,font=7,font.lab=7)
plotphase(nocatch,"Catch = 0",ymax=1300)
plotphase(catch50,"Catch = 50",ymax=1300)
plotphase(catch200,"Catch = 200",ymax=1300)
plotphase(catch300,"Catch = 300",ymax=1300)
mtext("nt",side=1,outer=T,line=0.0,font=7,cex=1.0)
mtext("nt+1",side=2,outer=T,line=0.0,font=7,cex=1.0)
```

Notice that as constant catches increase the maximum of the implied parabolic curve (strange attractor) drops. Catches ($\equiv$ predation) would appear to have a stabilizing influence on a stock's dynamics. Whether this would be the case in nature would depend on whether the biology of the particular stock reflected the assumptions underlying the model used to describe the dynamics (for example, the Schaefer model assumes linear density-dependence operates in the population).

Hopefully now, it should be clear that even simple models can exhibit complex behaviour. We have only lightly touched on this subject but equally clearly, using R, it is possible to explore these ideas to whatever depth one wishes.

3.1.6 Determinism

All the models in this chapter exhibit deterministic dynamics; even the chaotic dynamics will follow the same trajectory if the starting conditions are repeated. In nature, of course, such deterministic behaviour would be rare. The models are always an abstraction of what may really be going on. The natural processes relating to productivity (natural mortality, growth, and reproduction) will all be variable through time in response to food availability, ecological interactions, and environmental influences such as temperature). Such

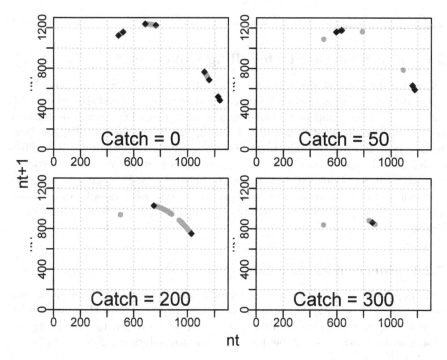

FIGURE 3.5 Schaefer model dynamics. Phase plots for each of the four constant catch scenarios.

variability within processes can be approximately included into models by including what is known as process error, which might be represented in equation form by:

$$N_{t+1} = N_t + (r + \varepsilon_t)N_t \left(1 - \left(\frac{N_t}{(K + \epsilon_t)}\right)\right) \tag{3.2}$$

where ε and ϵ represent Normal random errors specific to each model parameter, which vary through time t. Sometimes, in simulation, just one parameter will have variation added. As a population phenomenon, this would lead population numbers to follow trajectories that were not necessarily smooth, and, in a simulation, a single replicate would be insufficient to capture the potential range of the dynamics possible. The inclusion of such process errors in simulations would depend upon the purpose the modelling was being put to. Here, the intent is to explore the expectations of deterministic population dynamics, but one should always keep in mind that such models are only approximations and are intended to represent average behaviour.

3.2   Age-Structured Modelling Concepts

Of course the greatest simplification within the surplus production models we have just examined is that they ignore all aspects of age- and size-related differences in biology and behaviour. Such simplified models remain useful (see the chapter on *Surplus Production Models*), but in the face of acknowledged fishing gear selectivity (akin to predators having preferred prey) and differential growth and productivity at age, fisheries and other natural resource scientists often include age and or size in their models. Here we will only give such ideas an initial treatment as there are many other details about modelling dynamic processes required before embarking on such complexities in detail.

3.2.1   Survivorship in a Cohort

It is hopefully obvious that the survivorship of individuals, S, between two time periods is simply the ratio of the numbers at $t+1$ to the numbers at time t (keep in mind that numbers-at-age can be converted to mass-at-age or biomass-at-age when one know the weight-at-age relationship):

$$S_t = \frac{N_{t+1}}{N_t} \text{ or } N_{t+1} = S_t N_t \tag{3.3}$$

Such proportional reductions through time are suggestive of negative exponential growth. Given an instantaneous total mortality rate of Z (natural and fishing mortality combined), then the survivorship each year (that is the proportion surviving) is equal to e^{-Z}. This can be illustrated by setting up a simulated population with constant initial conditions and applying a vector of different instantaneous total mortality rates.

This would be equivalent to considering a single cohort recruiting as age 0+ individuals. Assuming no immigration, such a single-cohort population could only decline in numbers, which although seemingly trivially obvious is an important intuition when considering age-structured populations. The total mortality (usually denoted Z) is merely the sum of natural mortality (usually denoted M) and fishing mortality (usually denoted F). In the example below, notice that applying a $Z = 0.05$ leads to the population being greater than zero even after 50 years. Each of the cohort trajectories illustrated depicts an exponential or constant proportional decline in population size.

```
#Exponential population declines under different Z. Fig. 3.6
yrs <- 50;  yrs1 <- yrs + 1 # to leave room for B[0]
years <- seq(0,yrs,1)
B0 <- 1000        # now alternative total mortality rates
```

```
Z <- c(0.05,0.1,0.2,0.4,0.55)
nZ <- length(Z)
Bt <- matrix(0,nrow=yrs1,ncol=nZ,dimnames=list(years,Z))
Bt[1,] <- B0
for (j in 1:nZ) for (i in 2:yrs1) Bt[i,j] <- Bt[(i-1),j]*exp(-Z[j])
plot1(years,Bt[,1],xlab="Years",ylab="Population Size",lwd=2)
if (nZ > 1) for (j in 2:nZ) lines(years,Bt[,j],lwd=2,col=j,lty=j)
legend("topright",legend=paste0("Z = ",Z),col=1:nZ,lwd=3,
        bty="n",cex=1,lty=1:5)
```

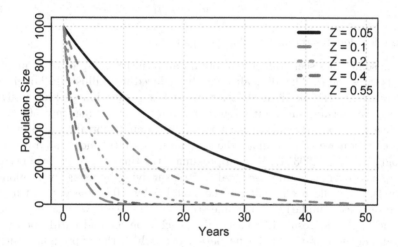

FIGURE 3.6 Exponential population declines under different levels of total mortality. The top curve is Z = 0.05 and the steepest curve is Z = 0.55.

3.2.2 Instantaneous vs Annual Mortality Rates

In the previous section we worked with what are known as instantaneous rates. They could be used to generate survivorship rates in terms of the proportion expected to survive after a year of a given instantaneous mortality rate. The idea behind an instantaneous mortality rate is that it is the mortality applied over infinitesimally small time periods.

We already know the annual proportion surviving:

$$S = e^{-Z} \tag{3.4}$$

Thus, the proportion that die (A) over the year is described by:

$$A = 1 - e^{-Z} \tag{3.5}$$

If we only consider the instantaneous fishing mortality rate, F and ignore the natural mortality rate, then the proportion dying is known as the harvest rate, H.

$$H = 1 - e^{-F} \qquad (3.6)$$

If we know the annual harvest rate we can back-transform that to estimate the instantaneous fishing mortality rate:

$$F = -\log(1 - H) \qquad (3.7)$$

where log denotes the natural log or logs to base e.

It can sometimes seem difficult to obtain an intuitive grasp of what instantaneous rates imply. This difficulty may be reduced if we illustrate how instantaneous rates can be derived. Starting with a population of 1000 individuals we can assume an annual removal rate of 0.5 meaning only 500 would survive the year. By substituting the total annual mortality rate into the last equation we can estimate the instantaneous total mortality rate to be $-\log(1-0.5) = 0.693147$. If we approximate the time steps this level needs to be applied over to obtain the total of 50% surviving by using two 6-monthly periods, we would need to divide the 0.693147 by 2.0 and apply that mortality rate twice. Similarly, if the time steps were monthly, we would divide by 12 and apply the result 12 times. While each month could hardly be termed instantaneous this would be the period over which the approximate instantaneous rate (0.693147/12) of mortality would be applied. Such a re-scaling could occur with smaller and smaller time intervals until the intervals were small enough that the result was not significantly different from the expected 500 individuals remaining at the end of a year, **Table 3.3**.

As becomes apparent, the shorter the time interval over which the approximate instantaneous rate is applied the closer the final value comes to the expected 0.5 of 1000 (i.e., 500). Hopefully, playing around with this code will assist you in gaining intuitions about the notion of instantaneous rates of mortality.

Annual rates can be plotted against instantaneous rates of mortality to illustrate the difference between the two. Note that up to about an instantaneous rate of 0.18 the annual harvest rate has approximately the same value. It should be clear that it is not possible to get a harvest rate greater than 1.0 (equivalent to capturing >100% of the exploitable biomass). High harvest rates begin to appear to be implausible and if they occur when fitting a fisheries model they should really need to be defended before being accepted. Instantaneous rates, on the other hand, can obviously be over 1.0, but ideally such an event should no longer be misunderstood.

```
#Prepare matrix of harvest rate vs time to appoximate F
Z <- -log(0.5)
timediv <- c(2,4,12,52,365,730,2920,8760,525600)
yrfrac <- 1/timediv
names(yrfrac) <- c("6mth","3mth","1mth","1wk","1d","12h",
                   "3h","1h","1m")
nfrac <- length(yrfrac)
columns <- c("yrfrac","divisor","yrfracH","Remain")
result <- matrix(0,nrow=nfrac,ncol=length(columns),
                dimnames=list(names(yrfrac),columns))
for (i in 1:nfrac) {
  timestepmort <- Z/timediv[i]
  N <- 1000
  for (j in 1:timediv[i]) N <- N * (1-timestepmort)
  result[i,] <- c(yrfrac[i],timediv[i],timestepmort,N)
}
```

TABLE 3.3: Outcome of applying a constant total mortality rate apportioned across shorter and shorter time periods each of which sum to a year. yrfrac is the fraction of a year, Approx is the instantaneous total mortality divided by the divisor, and Remain is the number of animals remaining out of 1000 after a year.

	yrfrac	divisor	yrfracH	Remain
6mth	0.5000000000	2	0.34657359	426.9661
3mth	0.2500000000	4	0.17328680	467.1104
1mth	0.0833333333	12	0.05776227	489.6953
1wk	0.0192307692	52	0.01332975	497.6748
1d	0.0027397260	365	0.00189903	499.6706
12h	0.0013698630	730	0.00094952	499.8354
3h	0.0003424658	2920	0.00023738	499.9589
1h	0.0001141553	8760	0.00007913	499.9863
1m	0.0000019026	525600	0.00000132	499.9998

```
#Annual harvest rate against instantaneous F, Fig. 3.7
Fi <- seq(0.001,2,0.001)
H <- 1 - exp(-Fi)
parset()  # a wrapper for simplifying defining the par values
plot(Fi,H,type="l",lwd=2,panel.first=grid(),
     xlab="Instantaneous Fishing Mortality F",
     ylab="Annual Proportion Mortality H")
lines(c(0,1),c(0,1),lwd=2,lty=2,col=2)
```

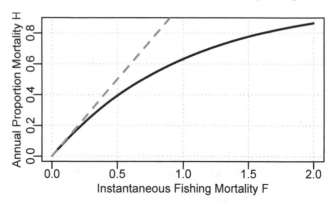

FIGURE 3.7 The relationship between the instantaneous fishing mortality rate (solid line) and the annual fishing mortality or annual harvest rate (dashed line). The values diverge at values of F greater than about 0.2.

3.3 Simple Yield per Recruit

We know that once a cohort recruits to a stock its numbers can only decrease (assuming no immigration). Take care over the concept of "recruitment", it can mean the post-larval arrival into the stock, however, it can also refer to when individuals become available or vulnerable to a fishery. Here we are concerned with post-larval arrival. We will deal with recruitment to a fishery when we consider the notion of selectivity and availability below and in *Static Models*. Interestingly, while numbers are expected to decrease through time, given that individuals can grow in size and weight, we do not yet know if the mass of the members of a cohort always decreases through time or if it increases first and then decreases as eventually all the individuals die.

The question of whether the mass of a cohort can increase for some years, even as its numbers decline was first explored because of a simplistic but common intuition about the relationship between catch and effort. It may seem obvious that the catch of a fish species landed will always increase with an increase in fishing effort, but that is short-term thinking. This seemingly obvious notion was refuted before 1930 and was illustrated nicely by some published work by Russell (1942), who demonstrated that the maximum on-going catch from a fishery need not necessarily arise from the maximum effort (or the maximum fishing mortality). This was not just an academic question but directly related to the management of fisheries. Once it was realized that it was possible to over-fish and even collapse a fish stock, which was only formally recognized at the start of the 20th century (Garstang, 1900), then how best to manage fisheries became an important question. Attempts were made to manage fisheries by attempting to manage the total effort applied. An argument in support of this derived from the notions of yield-per-recruit (YPR), which mathematically followed the fate, in terms of numbers and mass, of cohorts under different levels of imposed fishing mortality; fishing mortality was taken to be directly related to the effort applied. The notion of 'per-recruit' implied it followed individual cohorts and to begin with had the objective of maximizing the yield possible from each recruit.

We will repeat Russell's (1942) example to illustrate the ideas behind YPR. Russell's example only concerned itself with fishing mortality, which is the same as assuming no natural mortality (M was ignored). For this simple YPR we will also ignore natural mortality and just apply a series of constant harvest rates each year to calculate the numbers-at-age (equivalent to **Equ**(3.3)), then we will use a weight-at-age multiplied by the numbers-at-age series to calculate the weight of the catch-at-age, from which we can obtain the total expected catch.

```
# Simple Yield-per-Recruit see Russell (1942)
age <- 1:11;  nage <- length(age); N0 <- 1000  # some definitions
# weight-at-age values
WaA <- c(NA,0.082,0.175,0.283,0.4,0.523,0.7,0.85,0.925,0.99,1.0)
# now the harvest rates
H <- c(0.01,0.06,0.11,0.16,0.21,0.26,0.31,0.36,0.55,0.8)
nH <- length(H)
NaA <- matrix(0,nrow=nage,ncol=nH,dimnames=list(age,H)) # storage
CatchN <- NaA;  CatchW <- NaA      # define some storage matrices
for (i in 1:nH) {                 # Loop through the harvest rates
   NaA[1,i] <- N0 # start each harvest rate with initial numbers
   for (age in 2:nage) {  # Loop through over-simplified dynamics
      NaA[age,i] <- NaA[(age-1),i] * (1 - H[i])
      CatchN[age,i] <- NaA[(age-1),i] - NaA[age,i]
   }
   CatchW[,i] <- CatchN[,i] * WaA
}                                # transpose the vector of total catches to
totC <- t(colSums(CatchW,na.rm=TRUE))   # simplify later printing
```

The impact of the different harvest rates on the age-structure of the population is apparent in both the numbers-at-age and the weight-at-age in the catch. As the harvest rate increases the numbers-at-age in the older classes becomes more and more reduced, and similarly for the catch-at-age. For the higher harvest rates the stock becomes dependent upon the recent recruitment rather than any accumulation of numbers and biomass in the older classes (see Tables 3.4, 3.5, and 3.6).

TABLE 3.4: The population Numbers-at-Age for each of 10 annual harvest rates applied to the simple age-structured model to perform the yield-per-recruit analysis.

	0.01	0.06	0.11	0.16	0.21	0.26	0.31	0.36	0.55	0.8
1	1000	1000	1000	1000	1000	1000	1000	1000	1000.0	1000.0
2	990	940	890	840	790	740	690	640	450.0	200.0
3	980	884	792	706	624	548	476	410	202.5	40.0
4	970	831	705	593	493	405	329	262	91.1	8.0
5	961	781	627	498	390	300	227	168	41.0	1.6
6	951	734	558	418	308	222	156	107	18.5	0.3
7	941	690	497	351	243	164	108	69	8.3	0.1
8	932	648	442	295	192	122	74	44	3.7	0.0
9	923	610	394	248	152	90	51	28	1.7	0.0
10	914	573	350	208	120	67	35	18	0.8	0.0
11	904	539	312	175	95	49	24	12	0.3	0.0

TABLE 3.5: The population Catch Weight-at-Age for each of 10 annual harvest rates applied to the simple age-structured model to perform the yield-per-recruit analysis.

	0.01	0.06	0.11	0.16	0.21	0.26	0.31	0.36	0.55	0.8
2	0.82	4.92	9.02	13.12	17.22	21.32	25.42	29.52	45.10	65.60
3	1.73	9.87	17.13	23.52	29.03	33.67	37.43	40.32	43.31	28.00
4	2.77	15.00	24.66	31.95	37.09	40.29	41.77	41.73	31.52	9.06
5	3.88	19.93	31.02	37.93	41.42	42.14	40.74	37.75	20.05	2.56
6	5.02	24.50	36.10	41.66	42.78	40.78	36.75	31.59	11.80	0.67
7	6.66	30.82	43.00	46.84	45.23	40.39	33.94	27.06	7.10	0.18
8	8.00	35.18	46.47	47.78	43.39	36.29	28.44	21.03	3.88	0.04
9	8.62	35.99	45.01	43.67	37.30	29.22	21.35	14.65	1.90	0.01
10	9.14	36.21	42.87	39.26	31.54	23.15	15.77	10.03	0.92	0.00
11	9.14	34.38	38.54	33.31	25.17	17.30	10.99	6.49	0.42	0.00

TABLE 3.6: The total catches for each of 10 annual harvest rates applied to the simple age-structured model to perform the yield-per-recruit analysis.

0.01	0.06	0.11	0.16	0.21	0.26	0.31	0.36	0.55	0.8
55.8	246.8	333.8	359.1	350.2	324.5	292.6	260.2	166	106.1

If we use the **MQMF** `plot1()` function to plot out the total weight of catch against the harvest rate, we can immediately see that there is indeed an optimum harvest rate in terms of yield produced and that it is not the maximum used. Catches do not always increase if effort is allowed to increase. If the harvest rate is very low, then too few animals are caught to make up a significant catch. However, the yield rapidly increases with harvest rate and tails down more slowly as the harvest rate that generates the maximum yield is exceeded.

```
#Use MQMF::plot1 for a quick plot of the total catches. Fig. 3.8
plot1(H,totC,xlab="Harvest Rate",ylab="Total Yield",lwd=2)
```

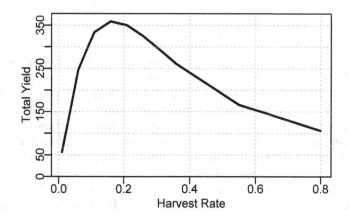

FIGURE 3.8 A simplified yield-per-recruit analysis that ignores natural mortality, selectivity-at-age, or any other influences. This analysis uses some weight-at-age data from Russell, 1942, although a wider array of harvest rates were examined than just the two by Russell.

As a quick illustration of the impact of different harvest rates this example was successful in 1942. A fuller treatment of yield-per-recruit needs to account for natural mortality in the dynamics and differential mortality rates by age brought about by gear selectivity.

3.3.1 Selectivity in Yield-per-Recruit

Usually, in a fully age-structured fisheries model one would implement a selectivity function to describe the different vulnerability of each age-class (or size-class) to a particular fishing gear. As the fish get older (or bigger), they recruit more and more to the fishery until they are completely vulnerable to fishing should it occur where they are. Normally, the parameters of such selectivity curves are obtained in the process of fitting the age-structured model to data from the fishery. Here, however, we will merely be attributing values (see more details of selectivity in the *Static Models* chapter). Early YPR analyses often used what is known as knife-edged selectivity, implying that fish became 100% vulnerable to gear at a particular age. Here however, we will implement a more realistic, yet simple, selectivity curve of vulnerability against age. There are many different equations used to describe the selectivity characteristics of different fishing gears but an extremely common one used for trawl gear is the standard logistic or S-shaped curve, which can also have a number of formulaic expressions. An equation commonly used to describe the logistic shape of selectivity with age, s_a, or size, and also maturity at age or size is defined as:

$$s_a = \frac{1}{1 + \left(e^{(\alpha+\beta a)}\right)^{-1}} = \frac{e^{(\alpha+\beta a)}}{1 + e^{(\alpha+\beta a)}} \tag{3.8}$$

where α and β are the logistic parameters, and $-\alpha/\beta$ is the age at a selectivity of 0.5 (50%). The inter-quartile distance (literally quantile 25% to quantile 75%; a measure of the gradient of the logistic curve) is defined as $IQ = 2\log(3)/\beta$ (see the **MQMF** function mature() for an implementation of this function). Generally, in age-structured modelling one needs length- or age-composition data so that the gear selectivity combined with fishery availability can be estimated directly. When working with yield-per-recruit calculations the usual reason to include a form of selectivity is to determine the optimum age at which to begin applying fishing mortality. In management terms this could be used to determine trawl or gillnet mesh sizes. This is the reason that knife-edged selectivity was often used, which would identify the age below which there is no selection and above which there is 100% selection. This is implemented in the **MQMF** function logist() (which uses a different formulation) but not in mature() (Figure 3.9). Knife-edged selectivity does not tend to be used in full age-structured stock assessment models.

```
#Logistic S shaped cureve for maturity
ages <- seq(0,50,1)
sel1 <- mature(-3.650425,0.146017,sizeage=ages) #-3.65/0.146=25
sel2 <- mature(-6,0.2,ages)
sel3 <- mature(-6,0.24,ages)
plot1(ages,sel1,xlab="Age Yrs",ylab="Selectivity",cex=0.75,lwd=2)
lines(ages,sel2,col=2,lwd=2,lty=2)
lines(ages,sel3,col=3,lwd=2,lty=3)
abline(v=25,col="grey",lty=2)
abline(h=c(0.25,0.5,0.75),col="grey",lty=2)
legend("topleft",c("25_15.04","30_10.986","25_9.155"),col=c(1,2,3),
       lwd=3,cex=1.1,bty="n",lty=1:3)
```

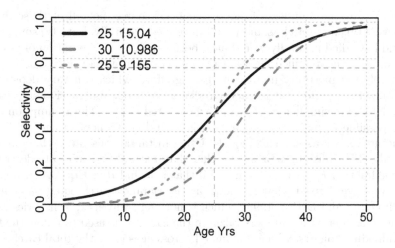

FIGURE 3.9 Examples of the logistic S-shaped curve using the `mature()` function. The legend consists of the L50 and the IQ for each curve with their parameters defined in the code.

3.3.2 The Baranov Catch Equation

Equ(3.4) described survivorship following the combination of fishing and natural mortality, but the simple Z for total mortality implied that all fish suffered the same fishing mortality. If we are to take account of selectivity then we need to consider fishing mortality at age:

$$S_{a,t} = e^{-(M+s_a F_t)} \tag{3.9}$$

where $S_{a,t}$ is the survivorship of fish aged a in year t, M is the natural mortality (assumed constant through time), s_a is the selectivity for age a, and F_t is known as the fully selected fishing mortality in year t.

Once we know the numbers surviving $S_t N_t$, then the numbers dying are obviously the difference between time periods. The total number dying within a cohort from one time period to the next can be described by:

$$N_z = N_t - N_{t+1} \tag{3.10}$$

If we replace the N_{t+1} with the survivorship equation, we get the complement of the survivorship equation, which describes the total numbers dying between time periods:

$$N_{a,z} = N_{a,t} - N_{a,t}e^{-(M+s_a F_t)} = N_{a,t}(1 - e^{-(M+s_a F_t)}) \tag{3.11}$$

Here $N_{a,z}$ is the numbers of age a dying. Of course, not all of those dying fish were taken as catch, some would have died naturally, but also some that would have died naturally would have been caught. The separation of mortality into M, the assumed constant across ages natural mortality, and F_t the fully selected mortality due to the fishing effort, which can vary depending in which time period t it is imposed allows the catch and survivorship to be calculated. For simplicity $s_a F_t$ can be designated as $F_{a,t}$. The assumption that it is possible to make this separation is a vital component of age- and size-structured stock assessment models. Using instantaneous rates, the numbers that die due to fishing mortality rather than natural mortality are described by the fraction $F_{a,t}/(M + F_{a,t})$, and if that is included in **Equ**(3.11), we find that we have derived what is known as the Baranov catch equation (Quinn, 2003; Quinn and Deriso, 1999), which is now generally used in fisheries modelling to estimate the numbers taken in the catch. It is used to follow the fate of individual cohorts, which if summed across ages gives the total catches:

$$\hat{C}_{a,t} = \frac{F_{a,t}}{M + F_{a,t}} N_{a,t} \left(1 - e^{-(M+F_{a,t})}\right) \tag{3.12}$$

where $\hat{C}_{a,t}$ is the expected catch at age a in year t, $F_{a,t}$ is the instantaneous fishing mortality rate at age a in year t $(s_a F_t)$, M is the natural mortality rate (assumed constant across age-classes), and $N_{a,t}$ is the numbers at age a in year t. This equation is implemented in the **MQMF** function bce().

```
# Baranov catch equation
age <- 0:12;  nage <- length(age)
sa <-mature(-4,2,age) #selectivity-at-age
H <- 0.2;  M <- 0.35
FF <- -log(1 - H)#Fully selected instantaneous fishing mortality
Ft <- sa * FF      # instantaneous Fishing mortality-at-age
N0 <- 1000
out <- cbind(bce(M,Ft,N0,age),"Select"=sa)  # out becomes Table 3.7
```

TABLE 3.7: The application of the Baranov catch equation to a population with an annual harvest rate of 0.2 and an instantaneous natural mortality rate of 0.3.

	Nt	N-Dying	Catch	Select
0	1000.000			0.018
1	686.191	291.645	22.164	0.119
2	432.501	192.368	61.322	0.500
3	250.395	116.618	65.488	0.881
4	141.728	66.827	41.840	0.982
5	79.943	37.766	24.018	0.998
6	45.071	21.298	13.574	1.000
7	25.409	12.007	7.655	1.000
8	14.325	6.769	4.316	1.000
9	8.075	3.816	2.433	1.000
10	4.553	2.151	1.372	1.000
11	2.567	1.213	0.773	1.000
12	1.447	0.684	0.436	1.000

3.3.3 Growth and Weight-at-Age

To obtain the catch as mass (i.e., the weight of catch) the numbers-at-age in the catch need to be multiplied by the weight-at-age. The weight-at-age can either be a vector of mean weight observations from a particular fishery for each particular year (Beverton and Holt, 1957) or, more commonly, a standard weight-at-age equation that derives from the von Bertalanffy growth curve, which starts as length-at-age:

$$L_a = L_\infty(1 - e^{(-K(a-t_0))})$$ (3.13)

where L_a is the length-at-age a, L_∞ is the mean maximum length-at-age, K is the coefficient, which determines how quickly L_∞ is achieved, and t_0 is the hypothetical age at length zero. This is implemented in the **MQMF** function vB(). The assumption is made that there is a power relationship between length-at-age and weight-at-age so two more parameters are added to give:

$$W_a = \alpha L_a{}^b = w_a(1 - e^{(-K(a-t_0))})^b$$
$$\log(W_a) = \log(\alpha) + bL_a$$ (3.14)

where α is a constant, w_a is (αL_∞), and b is the exponent, which usually approximates to a value of 3.0 (size measures are two-dimensional while weight measures are three-dimensional). The log-transformed version is, of course, linear. In the *Static Models* chapter we will see how parameters for such models can be estimated.

3.4 Full Yield-per-Recruit

Pulling all of this together means we can produce a much more complete yield-per-recruit (YPR) analysis. A standard YPR analysis assumes constant recruitment, which means we can follow the fate of a single cohort and still capture the required details. Of course, recruitment is not a constant so this approach is considered to be based upon long-run or equilibrium conditions. Ignoring variation and stochasticity in this way means that any conclusions drawn, with regard to potential productivity and resulting management decisions, need to be treated with caution (in fact, with great caution). But these approaches were developed when fisheries analyses were all deterministic and assumed to be at equilibrium, the implications of uncertainty had not yet been explored (Beverton and Holt, 1957).

It makes sense to do such analyses when the objective is to maximize yield. But the assumption of constant recruitment and the lack of attention paid to uncertainty means these analyses are generally not conservative. Attempts were made to improve on the recommendations that flowed from YPR analyses with one of the more useful being the advent of $F_{0.1}$ (pronounced F zero point one). This is defined as the harvest rate at which the rate of increase in yield is $1/10$ that at the origin of a yield curve (Hilborn and Walters, 1992). The advantage of $F_{0.1}$ is a relatively large reduction in fishing effort and can lead to only a small loss in yield. This can improve both the economics that the sustainability of a fishery. Nevertheless, essentially the use of $F_{0.1}$ remains an empirical rule which was found to be more sustainable in practice than either F_{max} (the point of maximum yield) and usually better than F_{msy} (the fishing mortality that, at equilibrium, would generate the *MSY*).

```
# A more complete YPR analysis
age <- 0:20;  nage <- length(age) #storage vectors and matrices
laa <- vB(c(50.0,0.25,-1.5),age) # Length-at-age
WaA <- (0.015 * laa ^ 3.0)/1000  # weight-at-age as kg
H <- seq(0.01,0.65,0.05);  nH <- length(H)
FF <- round(-log(1 - H),5)  # Fully selected fishing mortality
N0 <- 1000
M <- 0.1
numt <- matrix(0,nrow=nage,ncol=nH,dimnames=list(age,FF))
catchN <- matrix(0,nrow=nage,ncol=nH,dimnames=list(age,FF))
as50 <- c(1,2,3)
yield <- matrix(0,nrow=nH,ncol=length(as50),dimnames=list(H,as50))
```

```
for (sel in 1:length(as50)) {
   sa <- logist(as50[sel],1.0,age)  # selectivity-at-age
   for (harv in 1:nH) {
      Ft <- sa * FF[harv]        # Fishing mortality-at-age
      out <- bce(M,Ft,N0,age)
      numt[,harv] <- out[,"Nt"]
      catchN[,harv] <- out[,"Catch"]
      yield[harv,sel] <- sum(out[,"Catch"] * WaA,na.rm=TRUE)
   } # end of harv loop
} # end of sel loop
```

```
#A full YPR analysis  Fig. 3.10
plot1(H,yield[,3],xlab="Harvest Rate",ylab="Yield",cex=0.75,lwd=2)
lines(H,yield[,2],lwd=2,col=2,lty=2)
lines(H,yield[,1],lwd=2,col=3,lty=3)
legend("bottomright",legend=as50,col=c(3,2,1),lwd=3,bty="n",
       cex=1.0,lty=c(3,2,1))
```

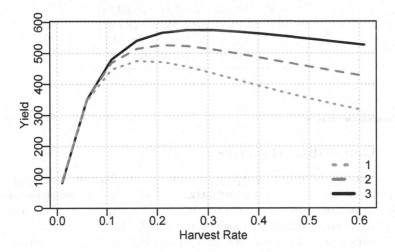

FIGURE 3.10 The effect on the total equilibrium yield of applying different harvest rates and different ages of first exploitation. The legend identifies the different age of first exploitation.

These detailed analyses, allowing for selectivity, weight-at-age, and natural mortality, provide a more accurate, albeit still deterministic, representation of the potential yield from a fishery under different conditions (Table 3.8). Once uncertainty and natural variation are taken into account, even just approximately by using $F_{0.1}$, then a YPR analysis can still provide some useful insights into the productive capacity of particular fisheries. Of course, these days, one would be more likely to conduct a profit-per-recruit analysis, but the principles remain the same (Table 3.8).

TABLE 3.8: The effect on the total equilibrium yield of applying different harvest rates (rows) and different ages of first exploitation (columns).

	1	2	3
0.01	85.82	84.152	81.136
0.06	347.12	352.465	350.567
0.11	447.07	469.250	480.134
0.16	475.15	514.595	540.249
0.21	471.99	526.522	565.880
0.26	455.84	522.962	574.288
0.31	434.88	512.360	573.986
0.36	412.72	498.720	569.223
0.41	390.96	483.969	562.163
0.46	370.25	469.031	553.935
0.51	350.83	454.348	545.146
0.56	332.72	440.100	536.113
0.61	315.81	426.329	526.995

3.5   Concluding Remarks

We have used relatively simple population models to illustrate numerous ideas from fisheries and ecology. Importantly, the focus was on simulation models rather than fitting models to data (see the next chapter on *Model Parameter Estimation*). Simulation models would usually include uncertainty in the model parameter values, however, as with many of the subjects given only an introductory treatment here, a more thorough treatment would entail a book in itself. Nevertheless, by exploring the implications of imposing a range of parameter values on the various models considered, the properties and implied dynamics of these models can be laid bare. As such, simulation studies are a vital tool in the modelling of any natural processes. Equally important, however, is using such simulations with plausible or realistic combinations of parameters. It is useful to understand the mathematical properties of all models but, naturally, the prime interest is when the models take on realistic values and might be found in nature.

The intuitions obtained when using any population model or their components have value in understanding any future work, but, of course, the context and assumptions in each of the models illustrated (especially the relatively simple models) have always to be kept in mind. The surprising thing is that even simple models can sometimes provide insight into various natural processes.

In this chapter we have only considered simple population models but we could equally well use R to explore interactions between species in processes such as competition, predation, parasitism, symbiosis, and others. All the early models (Lotka, 1925; Volterra, 1927; Gause, 1934) considered populations with no spatial structure. Using R it would be relatively straightforward to increase the complexity of such models and include spatial details such as explored experimentally by Huffaker (1958). Some of Huffaker's later experimental predator-prey arrangements lasted for 490 days. In such cases, simulation studies could become very useful for expanding on what explorations are possible. Such interesting work would need to be a part of a different book as here we still have many other areas to explore and discuss in the field of fisheries.

The use of R as the tool for conducting such simulations makes the analyses somewhat more abstract than laying the parameter manipulations out on a spreadsheet. Nevertheless, the advantages arising from the re-usability of code chunks and developed functions, as well as the potential for the incremental development of more complex models and functions, outweighs the more abstract nature of the programming environment. The use of R as a programming language in which to develop these different analyses is well suited to fisheries and ecological modelling work.

4

Model Parameter Estimation

4.1 Introduction

One of the more important aspects of modelling in ecology and fisheries science relates to the fitting of models to data. Such model fitting requires:

- data from a process of interest in nature (samples, observations),

- explicitly selecting a model structure suitable for the task in hand (model design and then selection—big subjects in themselves),

- explicitly selecting probability density functions to represent the expected distribution of how predictions of the modelled process will differ from observations from nature when compared (choosing residual error structures), and finally,

- searching for the model parameters that optimize the match between the predictions of the model and any observed data (the criterion of model fit).

Much of the skill/trickery/magic that is involved when fitting models to data centers on that innocent-looking word or concept **optimize** in the last requirement above. This is an almost mischievous idea that can occasionally lead one into trouble, though it is also a challenge and can often be fun. The different ways of generating a so-called *best-fitting* model is an important focus for this chapter. It centers around the idea of what criterion to use when describing the quality of model fit to the available data and how then to implement the explicitly selected criteria.

I keep using the term "explicit", and for good reason, but some clarification is needed. Very many people will have experienced fitting linear regressions to data but, I am guessing from experience, far fewer people realize that when they fit such a model they are assuming the use of additive normal random residual errors (normal errors) and that they are minimizing the sum of the square of those residuals. In terms of the four requirements above, when applying a linear regression to a data-set, the assumption of a linear relationship answers the second requirement, the use of normal errors (with the additional assumption of a constant variance) answers the third requirement, and the minimization of the sum-of-squares is the choice that meets the fourth requirement. As is generally the case, it is better to understand explicitly what one is doing rather than just operating out of habit or copying others. In order to make the most appropriate choice of these model fitting requirements (i.e., make selections that can be defended), an analyst also needs an understanding of the natural process being modelled. One can assume and assert almost anything but only so long as such selections can be defended validly. As a more general statement, if one cannot defend a set of choices then one should not make them.

4.1.1 Optimization

In Microsoft Excel, when fitting models to data, the optimum model parameters are found using the built-in Excel Solver. This involves setting up a spreadsheet so that the criterion of optimum fit (sum-of-squares, maximum likelihood, etc., see below) is represented by the contents of a single cell, and the model parameters and data used were contained in other inter-related cells. Altering a model's parameters would alter the predicted values generated, and this in turn would alter the value of the criterion of optimum fit. A 'best' parameter set can be found by searching for the parameters that optimize the match between the observed and the predicted values. It sounds straightforward but turns out to be quite an art form with many assumptions and decisions to be made along the way. In the Excel Solver, one identified the cells containing the model parameters and the Solver's internal code would then modify those values while monitoring the "criterion of best fit" cell until either a minimum (or maximum) value was found (or an exception was encountered). Effectively such a spreadsheet set-up constitutes the syntax for using the Solver within Excel. We will be using solver or optimization functions in R, and they too have a required syntax, but it is no more complex than setting up a spreadsheet, it is just rather more abstract.

Model fitting is usually relatively straightforward when modelling a non-dynamic process to a single set of data, such as length-at-age, with perhaps just two to six parameters. However, it can become much more involved when perhaps dealing with a population's dynamics through time involving recruitment, the growth of individuals, natural mortality, and fishing mortality from multiple fishing fleets. There may be many types of data, and possibly many 10's or even 100's of parameters. In such circumstances, to adjust the quality of the fit of predicted values to observed values some form of automated optimization or non-linear solver is a necessity.

Optimization is a very large subject of study and the very many options available are discussed at length in the CRAN Task View: *Optimization and Mathematical Programming* found at https://cran.r-project.org/. In the work here, we will mainly use the built-in function nlm() (try ?nlm), but there are many alternatives (including nlminb(), optim(), and others). If you are going to be involved in model fitting it is really worthwhile reading the Task View on optimization on R-CRAN and, as a first step, explore the help and examples given for the `nlm()` and the `optim()` functions.

It is sometimes possible to guess a set of parameters that generate what appears to be a reasonable visual fit to the available data, at least for simple static models. However, while such *fitting-by-eye* may provide usable starting points for estimating a model's parameters it does not constitute a defensible criterion for fitting models to data. This is so because my fitting-by-eye (or wild-stab-in-the-dark) may well differ from your fitting-by-eye (or educated-guess). Instead of using such opinions some more formally defined criteria of quality of model fit to data is required.

The focus here will be on how to set-up R code to enable model parameter estimation using either least-squares or maximum likelihood, especially the latter. Our later consideration of Bayesian methods will be focussed primarily on the characterization of uncertainty. We will illustrate the model fitting process through repeated examples and associated explanations. The objective is that reading this section should enable the reader to set up their own models to solve for the parameter values. We will attempt to do this in a general way that should be amenable to adaptation for many problems.

4.2 Criteria of Best Fit

There are three common options for determining what constitutes an optimum model fit to data.

In general terms, model fitting can involve the minimization of the sum of squared residuals (`ssq()`) between the values of the observed variable x and those predicted by candidate models proposed to describe the modelled process, $\hat{x}$:

$$ssq = \sum_{i=1}^{n} (x_i - \hat{x}_i)^2 \qquad (4.1)$$

where x_i is observation i from n observations, and $\hat{x}_i$ is the model predicted value for a given observation i (for example, if x_i was the length-at-age for fish i, then the $\hat{x}_i$ would be the predicted length-at-age for fish i derived from some candidate growth model). The use of the ˆ in $\hat{x}$ indicates a predicted value of x.

Alternatively, model fitting can involve minimizing the negative log-likelihood (in this book *-veLL* or *negLL*), which entails determining how likely each observed data point is given 1) a defined model structure, 2) a set of model parameters, and 3) the expected probability distribution of the residuals. Minimizing the negative log-likelihood is equivalent to maximizing either the product of all likelihoods or the sum of all log-likelihoods for the set of data points. Given a collection of observations x, a model structure that can predict $\hat{x}$, and a set of model parameters θ, then the total likelihood of those observations is defined as:

$$L = \prod_{i=1}^{n} L\left(x_i|\theta\right)$$

$$-veLL = -\sum_{i=1}^{n} \log\left(L\left(x_i|\theta\right)\right)$$

(4.2)

where L is the total likelihood, which is $\prod L\left(x|\theta\right)$ or the product of the probability density (likelihood) for each observation of x given the parameter values in θ (the further from the expected value in each case, $\hat{x}_i$, the lower the likelihood). $-veLL$ is the total negative log-likelihood of the observations x given the candidate model parameters in θ, which is the negative sum of the n individual log-likelihoods for each data point in x. We use log-likelihoods because most likelihoods are very small numbers, which when multiplied by many other very small numbers can become so small as to risk leading to floating-point overflow computer errors. Log-transformations change multiplications into additions and avoid such risks (the capital pi, Π turns into a capital sigma, Σ).

A third alternative is to use Bayesian methods that use prior probabilities, which are the initial relative weights given to each of the alternative parameter vectors being considered in the model fitting. Bayesian methods combine with and update any prior knowledge of the most likely model parameters (the prior probabilities) with the likelihoods of any new data that become available given the different candidate parameter vectors θ. The two key difference between Bayesian methods and maximum likelihood methods, for our purposes, is the inclusion of prior likelihoods and the re-scaling of values so that the posterior probabilities sum to a total of 1.0. An important point is that the likelihood of the data given a set of parameters is being converted into a true probability of the parameters given the data. The posterior probability of a particular parameter set θ given the data x is thus defined as:

$$P(\theta|x) = \frac{L(x|\theta)P(\theta)}{\sum\limits_{i=1}^{n}[L(x_i|\theta)P(\theta)]}$$

(4.3)

where $P\left(\theta\right)$ is the prior probability of the parameter set θ, which is updated by the likelihood of the data, x, given the parameters, θ, $L(x|\theta)P(\theta)$, and the divisor $\sum_{i=1}^{n} \left[L\left(x|\theta\right) P\left(\theta\right) \right]$ re-scales or normalizes the result so the sum of the posterior probabilities for all parameter vectors given the data, $\sum P(\theta|x)$, sums to 1.0. Formally, **Equ**(4.3) is an approximation as the summation in the divisor should actually be an integration across a continuous distribution, but in practice the approximation suffices and is the only practical option when dealing with the parameters of a complex fisheries model whose Posterior distribution has no simple analytical solution.

Here we will focus mainly on minimizing negative log-likelihoods (equivalent to the maximum likelihood). Although the other methods will also be given some attention. The Bayesian methods will receive much more attention when we explore the characterization of uncertainty.

Microsoft Excel is excellent software for many uses but implementing both maximum likelihood and particularly Bayesian methods tends to be slow and clumsy; they are much more amenable to implementation in R.

Identifying the sum-of-squared residuals, maximum likelihood, and Bayesian methods is not an exhaustive list of the criteria it is possible to use when fitting models to data. For example, it is possible to use the 'sum of absolute residuals' (SAR) which avoids the problem of combining positive and negative residuals by using their absolute values rather than squaring them (Birkes and Dodge, 1993). Despite the existence of such alternative criteria of optimal model fit we will focus our attention only on the three mentioned. Other more commonly used alternatives include what are known as robustified methods, which work to reduce the effect of outliers, or extreme and assumed to be atypical values, in available data. As stated earlier, optimization is a large and detailed field of study; I commend it to you for study, and good luck with it.

4.3 Model Fitting in R

While a grid search covering the parameter space might be a possible approach to searching for an optimum parameter set it would become more and more unwieldy as the number of parameters increased above two until eventually it would became unworkable. We will not consider that possibility any further. Instead, to facilitate the search for an optimum parameter set, a non-linear optimizer implemented in software is required.

The R system has a selection of different optimization functions, each using a different algorithm (do see the CRAN task view on Optimization). The solver function (nlm()), like the others, needs to be given an initial guess at the parameter values and then those initial parameter values are varied by the optimizer function and at each change the predicted values are recalculated as in the ssq() or negLL(). Optimization functions, such as nlm(), continue to vary the parameter values (how they do this is where the algorithms vary) until a combination is found that is defined to be the 'best fit' according to whatever criterion has been selected (or further improvements cannot be found). Fisheries stock assessment models can often have numerous parameters in the order of 10s or 100s (some have many more and require more highly specialized software, e.g., Fournier *et al*, 1998; Fournier *et al*, 2012; Kristensen *et al*, 2016). In this book we will not be estimating large numbers of parameters, but the principles remain similar no matter the number.

4.3.1 Model Requirements

Discussing the theory of model fitting is helpful but does not clarify how one would implement fitting a model in R in practice. The optimizing software is used to vary the values within a parameter vector but we need to provide it with the means of repeatedly calculating the predicted values that will be used to compare with the observed values as many times as required to find the optimum solution (if successful). We need to develop chunks of code that can be repeatedly called, which is exactly the purpose for which R functions are designed. To implement model fitting to real world problems within R we need to consider the four formal requirements:

- observations (data) from the system under study. This may be a fishery with observed catches, cpue, age- and length-composition of catches, etc., or it may be something simpler, such as the observed lengths and associated ages for a sample of fish (but how to get this into R?),

- an R function representing a candidate model of the system that when provided with a vector of parameters is used to calculate predicted values for comparison with whatever observed values are available,

- an R function calculating the selected criterion of best fit, the minimum least-squares or minimum negative log-likelihood, to enable a comparison of the observed with the predicted values. This needs to be able to return a single value, reflecting the input parameters and data, that can then be minimized by the final required function, which is

- an R function (we will tend to use nlm()) for automatically optimizing the value of the selected criterion of best fit.

Thus, input data and three functions are needed, **Figure** 4.1, but because we can use built-in functions to conduct the optimization, model fitting usually entails writing at most two functions, one to calculate the predicted values from whatever model one is using and the other to calculate the criterion of fit (sometimes, in simpler exercises, these two can be combined into one function).

We assume in this book that the reader is at least acquainted with the concepts behind model fitting, as in fitting a linear regression, so we will move straight to non-linear model fitting. The primary aim of these relatively simple examples will be to introduce the use and syntax of the available solvers within R.

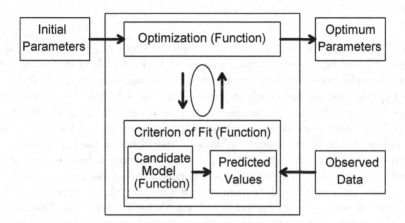

FIGURE 4.1 Inputs, functional requirements, and outputs, when fitting a model to data. The optimization function (here `nlm()`) minimizes the negative log-likelihood (or sum-of squares) and requires an initial parameter vector to begin. In addition, the optimizer requires a function (perhaps `negLL()`) to calculate the corresponding negative log-likelihood for each vector of parameters it produces in its search for the minimum. To calculate the negative log-likelihood requires a function (perhaps `vB()`) to generate predicted values for comparison with the input observed values.

4.3.2 A Length-at-Age Example

Fitting a model to data simply means estimating the model's parameters so that its predictions match the observations as well as it can according to the criterion of best-fit chosen. As a first illustration of fitting a model to data in R we will use a simple example of fitting the well-known von Bertalanffy growth curve (von Bertalanffy, 1938), to a set of length-at-age data. Such a data set is included in the R package **MQMF** (try `?LatA`). To use your own data one option is to produce a comma separated variable (csv) file with a minimum of columns of ages and lengths each with a column name (*LatA* has columns of just age and length; see its help page). Such csv files can be read into R using `laa <- read.csv(file="filename.csv", header=TRUE)`.

The von Bertalanffy length-at-age growth curve is described by:

$$\hat{L}_t = L_\infty \left(1 - e^{(-K(t-t_0))}\right)$$
$$L_t = L_\infty \left(1 - e^{(-K(t-t_0))}\right) + \varepsilon \qquad (4.4)$$
$$L_t = \hat{L}_t + \varepsilon$$

where $\hat{L}_t$ is the expected or predicted length-at-age t, L_∞ (pronounced L-infinity) is the asymptotic average maximum body size, K is the growth rate coefficient that determines how quickly the maximum is attained, and t_0 is the hypothetical age at which the species has zero length (von Bertalanffy, 1938). This non-linear equation provides a means of predicting length-at-age for different ages once we have estimates (or assumed values) for L_∞, K, and t_0. When we are fitting models to data we use either of the the bottom two equations in **Equ**(4.4), where the L_t are the observed values and these are equated to the predicted values plus a normal random deviate $\varepsilon = N(0, \sigma^2)$, each value of which could be negative or positive (the observed value can be larger or smaller than the predicted length for a given age). The bottom equation is really all about deciding what residual errors structure to use. In this chapter we will describe an array of alternative error structures used in ecology and fisheries. Not all of them are additive and some are defined using functional relationships rather than constants (such as σ).

The statement about **Equ**(4.4) being non-linear was made explicitly because earlier approaches to estimating parameter values for the von Bertalanffy (vB()) growth curve involved various transformations that aimed approximately to linearize the curve (e.g., Beverton and Holt, 1957). Fitting the von Bertalanffy curve was no minor undertaking in the late 1950s and 1960s. Happily, such transformations are no longer required, and such curve fitting has become straightforward.

4.3.3 Alternative Models of Growth

There is a huge literature on the study of individual growth and many different models are used to describe the growth of organisms (Schnute and Richards, 1990). The von Bertalanffy (vB()) curve has dominated fisheries models since Beverton and Holt (1957) introduced it more widely to fisheries scientists. However, just because it is very commonly used does not necessarily mean it will always provide the best description of growth for all species. Model selection is a vital if often neglected aspect of fisheries modelling (Burnham and Anderson, 2002; Helidoniotis and Haddon, 2013). Two alternative possibilities to the vB() for use here are the Gompertz growth curve (Gompertz, 1925):

$$\hat{L}_t = ae^{-be^{ct}} \text{ or } \hat{L}_t = a\exp(-b\exp(ct)) \qquad (4.5)$$

and the generalized Michaelis-Menten equation (Maynard Smith and Slatkin, 1973; Legendre and Legendre, 1998):

$$\hat{L}_t = \frac{at}{b + t^c} \tag{4.6}$$

each of which also have three parameters a, b, and c, and each can provide a convincing description of empirical data from growth processes. Biological interpretations (e.g., maximum average length) can be given to some of the parameters but in the end these models provide only an empirical description of the growth process. If the models are interpreted as reflecting reality this can lead to completely implausible predictions such as non-existent meters-long fish (Knight, 1968). In the literature the parameters can have different symbols (for example, Maynard Smith and Slatkin (1973), use *R0* instead of a for the Michaelis-Menton), but the underlying structural form remains the same. After fitting the von Bertalanffy growth curve to the fish in **MQMF's** length-at-age data-set *LatA*, we can use these different models to illustrate the value of trying such alternatives and maintaining an open mind with regard to which model one should use. This issue will arise again when we discuss uncertainty because we can get different results from different models. Model selection is one of the big decisions to be made when modelling any natural process. Importantly, trying different models in this way will also reinforce the processes involved when fitting models to data.

4.4 Sum of Squared Residual Deviations

The classical method for fitting models to data is known as least sum of squared residuals (see **Equs**(4.1) and (4.7)), or more commonly least-squares or least-squared. The method has been attributed to Gauss (Nievergelt, 2000, refers to a translation of Gauss' 1823 book written in Latin: *Theoria combinationis observationum erroribus minimis obnoxiae*). Whatever the case, the least-squares approach fits into a strategy used for more than two centuries for defining the best fit of a set of predicted values to those observed. This strategy is to identify a so-called *objective* function (our criterion of best-fit), which can either be minimized or maximized depending on how the function is structured. In the case of the sum-of-squared residuals one subtracts each predicted value from its associated observed value, square the separate results (to avoid negative values), sum all the values, and use mathematics (the analytical solution) or some other approach to minimize that summation:

$$ssq = \sum_{i=1}^{n} \left(O_i - \hat{E}_i \right)^2 \tag{4.7}$$

where ssq is the sum-of-squared residuals of n observations, O_i is observation i, and $\hat{E}_i$ is the expected or predicted value for observation i. The function ssq() within **MQMF** is merely a wrapper surrounding a call to whatever function is used to generate the predicted values and then calculates and returns the sum-of-squared deviations. It is common that we will have to create new functions as wrappers for different problems depending on their complexity and data inputs. ssq() illustrates nicely the fact that in among the arguments that one might pass to a function it is also possible to pass other functions (in this case, within ssq(), we have called the passed a function to *funk*, but of course when using ssq() we input the actual function relating to the problem at hand, perhaps vB, note with no brackets when used as a function argument).

4.4.1 Assumptions of Least-Squares

A major assumption of least-squares methodology is that the residual error terms exhibit a Normal distribution about the predicted variable with equal variance for all values of the observed variable; that is, in the $\varepsilon = N(0, \sigma^2)$ the σ^2 is a constant. If data are transformed in any way, the transformation effects on the residuals may violate this assumption. Conversely, a transformation may standardize the residual variances if they vary in a systematic way. Thus, if data are log-normally distributed then a log-transformation will normalize the data and least-squares can then be used validly. As always, a consideration or visualization of the form of both the data and the residuals, resulting from fitting a model, is good practice.

4.4.2 Numerical Solutions

Most of the interesting problems in fisheries science do not have an analytical solution (e.g., as available for linear regression), and it is necessary to use numerical methods to search for an optimum model fit using a defined criterion of 'best fit', such as the minimum sum of squared residuals (least-squares). This will obviously involve a little R programming, but a big advantage of R is that once you have a set of analyses developed it can become straightforward to apply them to a new set of data.

In the examples below we use some of the utility functions from **MQMF** to help with the plotting. But for fitting and comparing the three different growth models defined above we need five functions, four of which we need to write. The first three are used to estimate the predicted values of length-at-age used to compare with the observed data. This example has three alternative model functions, vB(), Gz(), and mm(), one each for the three different growth curves. The fourth function is needed to calculate the sum-of-squared residuals from the predicted values and their related observations. Here we are going to be using the **MQMF** function ssq() (whose code you should examine and understand). This function returns a single value, which is to be minimized by the final function, nlm(), which is needed to do the minimization in an auto-mated manner. The R function nlm() uses a user-defined generalized function, which it refers to as f (try args(nlm), or formals(nlm) to see the full list of arguments), for calculating the minimum (in this case of ssq()), which, in turn, uses the function defined for predicted lengths-at-age from the growth curve (e.g., vB()). Had we used a different growth curve function (e.g., Gz()) we only have to change everywhere the nlm() calling code points to vB() to Gz() and modify the parameter values to suit the Gz() function for the code to produce a usable result. Fundamentally, nlm() minimizes ssq() by varying the input parameters (which it refers to as p) that alter the outcome of the growth function vB(), Gz(), or mm(), whichever is selected.

nlm() is just one of the functions available in R for conducting non-linear opti-mization, alternatives include optim() and nlminb() (do read the documen-tation in ?nlm, and the task view on optimization on CRAN lists packages aimed at solving optimization problems).

```
#setup optimization using growth and ssq
data(LatA)      # try ?LatA   assumes library(MQMF) already run
#convert equations 4.4 to 4.6 into vectorized R functions
#These will over-write the same functions in the MQMF package
vB <- function(p, ages) return(p[1]*(1-exp(-p[2]*(ages-p[3]))))
Gz <- function(p, ages) return(p[1]*exp(-p[2]*exp(p[3]*ages)))
mm <- function(p, ages) return((p[1]*ages)/(p[2] + ages^p[3]))
 #specific function to calc ssq. The ssq within MQMF is more
ssq <- function(p,funk,agedata,observed) {       #general and is
  predval <- funk(p,agedata)        #not limited to p and agedata
  return(sum((observed - predval)^2,na.rm=TRUE))
} #end of ssq
 # guess starting values for Linf, K, and t0, names not needed
pars <- c("Linf"=27.0,"K"=0.15,"t0"=-2.0) #ssq should=1478.449
ssq(p=pars, funk=vB, agedata=LatA$age, observed=LatA$length)
```

```
# [1] 1478.449
```

```
# try misspelling LatA$Length with a capital. What happens?
```

The ssq() function replaces the MQMF::ssq() function in the global environ-ment, but it also returns a single number, e.g., the 1478.449 above, which is the first input to the nlm() function and is what gets minimized.

4.4.3 Passing Functions as Arguments to Other Functions

In the last example we defined some of the functions required to fit a model to data. We defined the growth models we were going to compare and we defined a function to calculate the sum-of-squares. A really important aspect of what we just did was that to calculate the sum-of-squares we passed the

vB() function as an argument to the ssq() function. What that means is
that we have passed a function that has arguments as one of the arguments
to another function. You can see the potential for confusion here so it is
necessary to concentrate and keep things clear. Currently, the way we have
defined ssq() this all does not seem so remarkable because we have explicitly
defined the arguments to both functions in the call to ssq(). But R has some
tricks up its sleeve that we can use to generalize such functions that contain
other functions as arguments and the main one uses the magic of the ellipsis
.... With any R function, unless an argument has a default value set in its
definition, each argument must be given a value. In the ssq() function above
we have included both the arguments used only by ssq() (*funk* and *observed*),
and those used only by the function *funk* (*p* and *agedata*). This works well
for the example because we have deliberately defined the growth functions to
have identical input arguments, but what if the *funk* we wanted to use had
different inputs, perhaps because we were fitting a selectivity curve and not
a growth curve? Obviously we would need to write a different ssq() function.
To allow for more generally useful functions that can be re-used in many more
situations, the writers of R (R Core Team, 2019) included this concept of ...,
which will match any arguments not otherwise matched, and so can be used
to input the arguments of the *funk* function. So, we could re-define ssq() thus:

```
# Illustrates use of names within function arguments
vB <- function(p,ages) return(p[1]*(1-exp(-p[2] *(ages-p[3])))) 
ssq <- function(funk,observed,...) { # only define ssq arguments
  predval <- funk(...) # funks arguments are implicit
  return(sum((observed - predval)^2,na.rm=TRUE))
} # end of ssq
pars <- c("Linf"=27.0,"K"=0.15,"t0"=-2.0) # ssq should = 1478.449
ssq(p=pars, funk=vB, ages=LatA$age, observed=LatA$length) #if no
ssq(vB,LatA$length,pars,LatA$age) # name order is now vital!
```

```
# [1] 1478.449
# [1] 1478.449
```

This means the ssq() function is now much more general and can be used with
any input function that can be used to generate a set of predicted values for
comparison with a set of observed values. This is how the ssq() function within
MQMF is implemented; read the help ?ssq. The general idea is that you must
define all arguments used within the main function but any arguments only
used within the called function (here called *funk*), can be passed in the .... It is
always best to explicitly name the arguments so that their order does not
matter, and you need to be very careful with typing as if you misspell the
name of an argument passed through the ... this will not always throw an
error! For example, using a uppercase *LatA$Age* rather than *LatA$age* does
not throw an error but leads to a result of zero rather than 1478.440. This is

because *LatA\$Age = NULL*, which is a valid if incorrect input. Clearly the ... can be very useful, but it is also inherently risky if you type as badly as I do.

```
# Illustrate a problem with calling a function in a function
# LatA$age is typed as LatA$Age but no error, and result = 0
ssq(funk=vB, observed=LatA$length, p=pars, ages=LatA$Age) # !!!
```

```
# [1] 0
```

And if you were rushed and did not bother naming the arguments that too will fail if you get the arguments out of order. If, for example, you were to input ssq(LatA\$length, vB, pars, LatA\$age) instead of ssq(vB, LatA\$length, pars, LatA\$age) then you will get an error: *Error in funk(...) : could not find function "funk"*. You might try that yourself just to be sure. It rarely hurts to experiment with your code; you cannot break your computer, and you might learn something.

4.4.4   Fitting the Models

If we plot up the *LatA* data-set, **Figure** 4.2, we see some typical length-at-age data. There are 358 points (try dim(LatA)) and many lie on top of each other, but the relative sparseness of fish in the older age classes becomes apparent when we use the rgb() option within the plot() function to vary the transparency of colour in the plot. Alternatively, we could use jitter() to add noise to each plotted point's position to see the relative density of data points. Whenever you are dealing with empirical data it is invariably worth your time to at least plot it up and otherwise explore its properties.

```
#plot the LatA data-set     Fig. 4.2
parset()         # parset and getmax are two MQMF functions
ymax <- getmax(LatA$length) # simplifies use of base graphics. For
# full colour, with the rgb as set-up below, there must be >= 5 obs
plot(LatA$age,LatA$length,type="p",pch=16,cex=1.2,xlab="Age Years",
     ylab="Length cm",col=rgb(1,0,0,1/5),ylim=c(0,ymax),yaxs="i",
     xlim=c(0,44),panel.first=grid())
```

Rather than continuing to guess parameter values and modifying them by hand we can use nlm() (or optim(), or nlminb(), which have different syntax for their use) to fit growth models or curves to the selected *LatA* data. This will illustrate the syntax of nlm() but also the use of two more **MQMF** utility R functions magnitude() and outfit() (check out ?nlm, ?magnitude, and ?outfit). You might also look at the code in each function (enter each function's name into the console without arguments or brackets). From now on I will prompt you less often to check out the details of functions that get used, but if you see a function that is new to you, hopefully, by now, it makes sense to review its help, its syntax, and especially its code, just as you might look at the contents of each variable that gets used.

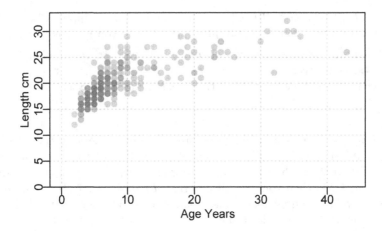

FIGURE 4.2 Simulated female length-at-age data for 358 Redfish, *Centroberyx affinis*, based on samples from eastern Australia. Full-intensity colour means >= 5 points.

Each of the three growth models require the estimation of three parameters, and we need initial guesses for each to start off the nlm() solver. All we do is provide values for each of the nlm() function's parameters/arguments. In addition, we have used two extra arguments, *typsize* and *iterlim*. The *typsize* is defined in the nlm() help as "an estimate of the size of each parameter at the minimum". Including this often helps to stabilize the searching algorithm used as it works to ensure that the iterative changes made to each parameter value are of about the same scale. A very common alternative approach, which we will use with more complex models, is to log-transform each parameter when input into nlm() and back-transform them inside the function called to calculate the predicted values. However, this could only work where the parameters are guaranteed to always stay positive. With the von Bertalanffy curve, for example, the t_0 parameter is often negative and so magnitude() should be used instead of the log-transformation strategy. The default iterlim=100 means a maximum of 100 iterations, which is sometimes not enough, so if 100 are reached you should expand this number to, say, 1000. You will quickly notice that the only thing changed between each model fitting effort is the function that *funk*, inside ssq(), is pointed at and the starting parameter values. This is possible by consciously designing the growth functions to use exactly the same arguments (passed via ...). You could also try running one or two of these without setting the *typsize* and *interlim* options. Notice also that we have run the Michaelis-Menton curve with two slightly different starting points.

```
 # use nlm to fit 3 growth curves to LatA, only p and funk change
ages <- 1:max(LatA$age) # used in comparisons
```

```
pars <- c(27.0,0.15,-2.0) # von Bertalanffy
bestvB <- nlm(f=ssq,funk=vB,observed=LatA$length,p=pars,
              ages=LatA$age,typsize=magnitude(pars))
outfit(bestvB,backtran=FALSE,title="vB"); cat("\n")
pars <- c(26.0,0.7,-0.5) # Gompertz
bestGz <- nlm(f=ssq,funk=Gz,observed=LatA$length,p=pars,
              ages=LatA$age,typsize=magnitude(pars))
outfit(bestGz,backtran=FALSE,title="Gz"); cat("\n")
pars <- c(26.2,1.0,1.0) # Michaelis-Menton - first start point
bestMM1 <- nlm(f=ssq,funk=mm,observed=LatA$length,p=pars,
              ages=LatA$age,typsize=magnitude(pars))
outfit(bestMM1,backtran=FALSE,title="MM"); cat("\n")
pars <- c(23.0,1.0,1.0) # Michaelis-Menton - second start point
bestMM2 <- nlm(f=ssq,funk=mm,observed=LatA$length,p=pars,
              ages=LatA$age,typsize=magnitude(pars))
outfit(bestMM2,backtran=FALSE,title="MM2"); cat("\n")
```

```
# nlm solution:  vB
# minimum      :  1361.421
# iterations   :  24
# code         :  2 >1 iterates in tolerance, probably solution
#           par      gradient
# 1 26.8353971 -1.133838e-04
# 2  0.1301587 -6.195068e-03
# 3 -3.5866989  8.326176e-05
#
# nlm solution:  Gz
# minimum      :  1374.36
# iterations   :  28
# code         :  1 gradient close to 0, probably solution
#           par      gradient
# 1 26.4444554  2.724757e-05
# 2  0.8682518 -6.455226e-04
# 3 -0.1635476 -2.046463e-03
#
# nlm solution:  MM
# minimum      :  1335.961
# iterations   :  12
# code         :  2 >1 iterates in tolerance, probably solution
#           par   gradient
# 1 20.6633224 -0.02622723
# 2  1.4035207 -0.37744316
# 3  0.9018319 -0.05039283
#
# nlm solution:  MM2
```

```
# minimum      :  1335.957
# iterations   :  25
# code         :  1 gradient close to 0, probably solution
#          par      gradient
# 1 20.7464274  8.465730e-06
# 2  1.4183164 -3.856475e-05
# 3  0.9029899 -1.297406e-04
```

These are numerical solutions and they do not guarantee a correct solution. Notice that the gradients in the first Michaelis-Menton solution (that started at 26.2, 1, 1) are relatively large, and yet its SSQ, at 1335.96, is very close to the second Michaelis-Menton model fit and smaller (better) than either the vB or Gz curves. However, the gradient values indicate that this model fit can, and should be, improved. If you alter the initial parameter estimate for parameter a (the first MM parameter) down to 23 instead of 26.2, as in the last model fit, we obtain slightly different parameter values, a slightly smaller SSQ, and much smaller gradients giving greater confidence that the result is a real minimum. As it happens, if one were to run cbind(mm(bestMM1$estimate,ages), mm(bestMM2$estimate,ages)), you can work out that the predicted values differ from −0.018 to 0.21% while, if you include the vB predictions, MM2 differs from the vB predictions from −6.15 to 9.88% (ignoring the very largest deviation of 40.6%). You could also try omitting the *typsize* argument from the estimate of the vB model, which will still give the optimal result but inspect the gradients to see why using *typsize* helps the optimization along. When setting up these examples, occasional runs of the Gz() model gave rise to a comment that the *steptol* might be too small and changing it from the default of 1e-06 to 1e-05 quickly fixed the problem. If it happens to you, add a statement ,steptol=1e-05 to the nlm() command and see if the diagnostics improve.

The obvious conclusion is that one should always read the diagnostic comments from nlm(), consider the gradients of the solution one obtains, and it is always a good idea to use multiple sets of initial parameter guesses to ensure one has a stable solution. Numerical solutions are based around software implementations and the rules used to decide when to stop iterating can sometimes be fooled by sub-optimal solutions. The aim is to find the global minimum not a local minimum. Any non-linear model can give rise to such sub-optimal solutions so automating such model fitting is not a simple task. Never assume that the first answer you get in such circumstances is definitely the optimum you are looking for, even if the plotted model fit looks acceptable.

Within a function call if you name each argument then the order does not strictly matter, but I find that consistent usage simplifies reading the code so would recommend using the standard order even when using explicit names. If we do not use explicit names the syntax for nlm() requires the function to be minimized (f) to be defined first. It also expects the f function, whatever it is, to use the initial parameter guess in the p argument, which, if unnamed must come second. If you type formals(nlm) or args(nlm) into the console one obtains the possible arguments that can be input to the function along with their defaults if they exist:

```
#The use of args() and formals()
args(nlm) # formals(nlm) uses more screen space. Try yourself.
```

```
# function (f, p, ..., hessian = FALSE, typsize = rep(1,
#    length(p)),fscale = 1, print.level = 0, ndigit = 12,
#    gradtol = 1e-06, stepmax = max(1000 *
#    sqrt(sum((p/typsize)^2)), 1000), steptol = 1e-06,
#    iterlim = 100, check.analyticals = TRUE)
```

As you can see the function to be minimized f (in this case ssq()) comes first, followed by the initial parameters, p, that must be the first argument required by whatever function is pointed to by f. Then comes the ellipsis (three dots) that generalizes the nlm() code for any function f, and then a collection of possible arguments all of which have default values. We altered *typsize* and *iterlim* (and *steptol* in Gz() sometimes); see the nlm() help for an explanation of each.

In R the ... effectively means whatever other inputs are required, such as the arguments for whatever function f is pointed at (in this case ssq). If you look at the *args* or code, or help for ?ssq, you will see that it requires the function, *funk*, that will be used to calculate the expected values relative to the next required input for ssq(), which is the vector of *observed* values (as in $O_i - E_i$). Notice there is no explicit mention of the arguments used by *funk*, which are assumed to be passed using the .... In each of our calls to ssq() we have filled in those arguments explicitly with, for example, nlm(f=ssq,funk=Gz, observed=LatA$length, p=pars, ages=LatA$age,). In that way all the requirements are filled and ssq() can do its work. If you were to accidentally omit, say, the ages=LatA$age, argument then, helpfully (in this instance), R will respond with something like *Error in funk(par, independent) : argument "ages" is missing, with no default* (I am sure you believe me but it does not hurt to try it for yourself!).

In terms of the growth curve model fitting, plotting the results provides a visual comparison that illustrates the difference between the three growth curves (Murrell, 2011).

```
#Female length-at-age + 3 growth fitted curves Fig. 4.3
predvB <- vB(bestvB$estimate,ages) #get optimumpredicted lengths
predGz <- Gz(bestGz$estimate,ages) # using the outputs
predmm <- mm(bestMM2$estimate,ages) #from the nlm analysis above
ymax <- getmax(LatA$length) #try ?getmax or getmax [no brackets]
xmax <- getmax(LatA$age)  #there is also a getmin, not used here
parset(font=7)   # or use parsyn() to prompt for the par syntax
```

```
plot(LatA$age,LatA$length,type="p",pch=16, col=rgb(1,0,0,1/5),
     cex=1.2,xlim=c(0,xmax),ylim=c(0,ymax),yaxs="i",xlab="Age",
     ylab="Length (cm)",panel.first=grid())
lines(ages,predvB,lwd=2,col=4)          # vB      col=4=blue
lines(ages,predGz,lwd=2,col=1,lty=2)    # Gompertz  1=black
lines(ages,predmm,lwd=2,col=3,lty=3)    # MM          3=green
#notice the legend function and its syntax.
legend("bottomright",cex=1.2,c("von Bertalanffy","Gompertz",
       "Michaelis-Menton"),col=c(4,1,3),lty=c(1,2,3),lwd=3,bty="n")
```

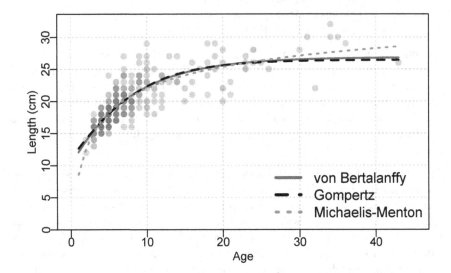

FIGURE 4.3 Female Length-at-Age data from 358 simulated female redfish with three optimally fitted growth curves drawn on top.

The rgb() function used in the plot implies that the intensity of colour represents the number of observations, with the most intense implying at least five observations. It is clear that with this data the vB() and Gz() curves are close to coincident over much of the extent of the observed data, while the mm() curve deviates from the other two but mainly at the extremes. The Michaelis-Menton curve is forced to pass through the origin, while the other two are not constrained in this way (even though the idea may appear to be closer to reality). One could include the length of post-larval fish to pull the Gompertz and von Bertalanffy curves downwards. But with living animals growth is complex. Many sharks and rays have live-births and do start post-development life at sizes significantly above zero. Always remember that these curves are empirical descriptions of the data and have limited abilities to reflect reality.

Most of the available data is between ages 3 and 12 (try `table(LatA$age)`), and then only single occurrences above age 24. Over the ages 3–24 the Gompertz and von Bertalanffy curves follow essentially the same trajectory and the Michaelis-Menton curve differs only slightly (you could try the following code to see the actual differences `cbind(ages, predvB, predGz, predmm)`. Outside of that age range bigger differences occur, although the lack of younger animals suggests the selectivity of the gear that caught the samples may underrepresent fish less than 3 years old. In terms of the relative quality of fit (the sum-of-squared residuals), the final Michaelis-Menton curve has the smallest `ssq()`, followed by the von Bertalanffy, followed by the Gompertz. But each provides a plausible mean description of the growth between the range of 3–24 years, where the data is most dense. When data is as sparse as it is in the older age classes there is also the question of whether or not this sample is representative of the population for those ages. Questioning one's data, the model used, and consequent interpretations, is an important aspect of producing useful models of natural processes.

4.4.5 Objective Model Selection

In the three growth models above the optimum model fit was defined as that which minimized the sum-of-squared residuals between the predicted and observed values. By that criterion the second Michaelis-Menton curve provided a 'better' fit than either von Bertalanffy or the Gompertz curves. But can we really say that the second Michaelis-Menton curve was 'better' fitting than the first? In terms of the gradients of the final solution things are clearly better with the second curve but strictly the criterion of fit was only the minimum SSQ and the difference was less than 0.01 units. Model selection is generally a trade-off between the number of parameters used and the quality of fit according to the criterion chosen. If we devise a model with more parameters this generally leads to greater flexibility and improved capacity to be closer to the observed data. In the extreme if we had as many parameters as we had observations we could have a perfect model fit, but, of course, would have learned nothing about the system we are modelling. With 358 parameters for the *LatA* data-set that would clearly be a case of over-parameterization, but what if we had only increased the number of parameters to, say, 10? No doubt the curve would have been oddly shaped but would likely have a lower SSQ. Burnham and Anderson (2002) provide a detailed discussion of the trade-off between number of parameters and quality of model fit to data. In the 1970s there was a move to using Information Theory to develop a means of quantifying the trade-off between model parameters and quality of model fit. Akaike (1974) described his Akaike's Information Criterion (AIC), which was based on maximum likelihood and information theoretic principles (more of which later) but fortunately, Burnham and Anderson (2002) provide an alternative when using the minimum sum-of-squared residuals, which is a variant of one included in Atkinson (1980):

$$AIC = N\left(log\left(\hat{\sigma}^2\right)\right) + 2p \tag{4.8}$$

where N is the number of observations, p is the "number of independently adjusted parameters within the model" (Akaike, 1974, p716), and $\hat{\sigma}^2$ is the maximum likelihood estimate of the variance, which just means the sum-of-squared residuals is divided by N rather than $N - 1$:

$$\hat{\sigma}^2 = \frac{\Sigma\varepsilon^2}{N} = \frac{ssq}{N} \tag{4.9}$$

Even with the AIC it is difficult to determine, when using least-squares, whether differences can be argued to be statistically significantly different. There are ways related to the analysis of variance, but such questions are able to be answered more robustly when using maximum likelihood, so we will address that in a later section.

If one wants to obtain biologically plausible or defensible interpretations when fitting a model, then model selection cannot be solely dependent upon quality of statistical fit. Instead, it should reflect the theoretical expectations (for example, should average growth in a population involve a smooth increase in size through time, etc.). Such considerations, other than statistical fit to data, do not appear to gain sufficient attention, but only become important when biologically implausible model outcomes arise or implausible model structures are proposed. It obviously helps to have an understanding of the biological expectations of the processes being modelled.

4.4.6 The Influence of Residual Error Choice on Model Fit

In the growth model example, we used Normal random deviates, but we can ask the question of whether we would have obtained the same answer had we used, for example, Log-Normal deviates? All we need to do in that case would be to log-transform the observed and predicted values before calculating the sum-of-squared residuals (see below with reference to Log-Normal residuals.

$$ssq = \sum_{i=1}^{n}\left(log(O_i) - log(\hat{E}_i)\right)^2 \tag{4.10}$$

Here we continue to use the *backtran=FALSE* option within outfit() because we are log-transforming the data not the parameters so no back-transformation is required.

```
# von Bertalanffy
pars <- c(27.25,0.15,-3.0)
bestvBN <- nlm(f=ssq,funk=vB,observed=LatA$length,p=pars,
```

```
              ages=LatA$age,typsize=magnitude(pars),iterlim=1000)
outfit(bestvBN,backtran=FALSE,title="Normal errors"); cat("\n")
 # modify ssq to account for Log-normal errors in ssqL
ssqL <- function(funk,observed,...) {
  predval <- funk(...)
  return(sum((log(observed) - log(predval))^2,na.rm=TRUE))
} # end of ssqL
bestvBLN <- nlm(f=ssqL,funk=vB,observed=LatA$length,p=pars,
              ages=LatA$age,typsize=magnitude(pars),iterlim=1000)
outfit(bestvBLN,backtran=FALSE,title="Log-Normal errors")
```

```
# nlm solution:  Normal errors
# minimum       :  1361.421
# iterations   :  22
# code          :  1 gradient close to 0, probably solution
#          par        gradient
# 1 26.8353990 -3.649702e-07
# 2  0.1301587 -1.576574e-05
# 3 -3.5867005  3.198205e-07
#
# nlm solution:  Log-Normal errors
# minimum       :  3.153052
# iterations   :  25
# code          :  1 gradient close to 0, probably solution
#          par        gradient
# 1 26.4409587  8.906655e-08
# 2  0.1375784  7.537147e-06
# 3 -3.2946087 -1.124171e-07
```

In this case the curves produced by using Normal and Log-Normal residual errors barely differ, **Figure** 4.4, even though their parameters differ (use ylim=c(10,ymax) to make the differences clearer). More than differing visually, the different model fits are not even comparable. If we compare their respective sum-of-squared residuals one has 1361.0 and the other only 3.153. This is not surprising when we consider the effect of the log-transformations within the calculation of the sum-of-squared. But what this means is we cannot look at the tabulated outputs alone and decide which version fits the data better than the other. They are strictly incommensurate even though they are using exactly the same model. The use of the different residual error structure needs to be defended in ways other than considering the relative model fit. This example emphasizes that while the choice of model is obviously important, the choice of residual error structure is part of the model structure and equally important.

```
 # Now plot the resultibng two curves and the data Fig 4.4
predvBN <- vB(bestvBN$estimate,ages)
```

```
predvBLN <- vB(bestvBLN$estimate,ages)
ymax <- getmax(LatA$length)
xmax <- getmax(LatA$age)
parset()
plot(LatA$age,LatA$length,type="p",pch=16, col=rgb(1,0,0,1/5),
     cex=1.2,xlim=c(0,xmax),ylim=c(0,ymax),yaxs="i",xlab="Age",
     ylab="Length (cm)",panel.first=grid())
lines(ages,predvBN,lwd=2,col=4,lty=2)    # add Normal dashed
lines(ages,predvBLN,lwd=2,col=1)         # add Log-Normal solid
legend("bottomright",c("Normal Errors","Log-Normal Errors"),
       col=c(4,1),lty=c(2,1),lwd=3,bty="n",cex=1.2)
```

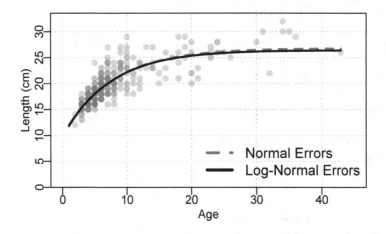

FIGURE 4.4 Female Length-at-Age data from 358 female redfish, *Centroberyx affinis*, with two von Bertalanffy growth curves fitted using Normal and Log-Normal residuals.

4.4.7 Remarks on Initial Model Fitting

The comparison of curves in the example above is itself interesting, however, what we have also done is illustrate the syntax of nlm() and how one might go about fitting models to data. The ability to pass functions as arguments into another function (as here we passed ssq as *f* into nlm(), and vB, Gz, and mm into ssq() as *funk*) is one of the strengths but also complexities of R. In addition, the use of the ... to pass extra arguments without having to name them explicitly beforehand helps produce re-usable code. It simplifies the re-use of functions like ssq(), where all we need do is change the input function, with potentially completely different input requirements, to get an entirely different answer from essentially the same code. The very best way of becoming familiar with these methods is to employ them with your own data-sets. Plot your data and any model fits because if the model fit appears unusual then it quite likely is, and deserves a second and third look.

One can go a long way using sum-of-squares but the assumptions requiring normally distributed residuals around the expected values, and a constant variance are constraining when dealing with the diversity of the real world. In order to use probability density distributions other than Normal and to use non-constant variances, one should turn to using maximum likelihood.

4.5 Maximum Likelihood

The use of likelihoods is relatively straightforward in R as there are built-in functions for very many Probability Density Functions (PDFs) as well as an array of packages that define other PDFs. To repeat, the aim with maximum likelihood methods is to use software to search for the model parameter set that maximizes the total likelihood of the observations. To use this criterion of optimal model fit requires the model to be defined so that it specifies probabilities or likelihoods for each of the observations (the available data) as a function of the parameter values and other variables within the model (**Equ**(4.2) and **Equ**(4.11)). Importantly, such a specification includes an estimate of the variability or spread of the selected PDF (the σ in the Normal distribution; which is only a by-product of the least-squares approach). A major advantage of using maximum likelihood is that the residual structure or the expected distribution of observations about the expected central tendency of the data need not be normally distributed. If the probability density function (PDF) can be defined, it can be used in a maximum likelihood framework; see Forbes *et al*, (2011) for the definitions of many useful probability density functions.

4.5.1 Introductory Examples

We will use the well-known Normal distribution to illustrate the methods but then extend the approach to include an array of alternative PDFs. For the purposes of model fitting to data the primary interest for each PDF is in the definition of the probability density or likelihood for individual observations. For a Normal distribution with a mean expected value of $\bar{x}$ or μ, the probability density or likelihood of a given single value x_i is defined as:

$$L\left(x_i|\mu_i, \sigma\right) = \frac{1}{\sigma\sqrt{2\pi}}e^{\left(\frac{-(x_i-\mu_i)^2}{2\sigma^2}\right)} \tag{4.11}$$

where σ is the standard deviation associated with μ_i. This identifies an immediate difference between least-squared methods and maximum likelihood methods; in the latter, one needs a full definition of the PDF, which in the case of the Normal distribution includes an explicit estimate of the standard deviation of the residuals around the mean estimates μ. Such an estimate is not required with least-squares, although it is easily derived from the SSQ value.

As an example, we can generate a sample of observations from a Normal distribution (see ?rnorm) and then calculate that sample's mean and stdev and compare how likely the given sample values are for these parameter estimates relative to how likely they are for the original mean and stdev used in the *rnorm* function, **Table** 4.1:

```
# Illustrate Normal random likelihoods. See Table 4.1
set.seed(12345)          # make the use of random numbers repeatable
x <- rnorm(10,mean=5.0,sd=1.0)          # pseudo-randomly generate 10
avx <- mean(x)                          # normally distributed values
sdx <- sd(x)          # estimate the mean and stdev of the sample
L1 <- dnorm(x,mean=5.0,sd=1.0)   # obtain likelihoods, L1, L2 for
L2 <- dnorm(x,mean=avx,sd=sdx)    # each data point for both sets
result <- cbind(x,L1,L2,"L2gtL1"=(L2>L1))        # which is larger?
result <- rbind(result,c(NA,prod(L1),prod(L2),1)) # result+totals
rownames(result) <- c(1:10,"product")
colnames(result) <- c("x","original","estimated","est > orig")
```

TABLE 4.1: An illustration of using Normal random values and related Normal likelihoods. The estimated column has the larger total likelihood. 1=TRUE, 0=FALSE.

	x	original	estimated	est > orig
1	5.5855	0.33609530	0.33201297	0
2	5.7095	0.31017782	0.28688183	0
3	4.8907	0.39656626	0.49010171	1
4	4.5465	0.35995784	0.45369729	1
5	5.6059	0.33204382	0.32465621	0
6	3.1820	0.07642691	0.05743702	0
7	5.6301	0.32711267	0.31586172	0
8	4.7238	0.38401358	0.48276941	1
9	4.7158	0.38315644	0.48191389	1
10	4.0807	0.26144927	0.30735328	1
product		0.00000475	0.00000892	1

The bottom line of **Table** 4.1 contains the product (obtained using the R function prod()) of each of the columns of likelihoods. Not surprisingly the maximum likelihood is obtained when we use the sample estimates of the mean and stdev (estimated, L2) rather than the original values of *mean=5* and *sd=1.0* (original, L1); that is $8.9201095 \times 10^{-6} > 4.7521457 \times 10^{-6}$. I can be sure of these values in this example because the set.seed() R function was used at the start of the code to begin the pseudo-random number generators at a specific location. If you commonly use set.seed() do not repeatedly use the same old sequences, such as 12345, as you risk undermining the idea of the pseudo-random numbers being a good approximation to a random number sequence, perhaps use getseed() to provide a suitable seed number instead.

So rnorm() provides pseudo-random numbers from the distribution defined by the mean and stdev, and dnorm() provides the probability density or likelihood of an observation given its mean and stdev (the equivalent of **Equ**(4.11)). The cumulative probability density function (cdf) is provided by the function pnorm(), and the quantiles are identified by qnorm(). The mean value will naturally have the largest likelihood. Note also that the Normal curve is symmetrical around the mean.

```
# some examples of pnorm, dnorm, and qnorm, all mean = 0
cat("x = 0.0           Likelihood =",dnorm(0.0,mean=0,sd=1),"\n")
cat("x = 1.95996395 Likelihood =",dnorm(1.95996395,mean=0,sd=1),"\n")
cat("x =-1.95996395 Likelihood =",dnorm(-1.95996395,mean=0,sd=1),"\n")
# 0.5 = half cumulative distribution
cat("x = 0.0           cdf = ",pnorm(0,mean=0,sd=1),"\n")
cat("x = 0.6744899   cdf = ",pnorm(0.6744899,mean=0,sd=1),"\n")
cat("x = 0.75          Quantile =",qnorm(0.75),"\n") # reverse pnorm
cat("x = 1.95996395 cdf = ",pnorm(1.95996395,mean=0,sd=1),"\n")
cat("x =-1.95996395 cdf = ",pnorm(-1.95996395,mean=0,sd=1),"\n")
cat("x = 0.975        Quantile =",qnorm(0.975),"\n") # expect ~1.96
 # try x <- seq(-5,5,0.2); round(dnorm(x,mean=0.0,sd=1.0),5)
```

```
# x = 0.0           Likelihood = 0.3989423
# x = 1.95996395 Likelihood = 0.05844507
# x =-1.95996395 Likelihood = 0.05844507
# x = 0.0           cdf =   0.5
# x = 0.6744899   cdf =   0.75
# x = 0.75          Quantile = 0.6744898
# x = 1.95996395 cdf =   0.975
# x =-1.95996395 cdf =   0.025
# x = 0.975        Quantile = 1.959964
```

As we can see, individual likelihoods can be relatively large numbers, however, when multiplied together they can quickly lead to relatively small numbers. Errors can arise when the number of observations increases. Even with only ten numbers when we multiply all the individual likelihoods together (using prod()) the outcome quickly becomes very small indeed. With another similar ten numbers to the ten used in **Table** 4.1, the overall likelihood could easily be down to 1e-11 or 1e-12. As the number of observations increases the chances of a rounding error (even on a 64-bit computer) begin to increase. Rather than multiply many small numbers to get an extremely small number the standard solution to multiplying these small numbers together is to natural-log-transform the likelihoods and then add them together. Maximizing the sum of the log-transformed likelihoods obtains the optimum parameters exactly as does maximizing the product of the individual likelihoods. In addition, many optimizers, in software, appear to have been designed to be most efficient at minimizing a function. The simple solution is instead of maximizing the product of the individual likelihoods we minimize the sum of the negative log-likelihoods (*-veLL* or negLL()).

4.6   Likelihoods from the Normal Distribution

Likelihoods can appear to be rather strange beasts. When from continuous PDFs they are not strictly probabilities although they share many of their properties (Edwards, 1972). Strictly they relate to the probability density at a particular point under a probability density function. The area under the full curve must, by the definition of probabilities, sum to 1.0, but the area under any single point of a continuous PDF becomes infinitesimally small. Normal likelihoods are defined using **Equ**(4.11), whereas the cumulative density function is:

$$cdf = 1 = \int_{x=-\infty}^{\infty} \frac{1}{\sigma\sqrt{2\pi}} e^{\left(\frac{-(x-\mu)^2}{2\sigma^2}\right)} \tag{4.12}$$

We can use dnorm() and pnorm() to calculate both the likelihoods and the *cdf* (**Figure** 4.5).

```
# Density plot and cumulative distribution for Normal    Fig. 4.5
x <- seq(-5,5,0.1)  # a sequence of values around a mean of 0.0
NL <- dnorm(x,mean=0,sd=1.0)    # normal likelihoods for each X
CD <- pnorm(x,mean=0,sd=1.0)    # cumulative density vs X
plot1(x,CD,xlab="x = StDev from Mean",ylab="Likelihood and CDF")
lines(x,NL,lwd=3,col=2,lty=3) # dashed line as these are points
abline(h=0.5,col=4,lwd=1)
```

That all sounds fine, but what does it mean to identify a specific value under such a curve for a specific value of the x variable? In **Figure** 4.5 we used a dotted line to suggest that the likelihoods in the plot are local estimates and do not make up a continuous line. Each represents the likelihood at exactly the given x value. We have seen above that the probability density at a value of 0.0 for a distribution with mean = 0.0 and stdev = 1.0 would be 0.3989423. Let us briefly examine this potential confusion between likelihood and probability. If we consider a small portion of the probability density function with a mean

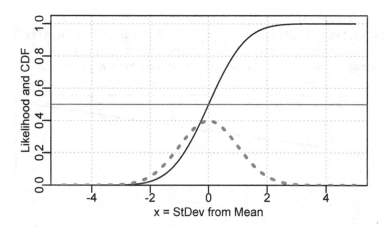

FIGURE 4.5 A dotted red curve depicting the expected Normal likelihoods for a mean = 0 and sd = 1.0, along with the cumulative density of the same Normal likelihoods as a black line. The blue line identifies a cumulative probability of 0.5.

= 5.0, and st.dev = 1.0 between the values of x = 3.4 to 3.6 we might see something like **Figure** 4.6:

```
#function facilitates exploring different polygons Fig. 4.6
plotpoly <- function(mid,delta,av=5.0,stdev=1.0) {
  neg <- mid-delta;   pos <- mid+delta
  pdval <- dnorm(c(mid,neg,pos),mean=av,sd=stdev)
  polygon(c(neg,neg,mid,neg),c(pdval[2],pdval[1],pdval[1],
                        pdval[2]),col=rgb(0.25,0.25,0.25,0.5))
  polygon(c(pos,pos,mid,pos),c(pdval[1],pdval[3],pdval[1],
                        pdval[1]),col=rgb(0,1,0,0.5))
  polygon(c(mid,neg,neg,mid,mid),
       c(0,0,pdval[1],pdval[1],0),lwd=2,lty=1,border=2)
  polygon(c(mid,pos,pos,mid,mid),
       c(0,0,pdval[1],pdval[1],0),lwd=2,lty=1,border=2)
  text(3.395,0.025,paste0("~",round((2*(delta*pdval[1])),7)),
       cex=1.1,pos=4)
  return(2*(delta*pdval[1])) # approx probability, see below
} # end of plotpoly, a temporary function to enable flexibility
 #This code can be re-run with different values for delta
x <- seq(3.4,3.6,0.05) # where under the normal curve to examine
pd <- dnorm(x,mean=5.0,sd=1.0) #prob density for each X value
mid <- mean(x)
delta <- 0.05   # how wide either side of the sample mean to go?
parset()        # a pre-defined MQMF base graphics set-up for par
```

```
ymax <- getmax(pd)      # find maximum y value for the plot
plot(x,pd,type="l",xlab="Variable x",ylab="Probability Density",
      ylim=c(0,ymax),yaxs="i",lwd=2,panel.first=grid())
approxprob <- plotpoly(mid,delta)   #use function defined above
```

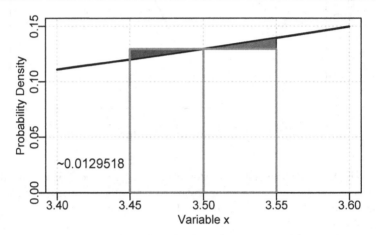

FIGURE 4.6 Probability densities for a normally distributed variate X, with a mean of 5.0 and a standard deviation of 1.0. The PDF value at x = 3.5 is 0.129518, so the area of the boxes outlined in red is 0.0129518, which approximates the total probability of a value between 3.45–3.55, which would really be the area under the curve.

The area under the complete PDF sums to 1.0, so the probability of obtaining a value between, say, 3.45 and 3.55, in **Figure** 4.6, is the sum of the areas of the oblongs minus the left triangle plus the right triangle. The triangles are almost symmetrical and so approximately cancel each other out, so an approximate solution is simply 2.0 times the area of one of the oblongs. With a delta (width of oblong on x-axis) of 0.05 that probability = 0.0129518. If you change the delta to become 0.01 then the approximate probability = 0.0025904, and as the delta value decreases so does the total probability although the probability density at 3.5 remains the same at 0.1295176. Clearly likelihoods are not identical to probabilities in continuous PDFs (see Edwards, 1972). The best estimate of the probability of the area under the curve is obtained with pnorm(3.55,5,1) - pnorm(3.45,5,1), which = 0.0129585.

4.6.1 Equivalence with Sum-of-Squares

When using Normal likelihoods to fit a model to data what we actually do is set things up so as to minimize the sum of the negative log-likelihoods for each available observation. Fortunately, we can use dnorm() to estimate those likelihoods. In fact, if we fit a model using maximum likelihood methods with

Normally distributed residuals, or log-transformed Log-Normal distributed data, then the parameter estimates obtained are the same as those obtained using the least-squared approach (see the derivation and form of **Equ**(4.19), below). Fitting a model would require generating a set of predicted values $\hat{x}$ (x-hat) as a function of some other independent variable(s) $\theta(x)$, where θ is the list of parameters used in the functional relationship. The log-likelihood for n observations would be defined as:

$$LL(x|\theta) = \sum_{i=1}^{n} log\left(\frac{1}{\hat{\sigma}\sqrt{2\pi}}e^{\left(\frac{-(x_i-\hat{x}_i)^2}{2\hat{\sigma}^2}\right)}\right) \tag{4.13}$$

$LL(x|\theta)$ is read as the log-likelihood of x, the observations, given θ the parameters (μ and $\hat{\sigma}$); the symbol $|$ is read as "given". This superficially complex equation can be greatly simplified. First, we could move the constant before the exponential term outside of the summation term by multiplying it by n, and the natural log of the remaining exponential term back-transforms the exponential:

$$LL(x|\theta) = nlog\left(\frac{1}{\hat{\sigma}\sqrt{2\pi}}\right) + \frac{1}{2\hat{\sigma}^2}\sum_{i=1}^{n}\left(-(x_i-\hat{x}_i)^2\right) \tag{4.14}$$

The value of $\hat{\sigma}^2$ is the maximum likelihood estimate of the variance of the data (note the division by n rather than $n-1$:

$$\hat{\sigma}^2 = \frac{\sum_{i=1}^{n}(x_i-\hat{y}_i)^2}{n} \tag{4.15}$$

If we replace the use of $\hat{\sigma}^2$ in **Equ**(4.14)) with **Equ**(4.15), the $(x_i-\hat{y}_i)^2$ cancels out leaving $-n/2$:

$$LL(x|\theta) = nlog\left(\left(\hat{\sigma}\sqrt{2\pi}\right)^{-1}\right) - \frac{n}{2} \tag{4.16}$$

simplifying the square root term means bringing the -1 out of the log term, which changes the n to $-n$, and we can change the square root to an exponent of $1/2$ and add $log(\hat{\sigma})$ to the log of the π term:

$$LL(x|\theta) = -n\left(log\left((2\pi)^{\frac{1}{2}}\right) + log(\hat{\sigma})\right) - \frac{n}{2} \tag{4.17}$$

then move the power of $1/2$ to outside the first *log* term:

$$LL(x|\theta) = -\frac{n}{2}\left(log\left(2\pi\right) + 2log\left(\hat{\sigma}\right)\right) - \frac{n}{2} \tag{4.18}$$

then simplify the $n/2$ and multiply throughout by -1 to convert to a negative log-likelihood to give the final simplification of the negative log-likelihood for normally distributed values:

$$-LL(x|\theta) = \frac{n}{2}\left(log\left(2\pi\right) + 2log\left(\hat{\sigma}\right) + 1\right) \tag{4.19}$$

The only non-constant part of this is the value of $\hat{\sigma}$, which is the square root of the sum of squared residuals divided by n, so now it should be clear why the parameters obtained when using maximum likelihood, if using Normal random errors, are the same as derived from a least-squares approach.

4.6.2   Fitting a Model to Data Using Normal Likelihoods

We can repeat the example using the simulated female Redfish data in the data-set *LatA*, **Figure** 4.7, which we used to illustrate the use of sum-of-squared residuals. Ideally, we should obtain the same answer but with an estimate of σ as well. The **MQMF** function plot1() is just a quick way of plotting a single graph (either *type="l"* or *type="p"*; see ?plot1), without too much white space. Edit plot1() if you like more white space than I do!

```
#plot of length-at-age data   Fig. 4.7
data(LatA) # load the redfish data-set into memory and plot it
ages <- LatA$age;  lengths <- LatA$length
plot1(ages,lengths,xlab="Age",ylab="Length",type="p",cex=0.8,
     pch=16,col=rgb(1,0,0,1/5))
```

Now we can use the **MQMF** function negNLL() (as in negative Normal log-likelihoods) to determine the sum of the negative log-likelihoods using Normal random errors (negLL() does the same but for log-normally distributed data). If you look at the code for negNLL() you will see that, just like ssq(), it is passed a function as an argument, which is then used to calculate the predicted mean values for each input age (in this case lengths-at-age using the **MQMF** function vB()), and then uses dnorm() to calculate the sum of the -veLL using the predicted values as the mean values and the observations of the length-at-age values in the data. The *ages* data is passed through the ellipsis (...) without being explicitly declared as an argument in negNLL(). The function is very similar in structure to ssq() in that it has identical input requirements, however, *pars* is passed explicitly rather than inside ..., because the last value of *pars* must be the *stdev* of the residuals, which is used inside negNLL() rather than just inside *funk*. The use of likelihoods means there is a

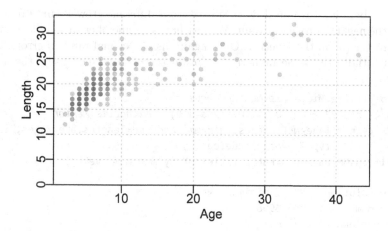

FIGURE 4.7 The length-at-age data contained within the LatA data-set for female Redfish *Centroberyx affinis*. Full colour means >= 5 points.

need to include an estimate of the standard deviation at the end of the vector of parameters. negNLL() thus operates in a manner very similar to ssq() in that it is a wrapper for calling the function that generates the predicted values and then uses dnorm() to return a single number every time it is called. Thus, nlm() minimizes negNLL(), which in turn calls vB().

```
# Fit the vB growth curve using maximum likelihood
pars <- c(Linf=27.0,K=0.15,t0=-3.0,sigma=2.5) # starting values
# note, estimate for sigma is required for maximum likelihood
ansvB <- nlm(f=negNLL,p=pars,funk=vB,observed=lengths,ages=ages,
             typsize=magnitude(pars))
outfit(ansvB,backtran=FALSE,title="vB by minimum -veLL")
```

```
# nlm solution:  vB by minimum -veLL
# minimum     :  747.0795
# iterations  :  26
# code        :  1 gradient close to 0, probably solution
#           par       gradient
# 1 26.8354474 4.490629e-07
# 2  0.1301554 1.659593e-05
# 3 -3.5868868 3.169513e-07
# 4  1.9500896 8.278354e-06
```

If you look back at the ssq() example for the von Bertalanffy curve, you will see we obtained an SSQ value of 1361.421 from a sample of 358 fish (try nrow(LatA)). Thus the estimate of σ from the ssq() approach was $\sqrt{(1361.421/358)} = 1.95009$, which, as expected, is essentially identical to the maximum likelihood estimate.

Just as before all we need do is substitute a different growth curve function into the negNLL() to get a result. We just have to remember to include a fourth parameter (σ) in the p vector. Once again, using Normal random errors leads to essentially the same numerical solution as that obtained using the ssq() approach.

```
#Now fit the Michaelis-Menton curve
pars <- c(a=23.0,b=1.0,c=1.0,sigma=3.0) # Michaelis-Menton
ansMM <- nlm(f=negNLL,p=pars,funk=mm,observed=lengths,ages=ages,
             typsize=magnitude(pars))
outfit(ansMM,backtran=FALSE,title="MM by minimum -veLL")
```

```
# nlm solution:  MM by minimum -veLL
# minimum      :  743.6998
# iterations   :  34
# code         :  1 gradient close to 0, probably solution
#           par       gradient
# 1 20.7464280  -6.195465e-06
# 2  1.4183165   1.601881e-05
# 3  0.9029899   2.461405e-04
# 4  1.9317669  -3.359816e-06
```

Once again the gradients of the solution are small, which increases confidence that the solution is not just a local minimum, so we should plot out the solution to see its relative fit to the data.

By plotting the fitted curves on top of the data points the data do not obscure the lines. The actual predictions that can now be produced from this analysis can also be tabulated along with the residual values. By including the individual squared residuals, it becomes more clear which points (see record 3) could have the most influence.

```
#plot optimum solutions for vB and mm. Fig. 4.8
Age <- 1:max(ages) # used in comparisons
predvB <- vB(ansvB$estimate,Age) #optimum solution
predMM <- mm(ansMM$estimate,Age) #optimum solution
parset()                      # plot the deata points first
plot(ages,lengths,xlab="Age",ylab="Length",type="p",pch=16,
     ylim=c(10,33),panel.first=grid(),col=rgb(1,0,0,1/3))
lines(Age,predvB,lwd=2,col=4)     # then add the growth curves
lines(Age,predMM,lwd=2,col=1,lty=2)
legend("bottomright",c("von Bertalanffy","Michaelis-Menton"),
       col=c(4,1),lwd=3,bty="n",cex=1.2,lty=c(1,2))
```

Generally, one would generate residual plots so as to check for patterns in the residuals, **Figure** 4.9.

```
# residual plot for vB curve   Fig. 4.9
predvB <- vB(ansvB$estimate,ages) # predicted values for age data
```

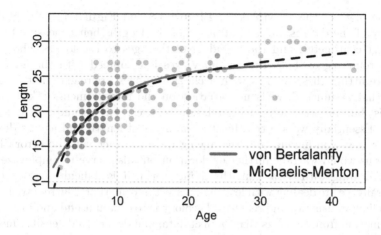

FIGURE 4.8 Optimum von Bertalanffy and Michaelis-Menton growth curves fitted to the female *LatA* Redfish data. Note the two curves are effectively coincident where the observations are most concentrated. Note the y-axis starts at 10.

```
resids <- lengths - predvB              # calculate vB residuals
plot1(ages,resids,type="p",col=rgb(1,0,0,1/3),xlim=c(0,43),
      pch=16,xlab="Ages Years",ylab="Residuals")
abline(h=0.0,col=1,lty=2)    # emphasize the zero line
```

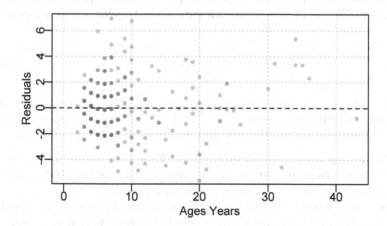

FIGURE 4.9 The residual values for von Bertalanffy curve fitted to the female *LatA* data. There is a clear pattern between the ages of 3–10, which reflects the nature of residuals when the mean expected length for a given age is constant and compared to these rounded length measurements.

In the plots of the growth data (**Figure** 4.8 and **Figure** 4.9) the grid-like nature of the data is a clear indication of the lengths being measured to the nearest centimeter, and the rounding of the ages to the lowest whole year. Such rounding on both the x- and y-axes is combined with the issue that we are fitting these models with a classical y-on-x approach (Ricker, 1973), and that assumes there is no variation on the measurements taken on the x-axis, but unfortunately, with relation to ages, this assumption is simply false. Essentially we are treating the variables of length and age as discrete rather than continuous, and the ageing data as exact with no error. These patterns would warrant further exploration but also serve to emphasize that when dealing with data from the living world, it is difficult to collect and generally we are dealing with less than perfect information. The real trick when modelling in fisheries and real world ecology is to obtain useful and interesting information from that less than perfect data, and do so in a defensible manner.

4.7　Log-Normal Likelihoods

The Normal distribution, with its additive residual errors that can be above or below an expected mean value, is generally well known and its properties form the basis for many intuitions concerning nature. However, it is the case that many variables found in exploited populations (CPUE, catch, effort, ...) exhibit highly skewed distributions where the central tendency does not lie in the center of a distribution (as it does in the Normal distribution). A very common PDF used to describe such data is the Log-Normal distribution :

$$L\left(x_i|m_i,\sigma\right) = \frac{1}{x_i\sigma\sqrt{2\pi}}e^{\left(\frac{-(log(x_i)-log(m_i))^2}{2\sigma^2}\right)} \tag{4.20}$$

where $L(x_i|m,\sigma)$ is the likelihood of data point x_i given the median m_i of the variable for instance i, where $m = e^\mu$ (and $\mu = log(m)$). μ is estimated as the mean of $log(x)$, and σ is the standard deviation of $log(x)$.

While the Log-Normal likelihood equation, **Equ**(4.20), has a visual similarity to that of the Normal distribution, it differs in that the Log-Normal distribution has multiplicative residual errors rather than additive. In detail, each likelihood is multiplied by the reciprocal of the observation and the observed and expected values are log-transformed. The log-transformation of the observed data and expected values imply that instead of using $x_i - m_i$ to calculate residuals we are using the equivalent of x_i/m_i. This means that the observations are divided by the expected values rather than subtracting the expected values from the observed. All such residuals will be positive and will vary about the value of 1.0. A residual of 1.0 would imply the data point exactly matches the expected median value for that data point.

4.7.1 Simplification of Log-Normal Likelihoods

As with Normal likelihoods, Log-Normal likelihoods can also be simplified to facilitate subsequent calculations:

$$-LL(x|\theta) = \frac{n}{2}\left(log\left(2\pi\right) + 2log\left(\hat{\sigma}\right) + 1\right) + \sum_{i=1}^{n} log\left(x_i\right) \qquad (4.21)$$

where the maximum likelihood estimate of $\hat{\sigma}^2$ is:

$$\hat{\sigma}^2 = \sum_{i=1}^{n} \frac{\left(log\left(x_i\right) - log\left(\hat{x}_i\right)\right)^2}{n} \qquad (4.22)$$

Once again note that the maximum likelihood estimate of a variance uses n rather than $n-1$. The $\sum log(x)$ term at the end of **Equ**(4.21) is a constant and is often omitted from calculations when fitting a model. As noted above **Equ**(4.21) appears identical to that for the Normal distribution (**Equ**(4.13)) assuming we omit the $\sum log(x)$ term. However, the σ now requires that the observations and predicted values are log-transformed, **Equ**(4.22). Hence we can use functions relating to Normal likelihoods (e.g., negNLL()) to fit models using Log-Normal residuals as long as we log-transform the data and predictions before the analysis. Generally, however, we will use negLL(), which expects log-transformed observations and a function for generating the log of predicted values (see ?negLL).

4.7.2 Log-Normal Properties

With the Normal distribution we know that the expected mean, median, and mode of the distribution are all identical, but this is not the case with the Log-Normal distribution. Given a set of values from a continuous variable x, the median is estimated as:

$$\text{median} = m = e^{\mu} \qquad (4.23)$$

where μ is the mean of $log(x)$. The mode of the Log-Normal distribution is defined as:

$$\text{mode} = \frac{m}{e^{\sigma^2}} = e^{(\mu - \sigma^2)} \qquad (4.24)$$

where σ is the standard deviation of $log(x)$. Finally, the mean or Expectation of the Log-Normal distribution is defined as:

$$\bar{x} = me^{(\sigma^2/2)} = e^{(\mu+\sigma^2/2)} \tag{4.25}$$

These equations (**Equs**(4.23) to (4.25)) imply that Log-Normal distributions are always skewed to the right (have a long tail to the right of the mode). In addition, in contrast to the Normal distribution, the Log-Normal distribution is only defined for positive values of x; see **Figure** 4.10).

```
 # meanlog and sdlog affects on mode and spread of lognormal Fig. 4.10
x <- seq(0.05,5.0,0.01)  # values must be greater than 0.0
y <- dlnorm(x,meanlog=0,sdlog=1.2,log=FALSE) #dlnorm=Likelihoods
y2 <- dlnorm(x,meanlog=0,sdlog=1.0,log=FALSE)#from log-normal
y3 <- dlnorm(x,meanlog=0,sdlog=0.6,log=FALSE)#distribution
y4 <- dlnorm(x,0.75,0.6)          #log=TRUE = log-likelihoods
parset(plots=c(1,2)) #MQMF shortcut plot formatting function
plot(x,y3,type="l",lwd=2,panel.first=grid(),
     ylab="Log-Normal Likelihood")
lines(x,y,lwd=2,col=2,lty=2)
lines(x,y2,lwd=2,col=3,lty=3)
lines(x,y4,lwd=2,col=4,lty=4)
legend("topright",c("meanlog sdlog","    0.0      0.6",
"    0.0       1.0","    0.0       1.2","    0.75     0.6"),
      col=c(0,1,3,2,4),lwd=3,bty="n",cex=1.0,lty=c(0,1,3,2,4))
plot(log(x),y3,type="l",lwd=2,panel.first=grid(),ylab="")
lines(log(x),y,lwd=2,col=2,lty=2)
lines(log(x),y2,lwd=2,col=3,lty=3)
lines(log(x),y4,lwd=2,col=4,lty=4)
```

In a similar manner it is possible to generate random numbers from a Log-Normal distribution and, just as before, a log transformation should generate a Normal distribution, as in **Figure** 4.11).

```
set.seed(12354) # plot random log-normal numbers as Fig. 4.11
meanL <- 0.7;   sdL <- 0.5  # generate 5000 random log-normal
x <- rlnorm(5000,meanlog = meanL,sdlog = sdL) # values
parset(plots=c(1,2)) # simplifies the plots par() definition
hist(x[x < 8.0],breaks=seq(0,8,0.25),col=0,main="")
meanx <- mean(log(x)); sdx <- sd(log(x))
outstat <- c(exp(meanx-(sdx^2)),exp(meanx),exp(meanx+(sdx^2)/2))
abline(v=outstat,col=c(4,1,2),lwd=3,lty=c(1,2,3))
legend("topright",c("mode","median","bias-correct"),
       col=c(4,1,2),lwd=3,bty="n",cex=1.2,lty=c(1,2,3))
outh <- hist(log(x),breaks=30,col=0,main="")   # approxnormal
```

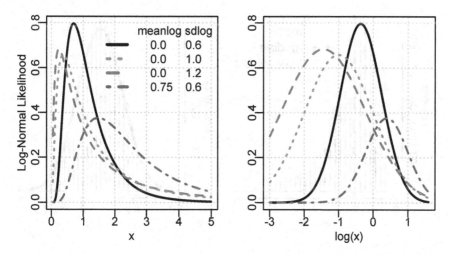

FIGURE 4.10 Two plots illustrating the Log-Normal probability density function. Left is a group of likelihood distributions for different parameter sets, while right is the log-transformed versions of these first four Log-Normal distributions.

```
hans <- addnorm(outh,log(x)) #MQMF function; try ?addnorm
lines(hans$x,hans$y,lwd=3,col=1) # type addnorm into the console
```

We can examine the expected statistics from the input parameters, and these can be compared with the parameter estimates made in the variable *outstat*.

```
#examine log-normal propoerties. It is a bad idea to reuse
set.seed(12345) #'random' seeds, use getseed() for suggestions
meanL <- 0.7;   sdL <- 0.5  #5000 random log-normal values then
x <- rlnorm(5000,meanlog = meanL,sdlog = sdL) #try with only 500
meanx <- mean(log(x)); sdx <- sd(log(x))
cat("             Original  Sample \n")
cat("Mode(x)    = ",exp(meanL - sdL^2),outstat[1],"\n")
cat("Median(x)  = ",exp(meanL),outstat[2],"\n")
cat("Mean(x)    = ",exp(meanL + (sdL^2)/2),outstat[3],"\n")
cat("Mean(log(x) = 0.7     ",meanx,"\n")
cat("sd(log(x)  = 0.5      ",sdx,"\n")
```

```
#              Original  Sample
# Mode(x)    = 1.568312 1.603512
# Median(x)  = 2.013753 2.052606
# Mean(x)    = 2.281881 2.322321
# Mean(log(x) = 0.7       0.7001096
# sd(log(x)  = 0.5       0.4944283
```

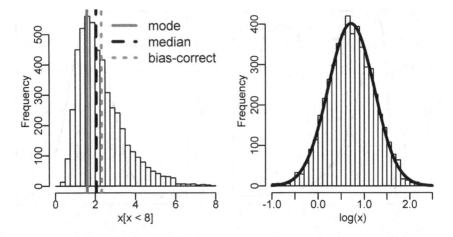

FIGURE 4.11 A Log-Normal distribution of 5000 random points with mean-log=0.7 and sdlog=0.5 showing the bias-corrected mean, the mode, and the median. On the right is the log-transformed version with a fitted Normal distribution.

The difference between the median and the mean, the $+(\sigma^2/2)$ term, is known as the bias-correction term and is attempting to account for the right-ward skew of the distribution by moving the measure of central tendency further to the right away from the mode. The mode appears to sit on the left side of the highest bar, but that merely reflects how R plots histograms (you could add half the bin width to fix this visual problem).

4.7.3 Fitting a Curve Using Log-Normal Likelihoods

We can use the spawning stock biomass and subsequent recruitment data from Penn and Caputi (1986), relating to the Australian Exmouth Gulf tiger prawns (*Penaeus semisulcatus*; **MQMF** data-set *tigers*). Stock recruitment relationships often assume Log-Normal residuals (sometimes a Gamma distribution, see later) and normally they would be derived inside a stock assessment model. Penn and Caputi (1986) used a Ricker curve but as an alternative we will attempt to fit a Beverton-Holt stock recruitment curve to these observations (we will examine stock recruitment relationships in more detail in the *Static Models* chapter). The Beverton-Holt stock recruitment relationship can have more than one form but for this example we will use **Equ**(4.26):

$$R_t = \frac{aB_t}{b + B_t}e^{N(0,\sigma^2)} \tag{4.26}$$

where R_t is the recruitment in year t, B_t is the spawning stock biomass that gave rise to R_t, a is the asymptotic maximum recruitment level, and b is the spawning stock biomass that gives rise to 50% of the maximum recruitment. The residual errors are Log-Normal with $\mu = 0$ and variance σ^2, which is estimated when fitting the model to the data. We will continue to use neg-NLL() but if you examine the code of negNLL() you will see we need to input log-transformed observed recruitment levels and write a short function to calculate the log of predicted recruitment levels. Once again we will use nlm() to minimize the output from negNLL(). When we consider the Log-Normal residuals (Observed Recruitment / Predicted recruitment) in **Table** 4.2, notice there are two exceptional residual values, one near 2.04 and the other near 0.44. The potential influence of the low point might bear further investigation as being relatively exceptional events (that appeared, in fact, to be influenced by the occurrence of cyclones, Penn and Caputi, 1986).

```
# fit a Beverton-Holt recruitment curve to tigers data Table 4.2
data(tigers)   # use the tiger prawn data-set
lbh <- function(p,biom) return(log((p[1]*biom)/(p[2] + biom)))
 #note we are returning the log of Beverton-Holt recruitment
pars <- c("a"=25,"b"=4.5,"sigma"=0.4)   # includes a sigma
best <- nlm(negNLL,pars,funk=lbh,observed=log(tigers$Recruit),
            biom=tigers$Spawn,typsize=magnitude(pars))
outfit(best,backtran=FALSE,title="Beverton-Holt Recruitment")
predR <- exp(lbh(best$estimate,tigers$Spawn))
 #note exp(lbh(...)) is the median because no bias adjustment
result <- cbind(tigers,predR,tigers$Recruit/predR)
```

```
# nlm solution:  Beverton-Holt Recruitment
# minimum      :  5.244983
# iterations   :  16
# code         :  1 gradient close to 0, probably solution
#          par        gradient
# 1 27.344523 -5.109267e-08
# 2  4.000166  1.270042e-07
# 3  0.351939  1.791807e-06
```

We can plot out the solution to compare the fit with the data visually:

```
# Fig. 4.12 visual examination of the fit to the tigers data
plot1(tigers$Spawn,predR,xlab="Spawning Biomass","Recruitment",
      maxy=getmax(c(predR,tigers$Recruit)),lwd=2)
points(tigers$Spawn,tigers$Recruit,pch=16,cex=1.1,col=2)
```

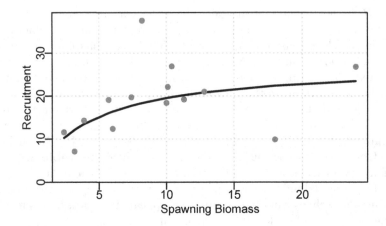

FIGURE 4.12 The optimum fit to the Exmouth Gulf tiger prawns Beverton-Holt stock recruitment relationship using Log-Normal likelihoods.

TABLE 4.2: The Exmouth Gulf tiger prawn data-set of spawning biomass with consequent recruitment levels, with the predicted recruitment level from the optimum model fit, along with the Log-Normal residuals.

SpawnB	Recruit	PredR	Residual
2.4	11.6	10.254	1.1313
3.2	7.1	12.153	0.5842
3.9	14.3	13.499	1.0593
5.7	19.1	16.068	1.1887
6.0	12.4	16.406	0.7558
7.4	19.7	17.750	1.1099
8.2	37.5	18.379	2.0404
10.0	18.4	19.532	0.9421
10.1	22.1	19.587	1.1283
10.4	26.9	19.749	1.3621
11.3	19.2	20.195	0.9507
12.8	21.0	20.834	1.0080
18.0	9.9	22.373	0.4425
24.0	26.8	23.438	1.1434

4.7.4 Fitting a Dynamic Model Using Log-Normal Errors

In this example, we will look ahead to an example from the chapter on *Surplus Production Models* (and previewed in the chapter on *Simple Population Models*). In particular we will be using what is known as the Schaefer (1954, 1957) surplus production model (Hilborn and Walters, 1992; Polacheck *et al*, 1993; Prager, 1994; Haddon, 2011). The simplest equations used to describe a Schaefer model involve two terms:

$$B_0 = B_{init}$$
$$B_{t+1} = B_t + rB_t \left(1 - \frac{B_t}{K}\right) - C_t \tag{4.27}$$

where r is the intrinsic rate of natural increase (a population growth rate term), K is the carrying capacity or unfished available biomass, commonly referred to as B_0 (confusingly, here B_0 is simply the initial biomass at $t = 0$ and B_{init} could possibly already be depleted below K). B_t represents the available biomass in year t, with B_{init} being the biomass in the first year of available data. If the population is unfished then $B_{init} = K$ but otherwise would constitute a separate model parameter. Finally, C_t is the total catch taken in year t. Of course, the time-step need not be in years and may need to be some shorter period depending on the biology of the species involved, however, the use of year is common. Notice that there are no error terms in **Equ**(4.27). This implies that the stock dynamics are deterministic and that the catches are known without error. Estimating the parameters for this model is an example of what is known as using an observation-error estimator.

The simple dynamic model in **Equ**(4.27), when given the required parameters $(B_{init}, r,$ and $K)$, projects the initial biomass forward to generate a time-series of population biomass levels. Such surplus production models can be compared with observations from nature if one has a time-series of relative abundance indices, which may be biomass estimates from surveys or standardized catch-per-unit-effort (CPUE) from fishery dependent data. In the following example we will use the catches and CPUE from a diver-caught invertebrate fishery contained in the **MQMF** data-set *abdat*. The assumption is made that there is a simple linear relationship between the index of relative abundance and the stock biomass:

$$I_t = \frac{C_t}{E_t} = qB_t e^\varepsilon \quad \text{or} \quad C_t = qE_t B_t e^\varepsilon \tag{4.28}$$

where I_t is the observed CPUE in year t, E_t is the effort in year t, q is known as the catchability coefficient (Arreguin-Sanchez, 1996), and e^ε represents the Log-Normal residual errors on the relationship between the CPUE and the stock biomass. This q could be estimated directly as a parameter although there is also what is known as a closed form estimate of the same value (Polacheck *et al*, 1993):

$$q = e^{\frac{1}{n}\sum log\left(\frac{I_t}{\hat{B}_t}\right)} = \exp\left(\frac{1}{n}\sum log\left(\frac{I_t}{\hat{B}_t}\right)\right) \tag{4.29}$$

which is basically the geometric mean of the vector of the observed CPUE divided by the predicted biomass levels in each year. An advantage of using such a closed form is that the model has fewer parameters to estimate when being fitted to data. It also emphasizes that the q parameter is merely a scaling factor that reflects the assumed linear relationship between exploitable biomass and the index of relative abundance. If a non-linear relationship is assumed a more complex representation of the catchability would be required.

The stock recruitment curve fitted in the previous example uses a relatively simple equation and related function as the model to be fitted, but there are no dynamics involved in the modelling. When trying to fit the surplus production model, as described in **Equ**(4.27) to **Equ**(4.29), to observed CPUE and catch data, the function that gives rise to the predicted CPUE will need to be more complex as it will need to include the stock dynamics. With simple, non-dynamic models the use of such things as the independent and the dependent variables are relatively straightforward to implement. Here we will illustrate (and reinforce) what is required when fitting a dynamic model to data.

We will use the **MQMF** data-set called *abdat*, named thus because it contains data on abalone.

```
data(abdat)  # plot abdat fishery data using a MQMF helper  Fig. 4.13
plotspmdat(abdat) # function to quickly plot catch and cpue
```

We need two functions to use within nlm() to find the optimum parameters. You should examine the code of each and understand how they relate to the equations used. The first is needed to calculate the log of the predicted cpue (we use simpspm()), while the second is used to calculate the -*veLL* where we are using Log-Normal residual errors to represent the residuals of the cpue data (so we use negLL(); compare its code with negNLL()). Note the expectation that the model parameters will be log-transformed, we do this as it often leads to greater stability than using the *typsize* option. You should experiment with this code by trying different starting points. You should carefully examine the code of simpspm and negLL() until you understand how they interact and believe you could repeat this analysis with a different data-set (see the *Surplus Production Models* chapter).

```
 # Use log-transformed parameters for increased stability when
 # fitting the surplus production model to the abdat data-set
param <- log(c(r= 0.42,K=9400,Binit=3400,sigma=0.05))
obslog <- log(abdat$cpue) #input log-transformed observed data
```

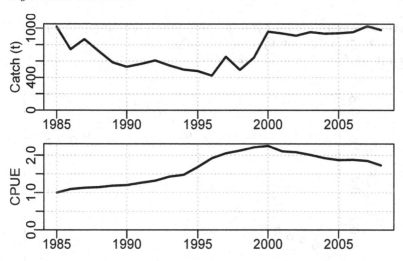

FIGURE 4.13 The *abdat* data-set plotting the catch and the cpue through time to illustrate their relationship.

```
bestmod <- nlm(f=negLL,p=param,funk=simpspm,indat=as.matrix(abdat),
                logobs=obslog)  # no typsize, or iterLim needed
 #backtransform estimates, outfit's default, as log-transformed
outfit(bestmod,backtran = TRUE,title="abdat")          # in param
```

```
# nlm solution:  abdat
# minimum      :  -41.37511
# iterations   :  20
# code         :  2 >1 iterates in tolerance, probably solution
#           par        gradient     transpar
# 1 -0.9429555  6.707523e-06     0.38948
# 2  9.1191569 -9.225209e-05  9128.50173
# 3  8.1271026  1.059296e-04  3384.97779
# 4 -3.1429030 -8.161433e-07     0.04316
```

It is always a good idea to plot the observed data against the predicted values from the optimum model. Here we have plotted them on their log-scale to illustrate exactly what was fitted and helps identify which points differed the most from their predicted values. As few have an intuitive grasp of the log-scale, it is also a good idea to plot the data and the fit on the nominal scale as well, and, in addition, one would usually plot the residuals to search for patterns.

```
# Fig. 4.14 Examine fit of predicted to data
predce <- simpspm(bestmod$estimate,abdat) #compare obs vs pred
ymax <- getmax(c(predce,obslog))
```

```
plot1(abdat$year,obslog,type="p",maxy=ymax,ylab="Log(CPUE)",
    xlab="Year",cex=0.9)
lines(abdat$year,predce,lwd=2,col=2)
```

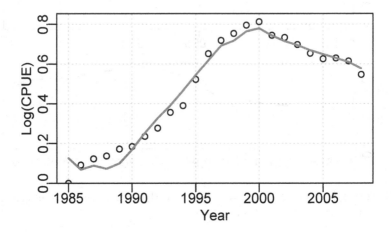

FIGURE 4.14 The optimum fit of the Schaefer surplus production model to the abdat data-set plotted in log-space on the y-axis.

4.8 Likelihoods from the Binomial Distribution

So far, with the Normal and Log-Normal distributions, we have been dealing with likelihoods of particular values of continuous variables (such as spawning biomass and cpue). But, of course, some observations and events are discrete in nature. Common examples would include an animal being either mature or not, has a tag or not, or similar yes/no situations. In contrast with continuous variables, when likelihoods are calculated for particular values such as the likelihood of seeing m tags in a sample of n observations, using discrete distributions such as the Binomial, they are true probabilities rather than just probability densities. Thus, the complication of understanding likelihoods that are not true probabilities is avoided; which might be why many texts describing maximum likelihood methods tend to start with examples that use the Binomial distribution.

In situations where an observation is true or false (a so-called Bernoulli trial; e.g., a captured fish either has or does not have a tag), and the probability of success out of n observations (trials) is the parameter p, then it would generally be best to use the Binomial distribution to describe observations.

The Binomial probability density function generates true probabilities and is characterized by two parameters, n, the number of trials (sample size), and p, the probability of success in a trial (an event/observation proving to be true):

$$P\{m|n,p\} = \left[\frac{n!}{m!\,(n-m)!}\right] p^m (1-p)^{(n-m)} \tag{4.30}$$

which is read as the probability of m events proving to be true given n trials (e.g., m tags from a sample of n observations), where p is the probability of an event being true. The term $(1-p)$ is often written as q, that is, $(1-p) = q$. The "!" symbol denotes a factorial. The term in the square brackets is the number of combinations that can be formed from n items taken m at a time, and is sometimes written as:

$$\binom{n}{m} = \frac{n!}{m!\,(n-m)!} \tag{4.31}$$

It is always the case that $n \geq m$ because one cannot have more successes than trials.

4.8.1 An Example Using Binomial Likelihoods

As a first example we might have a population from which a fishery focussed only on males (the Australian Queensland fishery for mud crabs, *Scylla serrata*, is an example). One might question whether this management strategy negatively affects the sex ratio of legal-sized animals from a particular population. In a hypothetical sample of 60 animals if we obtained 40 females and only 20 males would we conclude there had been an impact of the fishery on the sex-ratio? One way to answer such a question would be to examine the relative likelihood of finding such a result (while it is true that Binomial likelihoods are true probabilities we will continue to refer to them also as likelihoods). A typical sex-ratio of 1:1 would mean that we might expect 30 animals to be male from a sample of 60, so, if we declare, for the purposes of this example, that finding a male is a success, we should examine the likelihood of different values of m (the number of males in our sample of n) and determine the relative likelihood of finding $m = 20$ in our sample.

```
#Use Binomial distribution to test biased sex-ratio Fig. 4.15
n <- 60     # a sample of 60 animals
p <- 0.5    # assume a sex-ration of 1:1
m <- 1:60   # how likely is each of the 60 possibilites?
binom <- dbinom(m,n,p)    # get individual likelihoods
cumbin <- pbinom(m,n,p)   # get cumulative distribution
plot1(m,binom,type="h",xlab="Number of Males",ylab="Probability")
abline(v=which.closest(0.025,cumbin),col=2,lwd=2) # lower 95% CI
```

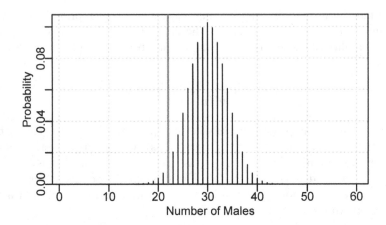

FIGURE 4.15 A graphical answer to the question of how likely is it to obtain only 20 males in a sample of 60 animals if the sex ratio is 1:1. The vertical red line is the lower bound of the 95% confidence intervals, which suggests that observing only 20 would be unlikely.

One can examine the specific values by examining the contents of *binom* (individual likelihoods) and *cumbin* (cumulative Binomial likelihoods). But clearly, if the sex ratio were 1:1 then obtaining fewer than 18 or greater than 42 are both extremely unlikely. Indeed, the chances of getting only 20 males is less that 95%. By examining *binom[20]* we can see that the likelihood of obtaining exactly 20 males (assuming 1:1 sex ratio) is just over a third of 1 percent (0.003636) and *cumbin[20]* tells us that the probability of 20 or less is only 0.006745. We would certainly have grounds for claiming there had been a depression in the sex ratio in the sample from this population. We have also placed a vertical line where the cumulative probability is approximately 0.025. As we are not really interested in the upper limit we could have used 0.05, which would have been more conservative. But such limits have an arbitrary element, and what really matters is what is the weight of evidence that the fishing has had no important impact on the sex ratio?

Notice that with relatively large samples the Binomial distribution becomes symmetrical. The tails, however, are shallower than a Normal distribution. For smaller samples, especially with low values of p, the distribution can be highly skewed with a value of zero successes having a specific probability.

Rather than determine whether a given sex ratio is plausible, we could search for the sex ratio (the p value) that maximized the likelihood of obtaining 20 males out of 60. We would expect it to be 20/60 (0.333...), but there remains interest in knowing the range of plausible values.

```
# plot relative likelihood of different p values Fig. 4.16
n <- 60  # sample size; should really plot points as each independent
m <- 20  # number of successes = finding a male
p <- seq(0.1,0.6,0.001) #range of probability we find a male
lik <- dbinom(m,n,p)    # R function for binomial likelihoods
plot1(p,lik,type="l",xlab="Prob. of 20 Males",ylab="Prob.")
abline(v=p[which.max(lik)],col=2,lwd=2) # try "p" instead of "l"
```

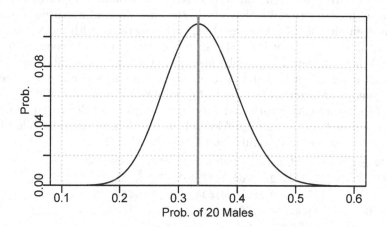

FIGURE 4.16 Representing the relative likelihood of the proportional sex-ratio when a sample exhibits only 20 males out of 60. Note that the likelihood of there having been a sex-ratio of 0.5 is confirmed as very low.

The use of optimize() (or optimise(), check out ?optimize) is much more efficient than using nlm() when we are only searching for a single parameter. It requires a function whose first term is the value to be minimized or maximized across the interval defined, in this case, by p, hence the simple wrapper function used around dbinom().

```
# find best estimate using optimize to finely search an interval
n <- 60; m <- 20  # trials and successes
p <- c(0.1,0.6) #range of probability we find a male
optimize(function(p) {dbinom(m,n,p)},interval=p,maximum=TRUE)
```

```
# $maximum
# [1] 0.3333168
#
# $objective
# [1] 0.1087251
```

4.8.2   Open Bay Juvenile Fur Seal Population Size

Rather than use a second hypothetical example it is more interesting to use data from a real world case. A suitable study was made on a population of New Zealand fur seal pups (*Arctocephalus forsteri*) on the Open Bay Islands off the west coast of the South Island of New Zealand (Greaves, 1992; York and Kozlof, 1987). The New Zealand fur seal has been recovering after having been heavily depleted in the 19 century, with new haul-out sites now found in both the South and North Island. Exploitation officially ceased in 1894, with complete protection within the New Zealand Exclusive Economic Zone beginning in 1978 (Greaves, 1992). In cooperation with the New Zealand Department of Conservation, Ms Greaves journeyed to and spent a week on one of these offshore islands. She marked 151 fur seal pups by clipping away a small patch of guard hairs on their heads, and then conducted a number of colony walk-throughs to re-sight tagged animals (Greaves, 1992). Each of these walk-throughs constituted a further sample, and differing numbers of animals were found to be tagged in each sample. The question is: What is the size of the fur seal pup population (X) on the island?

```
# Juvenile furseal data-set Greaves, 1992.  Table 4.3
furseal <- c(32,222,1020,704,1337,161.53,31,181,859,593,1125,
             135.72,29,185,936,634,1238,153.99)
columns <- c("tagged(m)","Sample(n)","Population(X)",
             "95%Lower","95%Upper","StErr")
furs <- matrix(furseal,nrow=3,ncol=6,dimnames=list(NULL,columns),
               byrow=TRUE)
```

TABLE 4.3: Three of the counts of New Zealand fur seal pups made by Greaves (1992) on Open Bay Island, West Coast, South Island, New Zealand. The Population estimates, Standard error, and confidence intervals were calculated using deterministic equations. The top two rows were independent counts while the bottom row averages six separate counts.

tagged(m)	Sample(n)	Population(X)	95%Lower	95%Upper	StErr
32	222	1020	704	1337	161.53
31	181	859	593	1125	135.72
29	185	936	634	1238	153.99

All the usual assumptions for tagging experiments are assumed to apply; i.e., we are dealing with a closed population—no immigration or emigration, with no natural or tagging mortality over the period of the experiment, no tags are lost, and tagging does not affect the recapture probability of the animals. Finally, the walk-throughs occurred on different days, which allowed the pups to move around, so the sightings were all independent of each other. Greaves

(1992) estimated all of these effects and accounted for them in her analysis. Having tagged and re-sighted tags, a deterministic answer was found with the Peterson estimator (Caughley, 1977; Seber, 1982):

$$\frac{n_1}{X} = \frac{m}{n} \quad \therefore \quad \hat{X} = \frac{n_1 n}{m} \tag{4.32}$$

where n_1 is the number of tags in the population, n is the subsequent sample size, m is the number of tags recaptured, and $\hat{X}$ is the estimated population size. An alternative estimator adjusts the counts on the second sample to allow for the fact that in such cases we are dealing with discrete events. This is Bailey's adjustment (Caughley, 1977):

$$\hat{X} = \frac{n_1 (n + 1)}{m + 1} \tag{4.33}$$

The associated estimate of standard error is used to calculate the approximate 95% confidence intervals using a Normal approximation, which leads to symmetrical confidence intervals when using the deterministic equations:

$$StErr = \sqrt{\frac{n_1^2 (n + 1) (n - m)}{(m + 1)^2 (m + 2)}} \tag{4.34}$$

These equations can be used to confirm the estimates in **Table** 4.3. However, instead of using the deterministic equations, a good alternative would be to use maximum likelihood to estimate the population size $\hat{X}$, using the Binomial probability density function.

We are only estimating a single parameter, $\hat{X}$, the population size, and this entails searching for the population size that maximizes the likelihood of the data. With the Binomial distribution, $P\{m|n, p\}$, **Equ**(4.30) provides the probability of observing m tagged individuals from a sample of n from a population with proportion p tagged (Snedecor and Cochran, 1967, 1989; Forbes *et al*, 2011). The proportion of fur seal pups that are marked relates to the population size $\hat{X}$ and the number of pups originally tagged, which was 151. Hence $p = 151/\hat{X}$. We will re-analyse the first two samples from **Table** 4.3.

```
# analyse two pup counts 32 from 222, and 31 from 181, rows 1-2 in
# Table 4.3.   Now set-up storage for solutions
optsol <- matrix(0,nrow=2,ncol=2,
                 dimnames=list(furs[1:2,2],c("p","Likelihood")))
X <- seq(525,1850,1) # range of potential population sizes
p <- 151/X  #range of proportion tagged; 151 originally tagged
m <- furs[1,1] + 1 #tags observed, with Bailey's adjustment
n <- furs[1,2] + 1 # sample size with Bailey's adjustment
```

```
lik1 <- dbinom(m,n,p) # individaul likelihoods
#find best estimate with optimize to finely search an interval
#use unlist to convert the output list into a vector
#Note use of Bailey's adjustment (m+1), (n+1) Caughley, (1977)
optsol[1,] <- unlist(optimize(function(p) {dbinom(m,n,p)},p,
                      maximum=TRUE))
m <- furs[2,1]+1;  n <- furs[2,2]+1 #repeat for sample2
lik2 <- dbinom(m,n,p)
totlik <- lik1 * lik2 #Joint likelihood of 2 vectors
optsol[2,] <- unlist(optimize(function(p) {dbinom(m,n,p)},p,
                      maximum=TRUE))
```

We could certainly tabulate the results but it is clearer to plot them as the likelihood (in this case probability) of each hypothesized population size. Then we can use the p column within the variable *optsol* to calculate the optimum population size in each case. The advantage of the plot is that one can immediately see the overlap of the likelihood curves for each sample, and gain an impression that any percentile confidence intervals would not be symmetric.

```
# Compare outcome for 2 independent seal estimates Fig. 4.17
# Should plot points not a line as each are independent
plot1(X,lik1,type="l",xlab="Total Pup Numbers",
      ylab="Probability",maxy=0.085,lwd=2)
abline(v=X[which.max(lik1)],col=1,lwd=1)
lines(X,lik2,lwd=2,col=2,lty=3)  # add line to plot
abline(v=X[which.max(lik2)],col=2,lwd=1) # add optimum
#given p = 151/X, then X = 151/p and p = optimum proportion
legend("topright",legend=round((151/optsol[,"p"])),col=c(1,2),lwd=3,
      bty="n",cex=1.1,lty=c(1,3))
```

4.8.3 Using Multiple Independent Samples

When one has multiple surveys, observations, or samples, or different types of data, and these are independent of one another, it is possible to combine the estimates to improve the overall estimate. Just as with probabilities, the likelihood of a set of independent observations is the product of the likelihoods of the particular observations. Thus, we can multiply the likelihoods from each population size for the two samples we have just examined to obtain a joint likelihood, which in the previous example was put into the variable *totlik*, **Figure** 4.18.

```
#Combined likelihood from 2 independent samples Fig. 4.18
totlik <- totlik/sum(totlik) # rescale so the total sums to one
cumlik <- cumsum(totlik) #approx cumulative likelihood for CI
plot1(X,totlik,type="l",lwd=2,xlab="Total Pup Numbers",
```

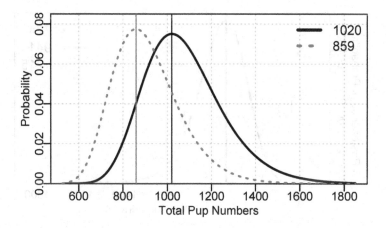

FIGURE 4.17 The population size versus likelihood distribution for two estimates of fur seal pup population size obtained using tagging experiments where 151 pups were tagged (Greaves, 1992). The right-hand black line is from a single count of 32 tags from 222 pups observed, while the left-hand dashed curve is from a count of 31 tags from 181 observed. The modes (optimum population estimates) are indicated by the vertical lines and in the legend.

```
        ylab="Posterior Joint Probability")
percs <- c(X[which.closest(0.025,cumlik)],X[which.max(totlik)],
          X[which.closest(0.975,cumlik)])
abline(v=percs,lwd=c(1,2,1),col=c(2,1,2))
legend("topright",legend=percs,lwd=c(2,4,2),bty="n",col=c(2,1,2),
       cex=1.2)  # now compare with averaged count
m <- furs[3,1];   n <- furs[3,2] # likelihoods for the
lik3 <- dbinom(m,n,p)            # average of six samples
lik4 <- lik3/sum(lik3)  # rescale for comparison with totlik
lines(X,lik4,lwd=2,col=3,lty=2) #add 6 sample average to plot
```

Notice that while the central tendency remains very similar, the 95% confidence intervals from the combination of likelihoods (thin vertical red lines) is tighter and asymmetric around the mean than would be the case from the six combined samples. As with all analyses, as long as the procedures used are defensible, then the analyses can proceed (i.e., in this case it could be argued that the samples were independent, meaning they were taken sufficiently far apart in time that there was no learning the locations of tagged pups, and the pups could move, etc.). There are similarities here to the Bayesian approach of updating a prior probability using likelihoods from new data. Details of actual Bayesian approaches will be given later. The averaging of six samples to obtain average count and sample sizes provides a very similar population

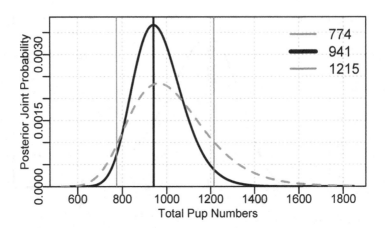

FIGURE 4.18 The combined likelihood from the first two fur seal pup samples (black line). These have tighter percentile confidence intervals (vertical thin red lines) than the individual samples. The average count from the other six combined samples (the curved green dashed line) remains similar in mean location to the single samples but would have much wider CI. The legend shows the optimum and the 95% percentile CI foir the combined likelihoods curve.

estimate but that remains associated with a wider range of uncertainty. Taking averages of such samples is not the optimum analytical strategy.

4.8.4 Analytical Approaches

Some biological processes, such as whether an animal is mature or not, or selected by fishing gear or not, are ideally suited to being analyzed using the Binomial distribution as the underlying basis to observations. However, such processes operate in a cumulative manner so that, for example, with maturity, the percent of a cohort within a population that become mature through time should eventually achieve 100%. Such processes can often be well described using the well-known logistic curve. While one could use numerical methods to estimate logistic model parameters it is also possible to use Generalized Linear Models with the Binomial distribution. We will describe these methods in the *Static Models* chapter that follows this chapter.

4.9 Other Distributions

In the base **stats** R package (use `sessionInfo()` to determine the seven base packages loaded) there are probability density functions available for numerous distributions. In fact by typing `?Distributions` into the console (with no brackets) you can obtain a list of those available immediately within R. In addition, there is a task view within the CRAN system on distributions that can be found at: https://CRAN.R-project.org/view=Distributions. That task view provides a detailed discussion but also points to a large array of independent packages that provide access to an even wider array of statistical distributions.

Within fisheries science a few of these other distributions are of immediate use. All have probability density functions used when calculating likelihoods (beginning with *d*, as in `dmultinom()`) and usually they all have random number generators (beginning with *r*, as in `rgamma()`). If you look at the help for a few you should get the idea and see the generalities. Here we will go through some of the details of a few of the more useful for fisheries and ecological work (see also Forbes *et al*, 2011).

4.10 Likelihoods from the Multinomial Distribution

As seen just above, we use the Binomial distribution when we have situations where there can be two possible outcomes to an observation (true/false, tagged/untagged, mature/immature). However, there are many situations where there are going to be more than two possible discrete outcomes to any observation, and in these situations, we can use the multinomial distribution. Such circumstances can arise when dealing with the frequency distribution within a sample of lengths or ages; e.g., given a random fish its age could lie in just one of a potentially large number of age-classes. In this multivariate sense, the multinomial distribution is an extension of the Binomial distribution. The multinomial is another discrete distribution that provides distinct probabilities and not just likelihoods.

With the Binomial distribution we used $P(m|n, p)$ to denote the likelihoods. With the multinomial, this needs to be extended so that instead of just two outcomes (one probability p), we have a probability for each of k possible outcomes (p_k) in our sample of n observations. The probability density function for the multinomial distribution is (Forbes *et al*, 2011):

$$P\{x_i|n, p_1, p_2, ..., p_k\} = n! \prod_{i=1}^{k} \frac{\hat{p}_i^{x_i}}{x!} \tag{4.35}$$

where x_i is the number of times an event of type i occurs in n samples (often referred to as trials in the statistical literature), and the p_i are the separate probabilities for each of the k types of events possible. The expectation of each type of event is $E(x_i) = np_i$, where n is the sample size and p_i the probability of event type i. Because of the presence of factorial terms in the PDF, which may lead to numerical overflow problems, a log-transformed version of the equation tends to be used:

$$LL\{x_i|n, p_1, ..., p_k\} = \sum_{j=1}^{n} log(j) + \sum_{i=1}^{k} \left[x_i log(\hat{p}_i) - \sum_{j=1}^{x} log(j) \right] \tag{4.36}$$

In practice, the log-transformed factorial terms $n!$ and $x!$, involving the sum of $log(j)$, where j steps from 1 to n, and 1 to x, will always be constant and are usually omitted from calculations. In addition the negative log-likelihood is used within minimizers:

$$-LL\{x_i|n, p_1, p_2, ..., p_k\} = -\sum_{i=1}^{k} [x_i log(\hat{p}_i)] \tag{4.37}$$

There are good reasons why we bother examining simplifications to the probability density functions when there are invariably very serviceable functions for their calculation already developed within R. It is always a good idea to understand what any function we use is doing and that follows from the fact that it is sensible to be aware of the properties of any statistical functions that we want to use in our analyses. In some cases, such as R's dmultinom() for the multinomial distribution, its help page tells us it is not currently vectorized and hence is a little clumsy in its use within fisheries. Instead we will use an R implementation of **Equ**(4.37) within the **MQMF** function mnnegLL(). In addition, while the difference in calculation speed is not that important when only dealing with a few hundred observations if one is repeating the calculation of the total log-likelihood for a set of data some millions of times, which is quite possible when using Markov-Chain-Monte-Carlo, or MCMC (sometimes McMC), on fisheries models (see the *On Uncertainty* chapter), then any way of saving time can be valuable. We will discuss such issues when we address Bayesian and other methods for characterizing uncertainty inherent in our models and data.

4.10.1   Using the Multinomial Distribution

When fitting a model to age- or size-composition data it is common to use
the Multinomial distribution to represent the expected distribution of observations among possible categories. We will use an example from a modal analysis
study of the growth of juvenile blacklip abalone (*Haliotis rubra*) from Hope
Island in the Derwent Estuary, Tasmania (Helidoniotis and Haddon, 2012). In
the sample collected in November 1992, two modes are clear when plotted in
2mm bins.

```
#plot counts x shell-length of 2 cohorts    Fig. 4.19
cw <- 2  # 2 mm size classes, of which mids are the centers
mids <- seq(8,54,cw) #each size class = 2 mm as in 7-9, 9-11, ...
obs <- c(0,0,6,12,35,40,29,23,13,7,10,14,11,16,11,11,9,8,5,2,0,0,0,0)
 # data from (Helidoniotis and Haddon, 2012)
dat <- as.matrix(cbind(mids,obs)) #xy matrix needed by inthist
parset() #set up par declaration then use an MQMF function
inthist(dat,col=2,border=3,width=1.8, #histogram of integers
    xlabel="Shell Length mm",ylabel="Frequency",xmin=7,xmax=55)
```

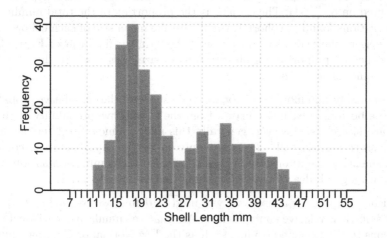

FIGURE 4.19 The length-frequency counts of a sample of juvenile abalone
from the south-east of Tasmania illustrating two modes taken in 1992.

It is assumed that the observable modes in the data relate to distinct cohorts
or settlements and we want to estimate the properties of each cohort. A Normal probability density function was used to describe the expected relative
frequency of each of the size classes within each mode/cohort, and these are
combined to generate the expected relative frequency in each of the k 2 mm
size classes found in the sample (Helidoniotis and Haddon, 2012). There are
five parameters required, the mean and standard deviation for each cohort
and the proportion of the total number of observations contained in the first

cohort (the proportion for the second is obtained through subtraction from 1.0). Thus, using $\theta = (\mu_c, \sigma_c, \varphi)$, where, in this case there are $c = 2$ cohorts, we can obtain the expected proportion of observations within each size class. One way would be to calculate the relative likelihood of the center of each 2mm-size class and multiply that by the total number in the sample as moderated by the proportion allocated to each cohort, **Equ**(4.38).

$$
\hat{N}_i = \varphi_1 n \sum_{S_i=6}^{56} \frac{1}{\sigma_1 \sqrt{2\pi}} \exp\left(\frac{-(S_i - \mu_1)}{2\sigma_1}\right)
$$

$$
+ (1 - \varphi_2) n \sum_{S_i=6}^{56} \frac{1}{\sigma_2 \sqrt{2\pi}} \exp\left(\frac{-(S_i - \mu_2)}{2\sigma_2}\right) \tag{4.38}
$$

$$
p_i = \hat{N}_i / \sum \hat{N}_i
$$

where $\hat{N}_i$ is the expected number of observations in each size class i, and i indexes the number of size classes (here from 7–55 mm in steps of 2mm, centered on 8, ..., 54). The φ (*phi*) is the proportion of the total number n of observations found in cohort 1, the μ_c are the mean size of each cohort c, and σ_c refers to their standard deviations. The p_i in the final row of **Equ**(4.38) is the expected proportion of observations in size-class i, where the S_i are the mid-points of each size-class.

Alternatively, and more precisely, we could subtract the cumulative probability for the bottom of each size class i from the cumulative probability of the top of each size class. However, generally this makes almost no difference in the analysis so here we will concentrate only on the first approach (you could try implementing the alternative yourself to make the comparison as sometimes it is not a good idea to trust everything that is put into print!).

Whichever approach is used, we will need to define two Normal distributions and sum their relative contributions. Alternative cumulative statistical distributions to the expected counts, such as the Log-Normal or Gamma, could be used in place of the Normal. If no compelling argument can be identified for using one or the other then ideally one would compare the relative fit to the available data obtained by using alternative distributions.

To obtain expected frequencies for comparison with the observed frequencies, it is necessary to constrain the total expected numbers to approximately the same as the numbers observed. The negative log-likelihoods from the multinomial are:

$$
-LL\{N|\mu_c, \sigma_c, \varphi\} = -\sum_{c=1}^{2}\sum_{i=1}^{k} N_i log\left(\hat{p}_i\right) = -\sum_{i=1}^{k} N_i log\left(\frac{\hat{N}_i}{\sum \hat{N}_i}\right) \tag{4.39}
$$

where μ_c and σ_c are the mean and standard deviation of the c cohorts being hypothesized as generating the observed distribution. There are k size classes and two cohorts c, and N_i is the observed frequency of size class i, while $\hat{p}_i$ is the expected proportion within size class i from the combined normal distributions. The objective would be to minimize the negative log-likelihood to find the optimum combination of the c Normal distribution parameters and φ.

```
#cohort data with 2 guess-timated normal curves Fig. 4.20
parset() # set up the required par declaration
inthist(dat,col=0,border=8,width=1.8,xlabel="Shell Length mm",
        ylabel="Frequency",xmin=7,xmax=55,lwd=2) # MQMF function
#Guess normal parameters and plot those curves on histogram
av <- c(18.0,34.5)    # the initial trial and error means and
stdev <- c(2.75,5.75) # their standard deviations
prop1 <- 0.55         # proportion of observations in cohort 1
n <- sum(obs) #262 observations, now calculate expected counts
cohort1 <- (n*prop1*cw)*dnorm(mids,av[1],stdev[1]) # for each
cohort2 <- (n*(1-prop1)*cw)*dnorm(mids,av[2],stdev[2])# cohort
#(n*prop1*cw) scales likelihoods to suit the 2mm class width
lines(mids,cohort1,lwd=2,col=1)
lines(mids,cohort2,lwd=2,col=4)
```

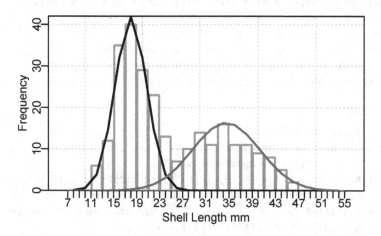

FIGURE 4.20 Two Normal distributions from initial parameter guesses imposed upon the length-frequency counts of juvenile abalone from southern Tasmania sampled in 1992.

The central estimates for the initial trial and error guesses for the two modes appear reasonable, but the spread in the left-hand cohort appears too small and the proportional allocation between the cohorts appears biased towards the first cohort. Being based upon proportions, which alter values in non-linear ways the search for an optimal likelihood can be sensitive to the starting values. We can search for a more optimal parameter set by applying nlm() to a wrapper function that is used to generate an estimate of the negative log-likelihood for the multinomial, as defined in **Equ**(4.39). As previously, we need a function to generate the predicted numbers of observations in each size class; in **MQMF** we have called it predfreq(). We also need a wrapper function that will calculate the negative log-likelihood using the predfreq() function, here we have called this wrapper() (see code chunk below). In developing the predfreq() function we have required that the parameters be ordered with the means of each cohort being fitted coming first, followed by their standard deviations, and then the proportions allocated to all but the last cohort coming at the end. This allows the algorithm to find the required parameters in their defined places. Thus, for a three cohort problem, in R we would have $pars = c(\mu_1, \mu_2, \mu_3, \sigma_1, \sigma_2, \sigma_3, \varphi_1, \varphi2)$. We have also included the option of using the cumulative Normal probability density.

```
#wrapper function for calculating the multinomial log-likelihoods
#using predfreq and mnnegLL, Use ? and examine their code
wrapper <- function(pars,obs,sizecl,midval=TRUE) {
  freqf <- predfreq(pars,sum(obs),sizecl=sizecl,midval=midval)
  return(mnnegLL(obs,freqf))
} # end of wrapper which uses MQMF::predfreq and MQMF::mnnegLL
mids <- seq(8,54,2) # each size class = 2 mm as in 7-9, 9-11, ...
av <- c(18.0,34.5)    # the trial and error means and
stdev <- c(2.95,5.75)  # standard deviations
phi1 <- 0.55       # proportion of observations in cohort 1
pars <-c(av,stdev,phi1)  # combine parameters into a vector
wrapper(pars,obs=obs,sizecl=mids) # calculate total -veLL
```

```
# [1] 708.3876
```

```
# First use the midpoints
bestmod <- nlm(f=wrapper,p=pars,obs=obs,sizecl=mids,midval=TRUE,
             typsize=magnitude(pars))
outfit(bestmod,backtran=FALSE,title="Using Midpts"); cat("\n")
#Now use the size class bounds and cumulative distribution
#more sensitive to starting values, so use best pars from midpoints
X <- seq((mids[1]-cw/2),(tail(mids,1)+cw/2),cw)
bestmodb <- nlm(f=wrapper,p=bestmod$estimate,obs=obs,sizecl=X,
             midval=FALSE,typsize=magnitude(pars))
outfit(bestmodb,backtran=FALSE,title="Using size-class bounds")
```

```
# nlm solution:  Using Midpts
# minimum     :  706.1841
# iterations  :  27
# code        :  1 gradient close to 0, probably solution
#          par     gradient
```

```
# 1 18.3300619  6.382071e-06
# 2 33.7907454  4.471337e-06
# 3  3.0390094 -2.835616e-05
# 4  6.0306017  6.975113e-06
# 5  0.5763628  7.515178e-05
#
# nlm solution:  Using size-class bounds
# minimum     :  706.1815
# iterations  :  24
# code        :  1 gradient close to 0, probably solution
#          par       gradient
# 1 18.3299573  2.363054e-06
# 2 33.7903327 -4.690083e-06
# 3  2.9831560  2.217978e-05
# 4  6.0030194 -2.880512e-05
# 5  0.5763426 -5.187824e-05
```

Now plot these optimal solutions against the original data. Plotting the data is simple enough but then we need to pull out the optimal parameters in each case and calculate the implied Normal distributions for each.

```
#prepare the predicted Normal distribution curves
pars <- bestmod$estimate # best estimate using mid-points
cohort1 <- (n*pars[5]*cw)*dnorm(mids,pars[1],pars[3])
cohort2 <- (n*(1-pars[5])*cw)*dnorm(mids,pars[2],pars[4])
parsb <- bestmodb$estimate # best estimate with bounds
nedge <- length(mids) + 1  # one extra estimate
cump1 <- (n*pars[5])*pnorm(X,pars[1],pars[3])#no need to rescale
cohort1b <- (cump1[2:nedge] - cump1[1:(nedge-1)])
cump2 <- (n*(1-pars[5]))*pnorm(X,pars[2],pars[4])  # cohort 2
cohort2b <- (cump2[2:nedge] - cump2[1:(nedge-1)])
```

```
#plot the alternate model fits to cohorts  Fig. 4.21
parset()  # set up required par declaration; then plot curves
pick <- which(mids < 28)
inthist(dat[pick,],col=0,border=8,width=1.8,xmin=5,xmax=28,
        xlabel="Shell Length mm",ylabel="Frequency",lwd=3)
lines(mids,cohort1,lwd=3,col=1,lty=2) # have used setpalette("R4")
lines(mids,cohort1b,lwd=2,col=4)      # add the bounded results
label <- c("midpoints","bounds")      # very minor differences .
legend("topleft",legend=label,lwd=3,col=c(1,4),bty="n",
        cex=1.2,lty=c(2,1))
```

In this case the parameters obtained when using the mid-points of the different size classes barely differ from the parameters obtained when using the upper and lower bounds of each size class with the resulting Normal curves almost completely overlapping. We can see the details of these fits by listing

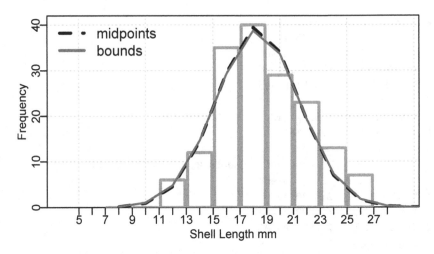

FIGURE 4.21 The optimum fit to the left-hand mode of the length-frequency information. The differences in parameter values between using size-class midpts or bounds occurs at the fourth decimal place. The difference in the right-hand mode was so small as not to be discernible.

the observed counts and the predicted counts, in each case, and the differences between the predicted and observed numbers in each size class. As in **Figure** 4.21, the numbers themselves illustrate more clearly the closeness of the two approaches, **Table** 4.4. The use of the upper and lower bounds is strictly more correct, but in practice it often makes little difference.

A different strategy for fitting modal distributions to mixed distributions can be found in Venebles and Ripley (2002, p436). Their approach is more sophisticated and elegant in that they use analytical gradients to assist the model fitting process. Hopefully it is clear that for almost any analytical problem there will be more than one way to solve it. Using numerical methods often entails exploring alternative methods in the search for a solution.

```
 # setup table of results for comparison of fitting strategies
predmid <- rowSums(cbind(cohort1,cohort2))
predbnd <- rowSums(cbind(cohort1b,cohort2b))
result <- as.matrix(cbind(mids,obs,predmid,predbnd,predbnd-predmid))
colnames(result) <- c("mids","Obs","Predmid","Predbnd","Difference")
result <- rbind(result,c(NA,colSums(result,na.rm=TRUE)[2:5]))
```

TABLE 4.4: A tabulation of the predicted counts for the two Normal distributions from the optimum model fit. The original sample had 262 observations.

mids	Obs	Predmid	Predbnd	Difference
8	0	0.1244	0.1487	0.0243
10	0	0.9323	1.0429	0.1106
12	6	4.5513	4.8236	0.2722
14	12	14.4344	14.6975	0.2631
16	35	29.7390	29.5255	-0.2135
18	40	39.8899	39.2110	-0.6789
20	29	35.1656	34.7612	-0.4044
22	23	21.2938	21.4733	0.1795
24	13	10.8868	11.2236	0.3368
26	7	8.0156	8.1948	0.1792
28	10	9.5120	9.5510	0.0390
30	14	12.0774	12.0505	-0.0268
32	11	14.0533	13.9954	-0.0579
34	16	14.6763	14.6094	-0.0669
36	11	13.7320	13.6777	-0.0543
38	11	11.5103	11.4832	-0.0270
40	9	8.6431	8.6454	0.0023
42	8	5.8142	5.8368	0.0226
44	5	3.5038	3.5337	0.0298
46	2	1.8916	1.9184	0.0268
48	0	0.9149	0.9339	0.0191
50	0	0.3964	0.4077	0.0113
52	0	0.1538	0.1596	0.0058
54	0	0.0535	0.0560	0.0025
	262	261.9656	261.9607	-0.0049

4.11 Likelihoods from the Gamma Distribution

The Gamma distribution is generally less well known than the statistical distributions we have considered in previous sections. Nevertheless, the Gamma distribution is becoming more commonly used in fisheries modelling and in simulation; practical examples can be found in the context of length-based population modelling (Sullivan *et al*, 1990; Sullivan, 1992). The probability density function for the Gamma distribution has two parameters: a scale parameter b ($b < 0$ (an alternative sometimes used is λ, where $\lambda < 1/b$)), and a shape parameter c ($c > 0$). The distribution extends over the range of $0 \leq x \leq \infty$

(i.e., no negative values). The expectation or mean of the distribution, $E(x)$, relates the two parameters, the scale b and the shape c. Thus:

$$E(x) = bc \quad \text{or} \quad c = \frac{E(x)}{b} \tag{4.40}$$

The probability density function for calculating the individual likelihoods for the Gamma distribution is (Forbes et al, 2011):

$$L\{x|b,c\} = \frac{\left(\frac{x}{b}\right)^{(c-1)} e^{\frac{-x}{b}}}{b\Gamma(c)} \tag{4.41}$$

where x is the value of the variate, b is the scale parameter, c is the shape parameter, and $\Gamma(c)$ is the gamma function for the c parameter. In cases where c takes on integer values the distribution is also known as the Erlang distribution, where the gamma function ($\Gamma(c)$) is replaced by factorial $(c-1)!$ (Forbes et al, 2011):

$$L\{x|b,c\} = \frac{\left(\frac{x}{b}\right)^{(c-1)} e^{\frac{-x}{b}}}{b\,(c-1)!} \tag{4.42}$$

As usual with likelihood calculations, to avoid computational issues of over- and under-flow it is standard to calculate the log-likelihoods, and more specifically negative log-likelihoods. The use of log-likelihoods is invariably more risk-averse:

$$-LL\{x|b,c\} = -\left[\left((c-1)log\left(\frac{x}{b}\right) - \frac{x}{b}\right) - \{log(b) + log(\Gamma(c))\}\right] \tag{4.43}$$

For multiple observations this enables their summation rather than product, but it still requires the calculation of the log-gamma function $log(\Gamma(c))$. Fortunately for us, there is both a gamma() and a lgamma() function defined in R.

The $Gamma$ distribution (not to be confused with the Gamma function, Γ) is extremely flexible, ranging in shape from effectively an inverse curve, through a right-hand skewed curve, to approximately Normal in shape. Its flexibility makes it a very useful function for simulations, **Figure** 4.22.

```
#Illustrate different Gamma function curves  Fig. 4.22
X <- seq(0.0,10,0.1) #now try different shapes and scale values
dg <- dgamma(X,shape=1,scale=1)
plot1(X,dg,xlab = "Quantile","Probability Density")
lines(X,dgamma(X,shape=1.5,scale=1),lwd=2,col=2,lty=2)
```

```
lines(X,dgamma(X,shape=2,scale=1),lwd=2,col=3,lty=3)
lines(X,dgamma(X,shape=4,scale=1),lwd=2,col=4,lty=4)
legend("topright",legend=c("Shape 1","Shape 1.5","Shape 2",
       "Shape 4"),lwd=3,col=c(1,2,3,4),bty="n",cex=1.25,lty=1:4)
mtext("Scale c = 1",side=3,outer=FALSE,line=-1.1,cex=1.0,font=7)
```

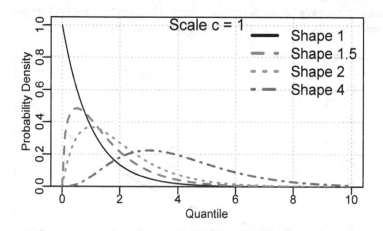

FIGURE 4.22 Different Gamma distributions all with a scale c = 1.0.

In the section on the Multinomial distribution we used Normal distributions to describe the expected distribution of lengths within a cohort. It is also possible that the growth pattern of a species might be better described using a Gamma distribution.

4.12   Likelihoods from the Beta Distribution

The Beta distribution is only defined for values of a variable between 0.0 and 1.0. This makes it another very useful distribution for use in simulations because it is relatively common that one is required to sample from a distribution of values between 0 to 1, without the possibility of obtaining values beyond those limits, **Figure** 4.23. It is recommended that you explore the possibilities available within R. In the chapter *On Uncertainty* we will be using the multivariate Normal distribution, which requires the use of an extra R package. The array of distributions available for analysis and simulation is very large. Exploration would be the best way of discovering their properties.

```
#Illustrate different Beta function curves. Fig. 4.23
x <- seq(0, 1, length = 1000)
```

```
parset()
plot(x,dbeta(x,shape1=3,shape2=1),type="l",lwd=2,ylim=c(0,4),
    yaxs="i",panel.first=grid(), xlab="Variable 0 - 1",
    ylab="Beta Probability Density - Scale1 = 3")
bval <- c(1.25,2,4,10)
for (i in 1:length(bval))
  lines(x,dbeta(x,shape1=3,shape2=bval[i]),lwd=2,col=(i+1),lty=c(i+1))
legend(0.5,3.95,c(1.0,bval),col=c(1:7),lwd=2,bty="n",lty=1:5)
```

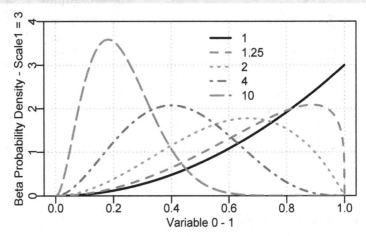

FIGURE 4.23 Different Beta probability density distributions all with a shape1 value = 3.0, and with values of shape2 ranging from 1 to 10.

4.13   Bayes' Theorem

4.13.1   Introduction

There has been an expansion in the use of Bayesian statistics in fisheries science (McAllister *et al*, 1994; McAllister and Ianelli, 1997; Punt and Hilborn, 1997; see Dennis, 1996, for an opposing view; more recently see Winker *et al*, 2018). An excellent book relating to the use of these methods was produced by Gelman *et al* (2014). Here we are not going to attempt a review of the methodology as it is used in fisheries; a good introduction can be found in Punt and Hilborn (1997), and there are many more recent examples. Instead, we will concentrate upon the foundation of Bayesian methods as used in fisheries and draw some comparisons with maximum likelihood methods. Details of how one can approach the use of these methods are provided in the chapter *On Uncertainty*.

Conditional probabilities are used to describe situations where one is interested in the probability that a particular event, say B_i, will occur given that an event A has already happened. The notation for this expression is $P(B_i|A)$, where the vertical line, "|", refers to the notion of *given*.

Bayes' theorem is based around a manipulation of such conditional probabilities. Thus, if a set of n events, labelled B_i, occur given that an event A has occurred, then we can formally develop Bayes' theorem. First, we consider the probability of observing a particular B_i given that A has occurred:

$$P(B_i|A) = \frac{P(A\&B_i)}{P(A)} \qquad (4.44)$$

That is, the probability of B_i occurring given the A has occurred is the probability of both A and B_i occurring divided by the probability that A occurs. In an equivalent manner we can consider the conditional probability of the event A given the occurrence of the B_i event:

$$P(A|B_i) = \frac{P(A\&B_i)}{P(B_i)} \qquad (4.45)$$

Those two conditional probabilities can be rearranged to give:

$$P(A|B_i)\,P(B_i) = P(A\&B_i) \qquad (4.46)$$

We can use this to replace the $P(A\&B_i)$ term in the **Equ**(4.44) to obtain the classical Bayes' Theorem:

$$P(B_i|A) = \frac{P(A|B_i)\,P(B_i)}{P(A)} \qquad (4.47)$$

If we translate this seemingly obscure theorem such that A represents data obtained from nature, and the various B_i as the separate hypotheses one might use to explain the data (where an hypothesis is a model plus a vector of parameters, θ), then we can derive the form of Bayes' Theorem as it is used in fisheries. Thus, the $P(A|B_i)$ is just the likelihood of the the data A given the model plus parameters B_i (the hypothesis); we are already familiar with this from maximum likelihood theory and practice. A new bit is the $P(B_i)$, which is the probability of the hypothesis before any analysis or consideration of the data A. This is known as the **prior** probability of the hypothesis B_i. The $P(A)$ in **Equ**(4.47) is the combined probability of all the possible combinations of data and hypotheses. The use of the word "all" needs emphasis as it really means a consideration of all possible outcomes. This completeness is why Bayesian statistics work so well for closed systems such as card games and other constrained games of chance. However, it is a strong assumption in the open world that all possibilities have been considered. It is best to interpret this as the array of alternative hypotheses (models plus particular parameters) considered constitute the domain of applicability of the analyses being undertaken. It means that for all B_i, $\sum P(B_i|A) = 0$ and hence **Equ**(4.48):

$$P(A) = \sum_{i=1}^{n} P(A|B_i)P(B_i) \qquad (4.48)$$

4.13.2 Bayesian Methods

As stated earlier, Bayes' theorem relates to conditional probabilities (Gelman *et al*, 2014), so that when we are attempting to determine which of a series of n discrete hypotheses (H_i = model + θ_i) is most probable, we use:

$$P\{H_i|data\} = \frac{L\{data|H_i\}\,P\{H_i\}}{\sum\limits_{i=1}^{n} [L\{data|H_i\}\,P\{H_i\}]} \qquad (4.49)$$

where H_i refers to hypothesis i out of the n being considered (a hypothesis would be a particular model with a particular set of parameter values), and the data are just the data to which the model is being fitted. Importantly, $P\{H_i|data\}$ is the **posterior** probability of the hypothesis H_i (meaning a strict probability between 0 and 1). This defines the divisor $\sum\limits_{i=1}^{n} [L\{data|H_i\}\,P\{H_i\}]$, which re-scales the posterior probability to sum to 1.0. $P\{H_i\}$ is the prior probability of the hypothesis (model plus particular parameter values), before the observed data are considered. Once again, this is a strict probability where the sum of the priors for all hypotheses being considered must be 1. Finally, $L\{data|H_i\}$ is the likelihood of the data given hypothesis i, just as previously discussed in the maximum likelihood section.

If the parameters are continuous variates (e.g., L_∞ and K from the von Bertalanffy curve), alternative hypotheses have to be described using a vector of continuous parameters instead of a list of discrete parameter sets, and the Bayesian conditional probability becomes continuous:

$$P\{H_i|data\} = \frac{L\{data|H_i\}\,P(H_i)}{\int\limits_{i=1}^{n} L\,[\{data|H_i\}\,P\{H_i\}]\,dH_i} \qquad (4.50)$$

In fisheries and ecology, to use Bayes' theorem to generate the required posterior distribution we need three things:

1. A list of hypotheses to be considered with the model under consideration (i.e., the combinations (or ranges) of parameters and models we are going to try);

2. A likelihood function required to calculate the probability density of the observed data given each hypothesis i, $L\{data|H_i\}$;

3. A prior probability for each hypothesis i, normalized so that the sum of all prior probabilities is equal to 1.0.

Apart from the requirement for a set of prior probabilities, this is identical to the requirements for determining the maximum likelihood. The introduction of prior probabilities is, however, a big difference, and is something we will focus on in our discussion. The essence of the approach is that the prior probabilities are updated using information from the data.

If there are many parameters being estimated in the model, the integration involved in determining the posterior probability in a particular problem can involve an enormous amount of computer time. There are a number of techniques used to determine the Bayesian posterior distribution, and Gelman *et al* (2014) introduce the more commonly used approaches. We will introduce and discuss one flexible approach (MCMC) to estimating the Bayesian posterior probability, which we use in the chapter dealing with the characterization of uncertainty. This is effectively a new method for model fitting but for convenience will be included in the section on uncertainty. The explicit objective of a Bayesian analysis is not just to discover the mode of the posterior distribution, which in maximum likelihood terms might be thought to represent the optimum model. Rather, the aim is to explicitly characterize the relative probability of the different possible outcomes from an analysis, that is, to characterize the uncertainty about each parameter and model output. There may be a most probable result, but it is presented in the context of the distribution of probabilities for all other possibilities.

4.13.3 Prior Probabilities

There are no constraints placed on how prior probabilities are determined. One may already have good estimates of a model's parameters from previous work on the same or a different stock of the same species, or at least have useful constraints on parameters (such as negative growth not being possible or survivorship > 1 being impossible). If there is insufficient information to produce informative prior probabilities, then commonly, a set of uniform or non-informative priors are adopted in which all hypotheses being considered are assigned equal prior probabilities. This has the effect of assigning each hypothesis an equal weight before analysis. Of course, if a particular hypothesis is not considered in the analysis, this is the same as assigning that hypothesis (model plus particular parameters) a weighting or prior probability of zero.

One reason why the idea of using prior probabilities is so attractive is that it is counterintuitive to think of all possible parameter values being equally likely. Any experience in fisheries and biology provides one with prior knowledge about the natural constraints on living organisms. Thus, for example, even before thorough sampling it should have been expected that a deep-water (>800 m depth) fish species, like orange roughy (*Hoplostethus atlanticus*), would likely be long-lived and slow-growing. This characterization is a reflection of the implications of living in a low-temperature and low-productivity environment. One of the great advantages of the Bayesian approach is that it permits one to move away from the counterintuitive assumption of all possibilities being equally likely. One can attempt to capture the relative likelihood of different values for the various parameters in a model in a prior distribution. In this way, prior knowledge can be directly included in analyses.

Where this use of prior information can lead to controversy is when moves are made to include opinions. For example, the potential exists for canvassing a gathering of stakeholders in a fishery for their belief on the state of such parameters as current biomass (perhaps relative to five years previously). Such a committee-based prior probability distribution for a parameter could be included into a Bayesian analysis as easily as could the results of a separate assessment (not previous assessment, which one would assume uses most of the same data, but one which used independent data). There is often debate about whether priors from such disparate sources should be equally acceptable in a formal analysis. In a nicely readable discussion on the problem of justifying the origin of priors, Punt and Hilborn (1997, p. 43) state:

We therefore strongly recommend that whenever a Bayesian assessment is conducted, considerable care should be taken to document fully the basis for the various prior distributions.... Care should be taken when selecting the functional form for a prior because poor choices can lead to incorrect inferences. We have also noticed a tendency to underestimate uncertainty, and hence to specify unrealistically informative priors—this tendency should be explicitly acknowledged and avoided.

The debate over the validity of using informative priors has been such that Walters and Ludwig (1994) recommended that non-informative priors be used as a default in Bayesian stock assessments. However, besides disagreeing with Walters and Ludwig, Punt and Hilborn (1997) highlighted a problem with our ability to generate non-informative priors (Box and Tiao, 1973). One problem with generating non-informative priors is that they are sensitive to the particular measurement system (**Figure** 4.24). Thus, a prior probability density that is uniform on a linear scale will not represent a uniform density on a log scale.

```
# can prior probabilities ever be uniniformative?  Fig. 4.24
x <- 1:1000
y <- rep(1/1000,1000)
```

```
cumy <- cumsum(y)
group <- sort(rep(c(1:50),20))
xlab <- seq(10,990,20)
par(mfrow=c(2,1),mai=c(0.45,0.3,0.05,0.05),oma=c(0.0,1.0,0.0,0.0))
par(cex=0.75, mgp=c(1.35,0.35,0), font.axis=7,font=7,font.lab=7)
yval <- tapply(y,group,sum)
plot(x,cumy,type="p",pch=16,cex=0.5,panel.first=grid(),
     xlim=c(0,1000),ylim=c(0,1),ylab="",xlab="Linear Scale")
plot(log(x),cumy,type="p",pch=16,cex=0.5,panel.first=grid(),
     xlim=c(0,7),xlab="Logarithmic Scale",ylab="")
mtext("Cumulative Probability",side=2,outer=TRUE,cex=0.9,font=7)
```

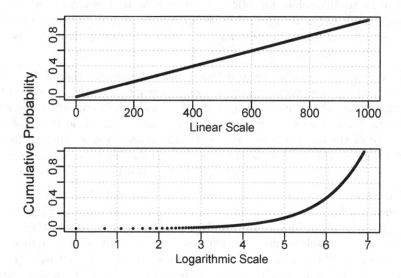

FIGURE 4.24 A constant set of prior probabilities accumulated on a linear scale and on a log scale.

As fisheries models tend to be full of nonlinear relationships, the use of non-informative priors is controversial because a prior that is non-informative with respect to some parameters will most likely be informative toward others. While such influences may be unintentional, they cannot be ignored. The implication is that information may be included into a model in a completely unintentional manner, which is one source of controversy when discussing prior probabilities. If priors are to be used, then Punt and Hilborn's (1997) exhortation to fully document their origin and properties is extremely sensible advice. We will examine the use of Bayesian methods in much more detail in the chapter *On Uncertainty*.

4.14   Concluding Remarks

The optimum method to use in any analytical situation depends largely on
the objectives of the analysis. If all one wants to do is to find the optimum
fit to a model, then it does not really matter whether one uses least squares,
maximum likelihood, or Bayesian methods. Sometimes it can be easier to fit a
model using least squares and then progress to using likelihoods or Bayesian
methods to create confidence intervals and risk assessments.

Confidence intervals around model parameters and outputs can be generated using traditional asymptotic methods (guaranteed symmetric and, with
strongly nonlinear models, only roughly approximate), using likelihood profiles
or by integrating Bayesian posteriors (the two are obviously strongly related),
or one can use bootstrapping or Monte Carlo techniques.

It is not the case that the more detailed areas of assessing the relative risks
of alternative management options are only possible using Bayesian methods.
Bootstrapping and Monte Carlo methods provide the necessary tools with
which to conduct such work. The primary concern should be to define the
objective of the analysis. However, it would be bad practice to fit a model
and not give some idea of the uncertainty surrounding each parameter and
the sensitivity of the model's dynamics of the various input parameters.

Because there is no clear winner among the methodological approaches, if one
has the time, it is a reasonable idea to use more than one approach (especially
a comparison of likelihood profiles, Bayesian posteriors, and bootstrapping).
Whether it is possible to have sufficient resources to enable a resource modeller
the time to conduct such an array of analyses is a different but equally real
problem. If significant differences are found, then it would be well to inves-
tigate the reasons behind them. If different procedures suggest significantly
different answers, it could be that too much is being asked of the available
data and different analyses would be warranted.

5

Static Models

5.1 Introduction

In this book we are focussing on relatively simple population models but will also begin to prepare the way towards understanding some of the requirements of more complex models. By "more complex" we are referring to the age-structured models commonly used to model the dynamics of harvested populations. Nevertheless, we can use components of these more complex models to illustrate fitting of what we can call "static models" to data. Static models are used to describe the functional form of such processes as the growth of individuals, how maturity changes with age or size, the "stock-recruitment" relationship between recruitment and mature or spawning biomass, and finally, the selectivity of fishing gear for a fished stock. Such static models exhibit a functional relationship between the variables of interest (length-at-age, maturity-at-age or -length, etc.) and these relationships are given the pretty big assumption of remaining fixed through time.

Such static models contrast with what we call dynamics models, which attempt to describe processes such as changes in population size, where the value being estimated at one time is a function of the population size at an earlier time. Such dynamics models invariably contain an element of self-reference, possibly along with other drivers, all of which contribute to the dynamics that we are attempting to describe or model.

In some situations, where clear changes have occurred in what are assumed to be static or stable processes, those would-be static relationships can be "time-blocked", which implies that such relationships are assumed to change in a step-like fashion between two different time periods. In reality, changes to such processes, if they occur, are likely to have occurred more smoothly over a period of time, but unfortunately, having sufficient data to describe such gradual changes well is generally rare. With the advent of directionally changing sea temperatures and acidity through the effects of climate change, such alterations in processes such as growth and maturity can be expected to occur. But, as always, tracking any such changes will be data dependent. Generally, such processes are assumed to be constant or static, but as long as you are aware of the possibility of such changes this can be monitored and looked for.

We begin by implementing more examples with such models because, being static, as we saw in the *Model Parameter Estimation* chapter, they tend to be simpler to implement than models with sufficient flexibility to describe dynamic processes.

5.2   Productivity Parameters

By definition, all biological populations are made up of collections of individuals and, as such, exhibit emergent properties that partially summarize the properties of those populations. With the many extinctions currently underway I agree that just a few or even single individuals of a species might constitute some sort of limit to what constitutes a 'population', but we will focus only on more abundant populations and leave aside these sad extreme cases. Here we are going to focus on individual growth, maturity, and recruitment, all of which relate to a population's productivity. A slow growing, late maturing, low fecundity species would tend to be less productive, though potentially more stable, than a fast growing, early maturing, highly fecund species. The evolution of life-history characteristics is a complex and fascinating subject, which I commend to your study (Beverton and Holt, 1959; Stearns, 1977, 1992). However, here we will focus on much simpler models. Even so, when modelling harvested populations a knowledge of the potential implications of the life-history characteristics of the species concerned should be very helpful in the interpretation of any observed dynamics. Do not forget that the modelling of biological processes benefits greatly from biological knowledge and understanding.

5.3   Growth

Ignoring any potential immigration and emigration, stock production is a combination of the recruitment of new individuals to a population and the growth of the individuals already in the population. Those are the positives, which are offset by the negatives of natural mortality and, in harvested populations, fishing mortality. The importance of individual growth is one reason there is a huge literature on the growth of individuals in fisheries ecology (Summerfelt and Hall, 1987). Much of that literature relates to the biology of growth, which we are going to ignore, despite what I just wrote about the importance of understanding biological processes if you are going to model them. Instead, we will concentrate on the mathematical description of individual growth. The emphasis is on description, which is an important point because some people appear to get overly fond of particular growth models that only describe growth and do not really explain the process.

In the previous chapter on *Model Parameter Estimation*, we have already introduced three alternative models that can be used to describe length-at-age (von Bertalanffy, Gompertz, and Michaelis-Menton), so we will not re-visit those. Instead we will consider two other aspects of the description of growth. In terms of length-at-age, we will examine ideas relating to seasonal growth models, though these tend to be of limited use, though important in freshwater systems. More commonly used, we will also examine how one might estimate individual growth from tagging data.

In this chapter we will concentrate primarily on the von Bertalanffy growth curve and extensions to it using the function vB():

$$L_t = L_\infty \left(1 - e^{-K(t-t_0)}\right) \tag{5.1}$$

5.3.1 Seasonal Growth Curves

Growth rates can be complex processes affected by the length or age of each animal, but also by the environmental conditions. As a result, in temperate and polar regions, where the temperature can vary greatly through the year, growth rates can alter seasonally. This can be expressed as literal growth rings in the ear-bones (otoliths) of fish and other hard parts that stem from changes in metabolism through the year. Of course, there are complications and obscurities brought about by other factors and events (for example, false rings in the hard parts), but again here we will focus on the description of the data assuming such complications have been dealt with before the analysis (reality often differs from this, do not under-estimate the difficulty of gathering valid fisheries data!).

Many different models of seasonal growth have been proposed but here we will use one proposed by Pitcher and Macdonald (1973), which modified the von Bertalanffy growth curve by including a sine wave into the growth rate term in an attempt to include cyclic water temperatures as a driver of growth rates such that they slowed in the winter and increased in the summer.

$$L_t = L_\infty \left(1 - e^{-\left[C\sin\left(\frac{2\pi(t-s)}{52}\right)+K(t-t_0)\right]}\right) \tag{5.2}$$

where L_∞, K, and t_0 are the usual von Bertalanffy parameters, t is the age in weeks at length L_t, C determines the magnitude of the oscillations around the non-seasonal growth curve, s is the starting point in the year of the sine wave, and the use of the constant 52 reflects the use of weeks as the units of within-year time-step. If we had used months or days as the units of time measured between sampling events then we would have used 12 or 365 respectively. We can use the data that Pitcher and Macdonald (1973) produced on English freshwater minnows (*Phoxinus phoxinus*; though the minnow data were read from their plots, and is thus only approximately correct, but suffices for a demonstration).

We can begin by plotting the available data and fitting a standard non-seasonal von Bertalanffy growth curve. The curve is fitted using Normal random errors rather than Log-Normal errors so we use the negNLL() function. It would be possible to use the ssq() function, but as we will be especially interested in the variation about the predicted curve we will stick with negNLL() in the following. If we ignore the seasonal trends then the expectation is that the variation around the curve (described by the *sigma* parameter), will be relatively large.

```
#vB growth curve fit to Pitcher and Macdonald derived seasonal data
data(minnow); week <- minnow$week; length <- minnow$length
pars <- c(75,0.1,-10.0,3.5); label=c("Linf","K","t0","sigma")
bestvB <- nlm(f=negNLL,p=pars,funk=vB,ages=week,observed=length,
              typsize=magnitude(pars))
predL <- vB(bestvB$estimate,0:160)
outfit(bestvB,backtran = FALSE,title="Non-Seasonal vB",parnames=label)
```

```
# nlm solution:  Non-Seasonal vB
# minimum     :  150.6117
# iterations  :  41
# code        :  3 Either ~local min or steptol too small
#                   par         gradient
# Linf    89.447640816   5.878836e-05
# K        0.009909338   2.705721e-01
# t0     -16.337065203  -4.717806e-05
# sigma    3.741419172   2.711341e-04
```

The standard von Bertalanffy curve **Equ**(5.1), fitted to the seasonal data, has a *sigma* parameter of about 3.7, which merely reflects the fact that the data oscillate about the mean growth curve and so the residuals can be expected to be relatively large. The expectation is that if we used a seasonally adjusted curve this variation would be greatly reduced; we can therefore predict that *sigma* should be much smaller. To fit the seasonal growth curve we need to define a modified vB() function that reflects the new model structure, **Equ**(5.2). Of course, when fitting any new model the requirement of a function to generate predicted values from a set of parameters would invariably be the first step. It would be possible to include the calculation of the negative log-likelihood in that new function but keeping the generation of predicted lengths in a separate function increases the flexibility of our code. Hence, we will stick to the strategy of having separate functions for the prediction and the calculation of the negative log-likelihood. In the vector of parameters, *pars*, we will also need to include initial estimates of C and s from **Equ**(5.2).

```
#plot the non-seasonal fit and its residuals.  Fig. 5.1
parset(plots=c(2,1),margin=c(0.35,0.45,0.02,0.05))
plot1(week,length,type="p",cex=1.0,col=2,xlab="Weeks",pch=16,
      ylab="Length (mm)",defpar=FALSE)
lines(0:160,predL,lwd=2,col=1)
 # calculate and plot the residuals
resids <- length - vB(bestvB$estimate,week)
plot1(week,resids,type="l",col="darkgrey",cex=0.9,lwd=2,
    xlab="Weeks",lty=3,ylab="Normal Residuals",defpar=FALSE)
points(week,resids,pch=16,cex=1.1,col="red")
abline(h=0,col=1,lwd=1)
```

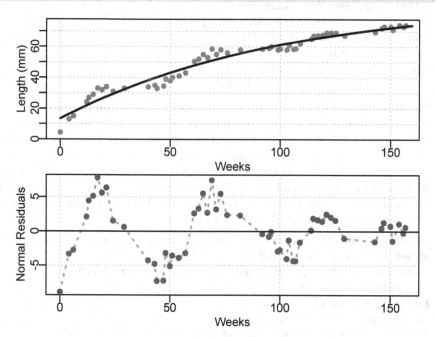

FIGURE 5.1 Length-at-age data derived from Pitcher and Macdonald (1973), illustrating strong seasonal growth fluctuations in growth rate when compared to the best-fitting von Bertalanffy curve. The seasonal pattern in the Normal random residuals in the bottom plot is very clear and reduces with age.

```
# Fit seasonal vB curve, parameters = Linf, K, t0, C, s, sigma
svb <- function(p,ages,inc=52) {
  return(p[1]*(1 - exp(-(p[4] * sin(2*pi*(ages - p[5])/inc) +
```

```
                          p[2] * (ages - p[3]))))))
} # end of svB
spars <- c(bestvB$estimate[1:3],0.1,5,2.0)  # keep sigma at end
bestsvb <- nlm(f=negNLL,p=spars,funk=svb,ages=week,observed=length,
               typsize=magnitude(spars))
predLs <- svb(bestsvb$estimate,0:160)
outfit(bestsvb,backtran = FALSE,title="Seasonal Growth",
       parnames=c("Linf","K","t0","C","s","sigma"))
```

```
# nlm solution:  Seasonal Growth
# minimum     :  105.2252
# iterations  :  21
# code        :  2 >1 iterates in tolerance, probably solution
#                   par        gradient
# Linf    89.06448058   7.259952e-05
# K        0.01040808   3.973411e-01
# t0     -13.46176840  -5.575708e-05
# C        0.10816263  -5.358108e-03
# s        6.96964772   2.209217e-05
# sigma    1.63926525   7.775261e-05
```

The seasonal adjustment to the growth rates only has minor effects on the L_∞ and K values, although it has a larger effect on the t_0 value as the seasonality can allow for the initial rapid rise rather better than the standard vB() function. As expected the impact on the *sigma* parameter is great (3.74 down to 1.64). The improvement to the model fit is also large (-veLL down from 150 to 105 for the cost of adding two parameters). This is also reflected by a reduction in the pattern across the residuals and the halving of their maximum and minimum values as in **Figure** 5.2.

```
#Plot seasonal growth curve and residuals    Fig. 5.2
parset(plots=c(2,1))  # MQMF utility wrapper function
plot1(week,length,type="p",cex=0.9,col=2,xlab="Weeks",pch=16,
      ylab="Length (mm)",defpar=FALSE)
lines(0:160,predLs,lwd=2,col=1)
 # calculate and plot the residuals
resids <- length - svb(bestsvb$estimate,week)
plot1(week,resids,type="l",col="darkgrey",cex=0.9,xlab="Weeks",
      lty=3,ylab="Normal Residuals",defpar=FALSE)
points(week,resids,pch=16,cex=1.1,col="red")
abline(h=0,col=1,lwd=1)
```

This is certainly not the only growth curve available that allows for the description of seasonal variations in growth rate, but this section constitutes only an introduction to the principles. Indeed, there are very many alternative growth curves in the literature, with special cases for taxa such as Crustacea that grow

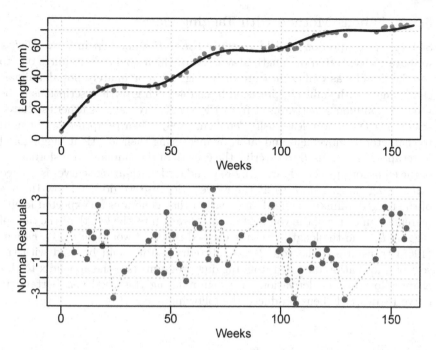

FIGURE 5.2 Approximate length-at-age data from Pitcher and Macdonald (1973) with a fitted seasonal version of the von Bertalanffy growth curve. The bottom plot is of the Normal random residuals, which still have series of runs above and below zero but are less patterned than the non-seasonal curve.

in size through moulting. Thus, strictly, describing their growth involves estimating the growth increments from initial starting sizes and the duration and frequency of growth increments when they do moult. At a population level, however, if there are sufficient numbers, very often, a continuous growth curve can provide an adequate approximation of such growth within a population dynamics model.

Pitcher and Macdonald (1973) objected to the appearance of negative growth predicted by their first seasonal growth curve, but as they were dealing with samples of fish from a population and fitting to the mean of those samples, there is no biological reason why the mean of those samples could not decline slightly at times during the year. They also pointed out that using a direct fitting process "... took several hours on-line to a PDP 8-E computer with interactive graphics: however, an efficient optimization method would reduce this time considerably" (Pitcher and Macdonald, 1973, p603). Hopefully, such a statement makes the reader more appreciate the now easy availability of non-linear optimizers and the remarkable computer speeds and their analytical capacity (freely available software, like R, is truly empowering).

5.3.2   Fabens Method with Tagging Data

So far in the chapter on *Model Parameter Estimation* and here in *Static Models*, we have concentrated on length-at-age data, which obviously requires one to age the samples as well as measure their length. Unfortunately, not all animals can be aged with sufficient precision to enable such growth descriptions to be used. A common practice used from the very early days of fisheries-related science is to tag and release fish (Petersen, 1896). Their recapture can be used to characterize movement and later, by measuring their length at tagging and at recapture it was used to describe the growth of the animals. Transformation of the equations used to describe more standard length-at-age curves is needed when using tagging data to fit the curves. Because we do not know the age of the tagged animals we need an equation that generates an expected length increment in terms of the von Bertalanffy parameters, the length at the time of tagging L_t, and the length after the elapsed time, Δt would be $L_{t+\Delta t}$. Fabens (1965) transformed the von Bertalanffy curve so it could be used with the sort of information obtained from tagging programs (see the appendix to this chapter for the full derivation). By manipulating the usual von Bertalanffy curve **Equ**(5.1) Fabens produced the following:

$$
\begin{aligned}
\Delta \hat{L} &= L_{t+\Delta t} - L_t \\
\Delta \hat{L} &= (L_\infty - L_t)\left(1 - e^{-K\Delta t}\right)
\end{aligned}
\tag{5.3}
$$

where, for an animal with an initial length of L_t, ΔL is the expected change in length through the duration of Δt. By using the minimum least squares or negative log-likelihood one can estimate values for L_∞ and K. To estimate a value for t_0, one would require the average length-at-age for a known age so, generally, no t_0 estimate can be made and the exact location of the growth curve along an age-axis is not determined. In such cases t_0 is often set to zero.

If one can obtain data that records at least the initial size at tagging, the time interval between tagging and recapture, and the growth increment that occurs over that time interval, then we can put together a function to generate the required predicted growth increments given initial lengths and the two parameters L_∞ and K, and then using maximum likelihood, we can obtain an optimum fit to whatever data we have.

This is best illustrated by plotting some tagging data, **Figure** 5.3; found in the **MQMF** data-set *blackisland* for an abalone population. This includes a number of zero growth increments for some larger abalone. The data cloud is scattered (noisy) but some form of trend is apparent, and it is this trend that the growth models will attempt to fit.

```
# tagging growth increment data from Black Island, Tasmania
data(blackisland);  bi <- blackisland # just to keep things brief
```

```
parset()
plot(bi$ll,bi$dl,type="p",pch=16,cex=1.0,col=2,ylim=c(-1,33),
    ylab="Growth Increment mm",xlab="Initial Length mm",
    panel.first = grid())
abline(h=0,col=1)
```

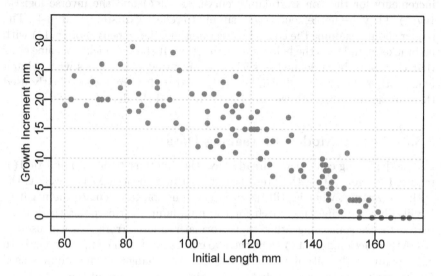

FIGURE 5.3 Tagging data collected for blacklip abalone from Black Island off the south-west corner of Tasmania. The time interval between tagging and recapture averaged 1.02 years.

As with the description of growth in the *Model Parameter Estimation* chapter, here we will fit and compare two growth curves to the example tagging data. These are the von Bertalanffy (Fabens, 1965; see **Equ**(5.3)), a standard for scalefish and many others species, and the inverse logistic (Haddon *et al*, 2008), which seems more suited to the indeterminate continuous growth exhibited by difficult to age invertebrates. The inverse logistic is more limited than the von Bertalanffy curve as it is designed to be used with time increments of one year, or at least a constant time increment

$$\Delta L = \frac{Max\Delta L}{1 + e^{log(19)(L_t - L_{50})/\delta}} + \varepsilon$$

$$L_{t+1} = L_t + \Delta L + \varepsilon$$

(5.4)

δ is equivalent to $L_{95} - L_{50}$, where L_{50} and L_{95} are the lengths where the growth increments are 50% and 5% of $Max\Delta L$, respectively, and ε represents the Normal random residual errors from the mean expected increment for a given length. In hindsight, L_{95} might have been better defined as L_5. Notice that because of the Fabens transformation the residual errors for the Fabens model differ from those for the length-at-age model even though both use the Normal probability density function (more on this later).

There are two functions defined in **MQMF** to generate the predicted growth increments for the von Bertalanffy curve, fabens() and the inverse logistic, invl(). They are implemented very simply to reflect **Equs**(5.3) and (5.4). The reader should examine the code of these functions (i.e. fabens() or invl() with no brackets) and read the help for each function. It should quickly become clear that they have been defined in such a manner that data.frames with column names other than *l1* and *dt* can be easily used. Below we have explicitly used these names even though we could have just used the default values.

5.3.3 Fitting Models to Tagging Data

In a model fitting context we already have the functions (fabens() and invl()) required for generating the predicted growth increments (ΔL) as well as the optimizer, nlm(), what is still required is a function to calculate the negative log-likelihood during the search for the minimum. At this point there is no reason to use something other than Normal random errors. We will use the **MQMF** function negNLL() to fit the two curves, and then compare them both visually and with a likelihood ratio test. If name changes to the columns used for the required data are needed this is inherent in the use of the ....

```
# Fit the vB and Inverse Logistic to the tagging data
linm <- lm(bi$dl ~ bi$l1) # simple linear regression
param <- c(170.0,0.3,4.0); label <- c("Linf","K","sigma")
modelvb <- nlm(f=negNLL,p=param,funk=fabens,observed=bi$dl,indat=bi,
               initL="l1",delT="dt") # could have used the defaults
outfit(modelvb,backtran = FALSE,title="vB",parnames=label)
predvB <- fabens(modelvb$estimate,bi)
cat("\n")
param2 <- c(25.0,130.0,35.0,3.0)
label2=c("MaxDL","L50","delta","sigma")
modelil <- nlm(f=negNLL,p=param2,funk=invl,observed=bi$dl,indat=bi,
               initL="l1",delT="dt")
outfit(modelil,backtran = FALSE,title="IL",parnames=label2)
predil <- invl(modelil$estimate,bi)

# nlm solution:  vB
# minimum      :  291.1691
# iterations   :  24
# code         :  1 gradient close to 0, probably solution
#                 par       gradient
# Linf   173.9677972 9.565398e-07
```

```
# K       0.2653003 2.657143e-04
# sigma   3.5861240 1.391815e-05
#
# nlm solution:  IL
# minimum     :  277.0122
# iterations  :  26
# code        :  1 gradient close to 0, probably solution
#            par       gradient
# MaxDL  21.05654 -2.021972e-06
# L50   130.92643  5.895934e-07
# delta  40.98771  2.218945e-07
# sigma   3.14555  4.553906e-06
```

The negative log-likelihood for the inverse-logistic is smaller than that for
the von Bertalanffy and the inverse-logistic exhibits a smaller σ value. If we
plot both growth curves against the original data one can see the differences
more clearly, **Figure** 5.4. Further an examination of their respective residuals
highlights differences between the curves, with the predicted von Bertalanffy
curve exhibiting a doming of its residuals consistent with its prediction of a
linear relationship between initial length and growth increment. In **Figure** 5.4
the linear regression is plotted over the top of the von Bertalanffy curve to
illustrate that it is effectively coincident.

```
#growth curves and regression fitted to tagging data Fig. 5.4
parset(margin=c(0.4,0.4,0.05,0.05))
plot(bi$ll,bi$dl,type="p",pch=16,cex=1.0,col=3,ylim=c(-2,31),
     ylab="Growth Increment mm",xlab="Length mm",panel.first=grid())
abline(h=0,col=1)
lines(bi$ll,predvB,pch=16,col=1,lwd=3,lty=1)  # vB
lines(bi$ll,predil,pch=16,col=2,lwd=3,lty=2)  # IL
abline(linm,lwd=3,col=7,lty=2) # add dashed linear regression
legend("topright",c("vB","LinReg","IL"),lwd=3,bty="n",cex=1.2,
                  col=c(1,7,2),lty=c(1,2,2))
```

The residuals of the two growth curves also exhibit the differences between
the respective model fits.

```
#residuals for vB and inverse logistic for tagging data Fig 5.5
parset(plots=c(1,2),outmargin=c(1,1,0,0),margin=c(.25,.25,.05,.05))
plot(bi$ll,(bi$dl - predvB),type="p",pch=16,col=1,ylab="",
     xlab="",panel.first=grid(),ylim=c(-8,11))
abline(h=0,col=1)
mtext("vB",side=1,outer=FALSE,line=-1.1,cex=1.2,font=7)
plot(bi$ll,(bi$dl - predil),type="p",pch=16,col=1,ylab="",
     xlab="",panel.first=grid(),ylim=c(-8,11))
abline(h=0,col=1)
mtext("IL",side=3,outer=FALSE,line=-1.2,cex=1.2,font=7)
```

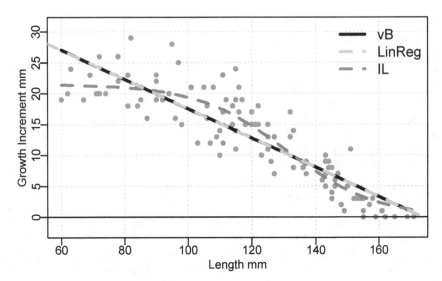

FIGURE 5.4 The von Bertalanffy (black), inverse logistic (curved dashed red), and linear regression (dashed yellow) fitted to blacklip abalone tagging data from Black Island. The time interval between tagging and recapture was 1.02 years. Obviously the vB and linear regression are identical.

```
mtext("Length mm",side=1,line=-0.1,cex=1.0,font=7,outer=TRUE)
mtext("Residual",side=2,line=-0.1,cex=1.0,font=7,outer=TRUE)
```

5.3.4   A Closer Look at the Fabens Methods

The growth curve described by the Fabens transformation of the von Bertalanffy equation uses the same parameters but has smuggled in differences that are not immediately obvious. You will remember the example where we fitted the vB curve using two different residual error structures and that made the outcomes incomparable or, more strictly, incommensurate. The Fabens transformation has altered the residual structure and how the parameters interact with each other. Instead of the L_∞ being the average asymptotic maximum size it becomes the absolute maximum size, which is the reason it can predict negative growth increments. Strictly the parameters of the Fabens model imply different things from that of the von Bertalanffy model.

We have compared the von Bertalanffy curve with the inverse-logistic but because we have used, as an example, data from an abalone stock, we have obtained a rather biased view of which curve might be more useful. In most scalefish fisheries, the von Bertalanffy curve has been found to be most useful as a general description of growth, although it does have a few further

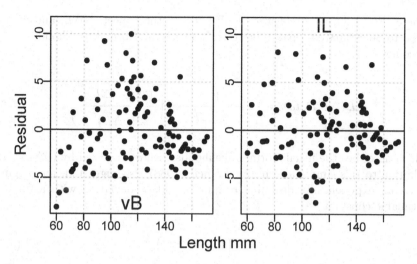

FIGURE 5.5 The von Bertalanffy (left hand) and inverse logistic (right hand) plots of the residuals from the curves fitted to blacklip abalone tagging data from Black Island. The time interval between tagging and recapture was 1.02 years.

problems. But a general problem with all growth models (not just tagging growth models) is the assumption that we have used so far, which is that the variation of the observations around the predicted growth trend will be constant. If we examine **Figure** 5.3 or **Figure** 5.4, and remember when we plotted the length-at-age data in the *Model Parameter Estimation* chapter, then in each case there are relatively clear trends in variation across either initial length or age. Instead of insisting on using a constant variance in the residuals (as required by SSQ), it is possible to use, for example, a constant coefficient of variation (σ/μ). If we used maximum likelihood methods it would be possible to have the variance of the residuals modelled separately allowing for a wide array of alternatives; Francis (1988) did exactly this. He fitted tagging models to data using maximum likelihood methods and assumed the residuals in each case were normally distributed, but he suggested a number of different functional forms for the relationship between residual variance and the expected ΔL (functional relationships with the initial length are also possible). The approach we have used to date has assumed a constant variance so we have estimated a constant σ parameter. Instead, Francis (1988) suggested that the variance could have a constant multiplier between σ and ΔL, which would allow for an inverse relationship for tagging data. These considerations also apply to length-at-age models where the constant multiplier would allow for a linear relationship between age and σ:

$$\sigma = v(\Delta \hat{L}) \tag{5.5}$$

where v (upsilon) is a constant multiplier on the expected ΔL, and would need to be estimated separately. In such a case the Normal likelihoods would become:

$$L\left(\Delta L | Data\right) = \sum_i \frac{1}{\sqrt{2\pi}v\Delta\hat{L}} \exp\left(\frac{\left(\Delta L - \Delta\hat{L}\right)^2}{2\left(v\Delta\hat{L}\right)^2}\right) \tag{5.6}$$

which describes the usual Normal likelihoods with σ^2 replaced by $(v\Delta\hat{L})^2$. In addition to this simple alternative Francis (1988) also suggested, for tagging data, an exponentially declining residual standard deviation with a further estimable constant τ:

$$\sigma = \tau\left(1 - e^{-v(\Delta\hat{L})}\right) \tag{5.7}$$

Finally, Francis (1988) suggested a residual standard deviation that would be described by a power law:

$$\sigma = v(\Delta\hat{L})^\tau \tag{5.8}$$

Francis (1988) also extended the Fabens approach to analyzing tagging data by considering the effect of bias, seasonal variation in growth rates, and the influence of outlier contamination. Often in the literature only a single aspect of a report is highlighted, which is one very good reason to always go back to the original literature rather than rely on a summary in another paper or book (like this one). Francis (1988) is an excellent place to start if you have to fit a growth curve to tagging data.

With each different formulation of the relationship between the variance of the residuals and the expected ΔL, the constants τ and v would change in their interpretation. As seen before, the parameter estimates obtainable from the same model can vary if different error structures are assumed and they become incomparable. Unfortunately, how we select which error structure is most appropriate is not a question that is simple to answer. At very best, using non-constant variances will provide another view of the uncertainty associated with whichever curve is selected.

5.3.5   Implementation of Non-Constant Variances

By including a relationship between the expected ΔL and the residual variance we would need to change the wrapper function around how the predicted values, $\Delta\hat{L}$, were used to calculate the negative log-likelihood. Previously, with constant σ we used negNLL() so we could start by modifying that. All we have done is include an extra function, funksig(), as an argument to negnormL()

(see its help and code) thereby including the calculation of the sigma values. By doing this we could include a funksig() that retained a constant σ, which would be a long version of the original negNLL(). But it also demonstrates that the function is behaving as expected. Then we can include an alternative funksig() that implements one of the three options above (or one of our own devising).

```
# fit the Fabens tag growth curve with and without the option to
# modify variation with predicted length. See the MQMF function
# negnormL. So first no variation and then linear variation.
sigfunk <- function(pars,predobs) return(tail(pars,1)) #no effect
data(blackisland)
bi <- blackisland # just to keep things brief
param <- c(170.0,0.3,4.0); label=c("Linf","K","sigma")
modelvb <- nlm(f=negnormL,p=param,funk=fabens,funksig=sigfunk,
               indat=bi,initL="l1",delT="dt")
outfit(modelvb,backtran = FALSE,title="vB constant sigma",
       parnames = label)
```

```
# nlm solution:  vB constant sigma
# minimum     :  291.1691
# iterations  :  24
# code        :  1 gradient close to 0, probably solution
#                  par       gradient
# Linf  173.9677972 9.565398e-07
# K       0.2653003 2.657143e-04
# sigma   3.5861240 1.391815e-05
```

```
sigfunk2 <- function(pars,predo) { # linear with predicted length
  sig <- tail(pars,1) * predo      # sigma x predDL, see negnormL
  pick <- which(sig <= 0)          # ensure no negative sigmas from
  sig[pick] <- 0.01                # possible negative predicted lengths
  return(sig)
} # end of sigfunk2
param <- c(170.0,0.3,1.0); label=c("Linf","K","sigma")
modelvb2 <- nlm(f=negnormL,p=param,funk=fabens,funksig=sigfunk2,
                indat=bi,initL="l1",delT="dt",
                typsize=magnitude(param),iterlim=200)
outfit(modelvb2,backtran = FALSE,parnames = label,
       title="vB inverse DeltaL, sigma < 1")
```

```
# nlm solution:  vB inverse DeltaL, sigma < 1
# minimum     :  298.9086
# iterations  :  46
# code        :  1 gradient close to 0, probably solution
#                  par       gradient
# Linf  170.9979114 -1.598948e-07
```

```
# K        0.2672788  1.488722e-06
# sigma    0.3941672  9.094947e-07
```

Remember that by altering the residual structure the likelihoods become incommensurate and so we cannot determine which provides the better fit. The constant variance allowed the von Bertalanffy curve to stay effectively above the data at lengths greater than about 148, the changing variance growth curve was still mostly above the data but was being pushed closer to the later data by the reducing variance with predicted length. One could experiment by re-writing `negnormL()` to work with the initial lengths rather than predicted lengths. As indicated earlier separating the mean predicted lengths from the associated variance leads to great flexibility.

```
#plot to two Faben's lines with constant and varying sigma Fig. 5.6
predvB <- fabens(modelvb$estimate,bi)
predvB2 <- fabens(modelvb2$estimate,bi)
parset(margin=c(0.4,0.4,0.05,0.05))
plot(bi$ll,bi$dl,type="p",pch=1,cex=1.0,col=1,ylim=c(-2,31),
     ylab="Growth Increment mm",xlab="Length mm",panel.first=grid())
abline(h=0,col=1)
lines(bi$ll,predvB,col=1,lwd=2)            # vB
lines(bi$ll,predvB2,col=2,lwd=2,lty=2)  # IL
legend("topright",c("Constant sigma","Changing sigma"),lwd=3,
       col=c(1,2),bty="n",cex=1.1,lty=c(1,2))
```

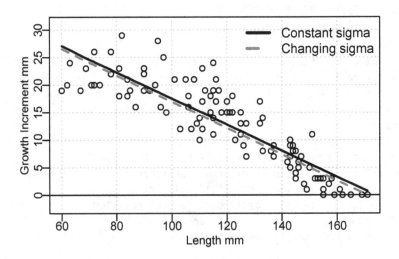

FIGURE 5.6 The von Bertalanffy (solid blue line) with constant sigma, and the same but with sigma related to the changing DeltaL (red dashed line) fitted to blacklip abalone tagging data from Black Island. The mean time interval between tagging and recapture was 1.02 years.

Our eyes are accustomed to appreciating symmetry and so the changing variation line in **Figure** 5.6 appears to be a relatively poor fit. But as a reflection of fitting a model assuming error only on the y-axis (the y-on-x issue) and with changing variances on the residuals, the lower dashed red line makes sense of data that has an intrinsic curve within its values. Using the more complex residual structure makes the model fitting more sensitive to initial conditions. You will have noticed that in the output the assumption of constant residual variance only took between 20 - 30 iterations, while with the functional relationship on the variance it took twice as many iterations (and that was with using the *typsize* optional argument to enhance stability, not used in the simpler model). If you modify those starting values you can easily obtain implausible answers.

Once again there needs to be good reasons for introducing such complexities into the modelling, although a changing variance tends to be clearly apparent when using length-at-age data, especially with very young ages.

5.4   Objective Model Selection

In the chapter on *Model Parameter Estimation* we saw a similarly named section where an information theoretical approximation for the Akaike's Information Criterion (AIC ; Akaike, 1974) based on sum-of-squares was presented (Burnham and Anderson, 2002). Now that we are using maximum likelihood we will introduce the original version of the AIC as well as a discussion of likelihood ratios.

5.4.1   Akiake's Information Criterion

The trade-off between the number of parameters used and the quality of a model's fit to a data-set is always an issue when selecting which model to use when attempting to describe a pattern in a data-set. The classical approach is to select the simplest model that does a sufficiently good job of that description (Occam's razor is simply an exhortation towards parsimony so if two models produce equivalent results select the simplest one). But how close is meant by "equivalent results" and what does "simplest" mean anyway. Akaike's (1974) solution was to use a combination of the likelihoods and the "number of independently adjusted parameters within the model". The AIC is a penalized likelihood-based criteria for model selection where the smallest AIC indicates the optimum model (in terms of likelihood balanced against the number of parameters):

$$AIC = -2\log(L) + \alpha p \qquad (5.9)$$

where $\log(L)$ is the total log-likelihood, p is the number of parameters explicitly varied when fitting the model, and α is a multiplier (the penalty in penalized likelihood) whose value AIC (1974) set to 2.0, using an argument from information theory. The emphasis on "independently adjusted parameters" that are "explicitly varied" is made to avoid confusion over model parameters which are held constant or which are derived from other parameters and model variables (we will see an example with the catchability parameter in surplus production models). For example, a common assumption in fishery models is that natural mortality, often depicted as M, is a constant. As such it would not count in the AIC's p value. The explicit identification of α for the penalty rather than just putting a constant value of 2.0 has been made here to emphasize that there has been debate (Bhansali and Downham, 1977; Schwartz, 1978; Akaike, 1979; Atkinson, 1980) over the exact value that should be used to ensure that the balance between model complexity (number of parameters) and quality of fit (maximum likelihood or minimum sum of squares) reflects reality in simulations. The admittedly relatively dense statistical arguments tend to boil down to how large should the value of α, the penalty on p, become? The outcome eventually became known as Schwartz's Bayesian Information Criterion or, more briefly, the BIC.

$$BIC = -2\log(L) + \log(n)p \qquad (5.10)$$

where n is the sample size or number of data points to which the model with p parameters is fitted. As long as the number of data points is 8 or greater ($log(8) = 2.079$) then the BIC will penalize complex models more than AIC. On the other hand, intuitively, it makes sense to take some account of the sample size. If we apply these statistics to the comparison of the von Bertalanffy and inverse-logistic model fits that used constant σ estimates, **Figure** 5.4, using the **MQMF** function `aicbic()`, the inverse-logistic has both a smaller AIC and BIC and so would be the preferred model in this case.

```
#compare the relative model fits of Vb and IL
cat("von Bertalanffy \n")
aicbic(modelvb,bi)
cat("inverse-logistic \n")
aicbic(modelil,bi)

# von Bertalanffy
#      aic      bic    negLL        p
# 588.3382 596.3846 291.1691   3.0000
# inverse-logistic
#      aic      bic    negLL        p
# 562.0244 572.7529 277.0122   4.0000
```

5.4.2 Likelihood Ratio Test

An alternative comparison between different model fits vs complexity can be made using what is known as a generalized likelihood ratio test (Neter *et al*, 1996). The method relies on the fact that likelihood ratio tests asymptotically approach the χ^2 distribution as the sample size gets larger. This means that with the usual extent of real fisheries data, this method is only approximate. Likelihood ratio tests are, exactly as their name suggests, the log of a ratio of two likelihoods, or if dealing directly with log-likelihoods, the subtraction of one from another, the two are equivalent (of course the likelihoods must be from the same PDF). We want to determine if either of two models, which use the same data and residual structure but different model structures (different parameters), provide for a significantly better model fit to the available data. As the likelihood ratio can be described by the χ^2 distribution it is possible to answer that question formally. Thus, for the likelihoods of two models, having either equal numbers of parameters or only one parameter difference, to be considered significantly different their ratio would need to be greater than the χ^2 distribution for the required number of degrees of freedom (the number of parameters that are different).

$$-2 \times log\left[\frac{L(\theta)_a}{L(\theta)_b}\right] \leq \chi^2_{1,1-\alpha}$$
$$-2 \times [LL(\theta)_a - LL(\theta)_b] \leq \chi^2_{1,1-\alpha}$$

(5.11)

where $L(\theta)_x$ is the likelihood of the θ parameters for model x, and $LL(\theta)_x$ is the equivalent log-likelihood. $\chi^2_{1,1-\alpha}$ is the $1-\alpha$th quantile of the χ^2 distribution with 1 degree of freedom (e.g., for 95% confidence intervals, $\alpha = 0.95$ and $1-\alpha = 0.05$, and $\chi^2_{1,1-\alpha} = 3.84$.

In brief, if we are comparing two models that only differ by one or no parameters then if their negative log-likelihoods differ by more than 1.92 (3.84/2) then they can be considered to be significantly different and one will provide a significantly better fit than the other. If there were two-parameters different between the two models, then the minimum difference between their respective negative log-likelihoods for the models to differ significantly would need to be 2.995 (5.99/2) $= \chi^2_{2,0.95}/2$; two degrees of freedom), and so on for greater numbers of parameter differences (Venzon and Moolgavkar, 1988).

With respect to the comparison of the von Bertalanffy and inverse-logistic curves, the inverse-logistic negative log-likelihood is $-2.0 \times (277.0122 - 291.1691) = 28.3138$, which is significantly smaller than that for the von Bertalanffy curve. We can use the **MQMF** function likeratio() to illustrate that this constitutes a highly significant difference between their respective fits to the data.

```
 # Likelihood ratio comparison of two growth models see Fig 5.4
vb <- modelvb$minimum # their respective -ve Log-likelihoods
il <- modelil$minimum
dof <- 1
round(likeratio(vb,il,dof),8)
```

```
#           LR         P      mindif           df
# 28.31380340  0.00000005  3.84145882  1.00000000
```

5.4.3 Caveats on Likelihood Ratio Tests

Given the prevalence of using weighted log-likelihoods in fishery models, when using a likelihood ratio test to compare alternative models one needs to be careful that the models are in fact comparable. When we are using the same data and the same probability density function to describe the residual distribution then we can use a likelihood ratio test. But remember the example where we compared the application of the same growth model to the same data only with a different assumption regarding the residual error structure (Log-Normal rather than Normal). Those curves were simply incomparable (incommensurate). Similarly with penalized likelihoods where different weightings may be given to particular data streams in different model versions, such changes make the models incommensurate. It sounds obvious that one should only compare comparable models, but sometimes, with some of the more complex models determining what constitutes a fair comparison is not always simple.

5.5 Remarks on Growth

Description of individual growth processes are important to fishery models because the growth in size and weight of individuals is a major aspect of the productivity of a given stock. This growth is obviously one of the positives in the balance between those processes that increase a stock's biomass and those that decrease it. But it is important to understand that invariably the descriptions of growth are only descriptions and are useful primarily across the range of sizes for which we have data. There are many instances where the weaknesses relating to the extrapolation of the von Bertalanffy curve lead to biologically ridiculous predictions (Knight, 1968). But such instances are mistaking a particular interpretation of the model's parameters for reality, whereas the curve is merely a description of the growth data and is strictly only valid across where there are data. This is one reason why growth model transformations have been produced that then have estimated parameters that can be located within the range of the available data (Schnute, 1981; Francis, 1988, 1995).

5.6 Maturity

5.6.1 Introduction

Fisheries management now tends to use what are known as biological reference points to form the foundation of what it hopes constitutes rational and defensible management decisions for harvested renewable natural resources (FAO, 1995, 1996, 1997; Restrepo and POwers, 1999; Haddon, 2007). Put very simply the idea is to have a target reference point, which is very often defined as a mature or spawning biomass level, or some proxy for this, that is deemed a desirable state for the stock. It is desirable because, generally, it ought to provide for good levels of subsequent recruitment and provide the stock with sufficient resilience to withstand occasional environmental shocks. Common default values can be B_{40}, meaning $40\%B_0$, or a depletion level of 40%, which is used as a proxy for B_{MSY}, or the spawning biomass that can sustainably produce the maximum sustainable yield. In Australia, the Commonwealth Harvest Strategy Policy defines the target as B_{48}, which is used as a proxy for B_{MEY}, or the spawning biomass that should be capable of producing the maximum economic yield (note it is the biomass level, B_{MEY}, that is the target not the MEY; DAFF, 2007; Rayns, 2007). In addition, one defines a limit reference point, again in terms of a level of mature biomass (or proxy) below which the stock concerned would be deemed to be at risk of its recruitment being compromised. Biological reference points are often couched in terms relative to an estimate of what the unfished spawning biomass would be (B_0). The concept of B_0 can be either an equilibrium notion or a more dynamic version that attempts to account for recruitment variation. Whatever the case this number tends to be uncertain within assessments and also often changes between assessments (Punt *et al*, 2018). To produce management advice one needs three things:

1) an assessment of what the current stock status is, in terms of spawning biomass depletion, and hence,

2) the current estimate of unfished spawning biomass, and finally,

3) a harvest control rule that determines what fishing mortality, effort, or catch will be allowed in the next season(s).

Perhaps you have noticed I am using the terms mature and spawning biomass inter-changeably. By doing this I am not trying to confuse you but rather trying to inure you to the varying terminology you will come across in the literature.

The emphasis on mature or spawning biomass is the reason that obtaining a reasonable estimate of the size- or age-at-maturity is so important. It is so that any estimate of how much mature biomass there is reflects the biology of the stock as well as the current state of the fishery. In this book we are trying to stay focussed on the modelling but it is always the case that we are trying to model an underlying biological reality. In fact, the biological reality, rather like the details of growth, is often rather complex. In fisheries theory many of the foundations derive from the northern hemisphere, and especially from working on highly productive stocks in the North Sea and the Atlantic (Smith, 1994). Many of the commercial species there have relatively simple reproductive histories. There are males and females that, as a population, mature to an average schedule, they spawn either once or each year until they die. That is, of course, an over-simplification, and the biological world is much more varied than that, even in the northern hemisphere. There are species that all start off as females with some of the larger females turning into males (protogynous), and vice versa (protandrous). There are many different reproductive strategies out there and many of them can influence things like the size and/or age at maturity. However, here we will focus only on the classical approach. Nevertheless, with your own examples, do not make assumptions about biology, it is always the case that representative evidence from a population is better than assumptions.

The size- or age-at-maturity is a population property. Individuals may undergo a process of maturation and take time to become capable of spawning, but the process we are describing is taken as an average across a population. Given a sample, hopefully a large sample that adequately bridges the size- or age-at-maturity, we are aiming to describe the proportion of fish that are mature at each size or age. The concept of size-at-maturity is often summarized as being the size at which 50% of a population are considered to be mature (capable of spawning). In fact, Lm_{50}, the size at 50% maturation, is only part of what is important; in addition, as we shall see, some measure of how quickly a species matures with respect to size or age is also important. Some simpler fisheries models still use an assumption of what is known as knife-edge maturity, which implies the strong assumption that 100% of animals become mature at a given age. We will discuss how the time taken for the population to mature affects the population dynamics later. How one determines maturity will not be considered here, though ideally one uses histology to confirm maturing gametes rather than simply using a visual inspection of fish gonads. Once again, there is a huge literature devoted to the details of determining maturity stages in a wide range of species, but we will not attempt to approach that. The task we have is finding a mathematical description of the expected population maturation curve. And the expectation is that the proportion of animals considered mature will increase with size and age until 100% are mature (though there are exceptions even to this, with the females of some species taking some years off from reproduction to recover from the large energy investment they make to the generation of eggs). Life history characteristics come into this in terms of whether a species is adapted to breed once (semelparity) or many times (iteroparity), or whether it generates vast numbers of small eggs or far fewer large eggs, or even produces a few live young. Always remember that biology can be complex and mathematical descriptions of biology are relatively simple abstractions.

5.6.2 Alternative Maturity Ogives

The general pattern of maturation within a population is that maturation will be found across a range of sizes and ages, with a proportion maturing both early and late, and most maturing around some average time. If we imagine some sort of parabolic curve of the proportion maturing through time, possibly close to normal but with shorter tails, what we want is the cumulative density function of that distribution running from 0 to 100% across some range of sizes or ages. The outcome is an S-shaped curve commonly called a logistic curve and in the literature there are a remarkable number of different mathematical representations of such curves. The statistics of interest are the Lm_{50} (the size/age at 50% maturity), and the IQ, the inter-quartile distance that measures the range of sizes/ages between the 25th and 75th quartiles of the maturity ogive (at what age are 25% and 75% mature). The use of the inter-quartile range is an arbitrary choice but reflects common practice (as seen in boxplots) but does not preclude using a wider or a narrower range if those are convenient.

Among the many different versions available there is a classical logistic curve, **Equ**(5.12), commonly used to describe maturation in many populations (see the **MQMF** function mature()). This formulation lends itself easily to being fitted using a Generalized Linear Model with Binomial errors. We have been using R as a programming environment but here is a chance to remember that its first purpose was to act as a tool for statistical analyses.

$$p_L = \frac{exp(a + bL)}{1 + exp(a + bL)} = \frac{1}{1 + exp(a + bL)^{-1}} \tag{5.12}$$

where p_L is the proportion mature at length L, and a and b are the exponential parameters. Notice I have not included an error term in the equation. Here we are attempting to make predictions about mature or not mature observations related to different sizes (for brevity I will stop writing "size or age" each time). This is what suggests that we should be using Binomial errors, as described in the *Model Parameter Estimation* chapter. Another nice thing about this formulation is that the $Lm50 = -a/b$, that is the size at 50% maturity is simply derived from the parameters. Similarly, the inter-quartile distance can also be derived from the parameters as $IQ = 2 \times log(3)/b = 2.197225 \times b$.

Maturity data is Binomial in nature, at a given time of sampling a population, the observations are whether each sampled fish, of size L, is mature or not. The book's R package, **MQMF** contains an example data-set for us to work with, *tasab*, which includes blacklip abalone (*Haliotis rubra*) data from two sites along a 16km stretch of Tasmanian coastline.

```
# The Maturity data from tasab data-set
data(tasab)         # see ?tasab for a list of the codes used
properties(tasab) # summarize properties of columns in tasab

#          Index isNA Unique     Class Min Max Example
# site         1    0      2   integer   1   2       1
# sex          2    0      3 character   0   0       I
# length       3    0     85   integer  62 160     102
# mature       4    0      2   integer   0   1       0
table(tasab$site,tasab$sex) # sites 1 & 2 vs F, I, and M

#
#        F   I   M
#  1   116  11 123
#  2   207  85 173
```

Abalone are notorious (infamous!) for being highly variable in their biological characteristics across relatively small spatial scales, so the fine detail of sampling site is important (Haddon and Helidoniotis, 2013). This data was all collected in the same month and year by the same people so the only factor other than shell length that we might think could influence maturation would be the specific site (the sexes appear to mature at the same time and sizes; **Figure** 5.7).

```
#plot the proportion mature vs shell length  Fig. 5.7
propm <- tapply(tasab$mature,tasab$length,mean) #mean maturity at L
lens <- as.numeric(names(propm))              # lengths in the data
plot1(lens,propm,type="p",cex=0.9,xlab="Length mm",
      ylab="Proportion Mature")
```

The data appear relatively noisy, although this is difficult to appreciate as some of the observed lengths whose proportion mature was neither 0 nor 1, may have had only a few observations. For example, there is a single point lined up on a proportion mature of 0.5 at length 100mm, and if you delve into the data you will find that the point is only made up of two observations, one mature the other not. One can find the lengths with pick <- which(tasab$length == 100), or which(propm == 0.5), and then use tasab[pick,] to see the data.

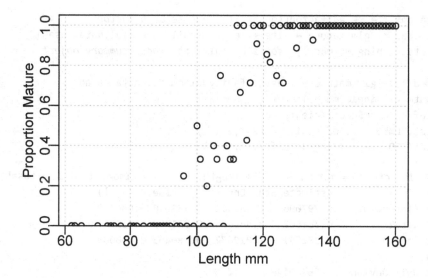

FIGURE 5.7 The proportion mature at length for the blacklip abalone maturity data in the tasab data-set.

Examining the data-set using `properties(tasab)` or `head(tasab,10)`, tells us what variable names we are dealing with. While we obviously want to examine the relationship between the variables *mature* and *length*, remember that initially we also want to examine the influence of *site* on maturation. Importantly, we need to convert the *site* variable into a categorical factor otherwise it will be treated as a vector of the integers 1 and 2, rather than as potentially different treatments. We cannot include *sex* as a factor as it has more than two outcomes in that all animals start out as Immature, although one can always analyse the male and female data separately. However, over many more sites, no repeatable differences in the rate or mean size at maturity x site have been found, so far, between the sexes in blacklip abalone. Once fitted we will see that *site* is uninformative to the analysis (it is insignificant) and so we repeat the analysis without *site*.

```
#Use glm to estimate mature logistic
binglm <- function(x,digits=6) { #function to simplify printing
  out <- summary(x)
  print(out$call)
  print(round(out$coefficients,digits))
  cat("\nNull Deviance  ",out$null.deviance,"df",out$df.null,"\n")
  cat("Resid.Deviance ",out$deviance,"df",out$df.residual,"\n")
  cat("AIC  = ",out$aic,"\n\n")
  return(invisible(out)) # retain the full summary
} #end of binglm
```

```
tasab$site <- as.factor(tasab$site) # site as a factor
smodel <- glm(mature ~ site + length,family=binomial,data=tasab)
outs <- binglm(smodel)  #outs contains the whole summary object

model <- glm(mature ~ length, family=binomial, data=tasab)
outm <- binglm(model)
cof <- outm$coefficients
cat("Lm50 = ",-cof[1,1]/cof[2,1],"\n")
cat("IQ   = ",2*log(3)/cof[2,1],"\n")
```

```
# glm(formula = mature ~ site + length, family = binomial, data = tasab)
#                Estimate Std. Error   z value Pr(>|z|)
# (Intercept) -19.797096    2.361561 -8.383056 0.000000
# site2        -0.369502    0.449678 -0.821703 0.411246
# length        0.182551    0.019872  9.186463 0.000000
#
# Null Deviance    564.0149 df 714
# Resid.Deviance   170.7051 df 712
# AIC  =   176.7051
#
# glm(formula = mature ~ length, family = binomial, data = tasab)
#                Estimate Std. Error   z value Pr(>|z|)
# (Intercept) -20.464131    2.265539 -9.032787        0
# length        0.186137    0.019736  9.431291        0
#
# Null Deviance    564.0149 df 714
# Resid.Deviance   171.3903 df 713
# AIC  =   175.3903
#
# Lm50 =   109.9414
# IQ   =   11.80436
```

In the first analysis, that included *site*, because I made sure that the sites included were similar when putting the data together, site 2 was not significantly different from site 1 (P=0.411), implying that, in this case, the *site* factor was not informative for this analysis. The analysis was thus repeated omitting *site* from the equation. Had the sites exhibited significant differences then we would require multiple curves to describe the model results. We will plot them as different curves anyway to illustrate the process. The extra *site* parameters are modifiers to the initial exponential intercept value. Thus, the *ab$length* parameter is the b parameter in both cases, but the curve for site 1 would be obtained by treating the intercept as the a parameter and the curve for site 2 would require us to add the parameter values for the intercept and *ab$site2* as the a parameter. Thus, site1 model $a = -19.797$ and $b = 0.18255$, while for site2 model $a = -19.797 - 0.3695$ and $b = 0.18255$. Adding the

final no-site model to the plot indicates that the combined model is closer
to site 2, which is a reflection of the fact that the number of observations at
site 2 was 465 while only 250 at site 1. Note that with one parameter less
the residual deviance is very slightly greater but despite this the AIC for the
simpler model is smaller than that for the model including *site* as a factor,
which again reinforces the idea that the simpler model provides an improved
balance between complexity and model fit. There are obvious analogies with
General Linear Models and their comparisons and manipulations. If one does
find a significant difference between the intercepts then it would be sensible
to test whether the difference also extended to the *b* parameter to determine
whether the curves needed to be treated completely separately. Sharing common parameters is often helpful as it leads to increases in sample size.

```
#Add maturity Logistics to the maturity data plot Fig. 5.8
propm <- tapply(tasab$mature,tasab$length,mean) #prop mature
lens <- as.numeric(names(propm))        # Lengths in the data
pick <- which((lens > 79) & (lens < 146))
parset()
plot(lens[pick],propm[pick],type="p",cex=0.9, #the data points
     xlab="Length mm",ylab="Proportion Mature",pch=1)
L <- seq(80,145,1) # for increased curve separation
pars <- coef(smodel)
lines(L,mature(pars[1],pars[3],L),lwd=3,col=3,lty=2)
lines(L,mature(pars[1]+pars[2],pars[3],L),lwd=3,col=2,lty=4)
lines(L,mature(coef(model)[1],coef(model)[2],L),lwd=2,col=1,lty=1)
abline(h=c(0.25,0.5,0.75),lty=3,col="grey")
legend("topleft",c("site1","both","site2"),col=c(3,1,2),lty=c(2,1,4),
       lwd=3,bty="n")
```

Large sample sizes often improve the quality of the model fit. With sufficient
data it may be the case that the variability currently exhibited by the data
would be sufficiently reduced that a statistically significant difference could be
found between these sites. However, if we consider that the difference between
the Lm_{50} for site 1 and site 2 is about 2mm (out of 110), then biologically, such
a difference may not be important as other factors such as growth between the
two sites may also vary. As with any model or statistical analyses the biological
implications should be taken into account in addition to being concerned about
statistical significance.

5.6.3 The Assumption of Symmetry

In very many instances the standard logistic curve does a fine job of describing
a transition from immature to mature (or maybe some other biological transition, such as cryptic to emergent, moult stage A to stage B, etc). However,
a major restriction of the classical logistic is that it is symmetric around the
L_{50} point, which may not be an optimum description for real-world events.

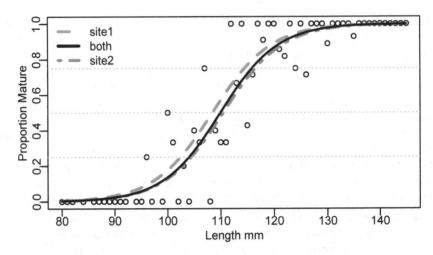

FIGURE 5.8 Proportion mature at length for blacklip abalone maturity data (*tasab* data-set). The combined analysis, without site (both) is closer to site 2 (dot-dash) than site 1 (dashes), which reflects site 2's larger sample size.

An array of alternative asymmetric curves have been suggested but fortunately, a general or unified model for the description of "fish growth, maturity, and survivorship data" was proposed by Schnute and Richards (1990), and this generalizes the classical logistic model used for maturity as well the growth models by Gompertz (1825), von Bertalanffy (1938), Richards (1959), Chapman (1961), and Schnute (1981), some of which can also be used to describe maturation.

The Schnute and Richards (1990) model has four parameters:

$$y^{-b} = 1 + \alpha \, exp(-a \, x^c) \qquad (5.13)$$

which can be re-arranged, **Equ**(5.14), to better illustrate the relationship with the classical logistic curve, **Equ**(5.12), which, if we set both $b = c = 1.0$ would be equivalent to the classical logistic (e.g., set $\alpha = 300$, and $a = 0.12$):

$$y = \frac{1}{(1 + \alpha \, exp(-a \, x^c)^{1/b}} \qquad (5.14)$$

If one of the special classical cases is used it may be that there are analytical solutions for determining the L_{50} and the IQ but otherwise they would need to be found numerically (see the **MQMF** functions bracket() and linter()). This curve can be fitted to maturity data using Binomial errors as before, although using nlm() or some other non-linear solver (Schnute and Richards,

1990, provide the required likelihoods). But it is likely that the special cases might provide more stable solutions to the full four-parameter model. The asymmetry of the possible curves from **Equ**(5.14) is easily demonstrated using the **MQMF** srug() function (Schnute and Richards Unified Growth). In the absence of an analytical solution to finding the L_{50} and IQ, we can use the two functions bracket() and linter(), which bracket the target values (in this case 0.25, 0.5, and 0.75) and then linearly interpolate to generate an approximate estimate of the required statistics (see their help files and examples).

```
#Asymmetrical maturity curve from Schnute and Richard's curve Fig.5.9
L = seq(50,160,1)
p=c(a=0.07,b=0.2,c=1.0,alpha=100)
asym <- srug(p=p,sizeage=L)
L25 <- linter(bracket(0.25,asym,L))
L50 <- linter(bracket(0.5,asym,L))
L75 <- linter(bracket(0.75,asym,L))
parset()
plot(L,asym,type="l",lwd=2,xlab="Length mm",ylab="Proportion Mature")
abline(h=c(0.25,0.5,0.75),lty=3,col="grey")
abline(v=c(L25,L50,L75),lwd=c(1,2,1),col=c(1,2,1))
```

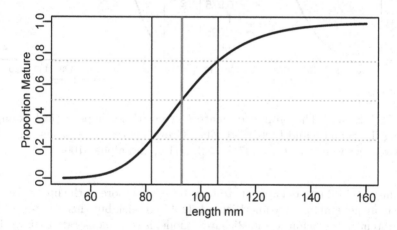

FIGURE 5.9 The proportion mature at length for a hypothetical example using the Schnute and Richards Unified Growth curve. The asymmetry of this logistic curve is illustrated by the difference between the left- and right-hand sides of the inter-quartile distance shown by the vertical thin lines.

Using the same parameters used in **Figure** 5.9 as a baseline, we can vary the individual parameters separately to determine the effect upon the resulting curves. The a and c parameters are relatively influential on the inter-quartile distance and all four are influential on the L_{50} (Figure 5.10).

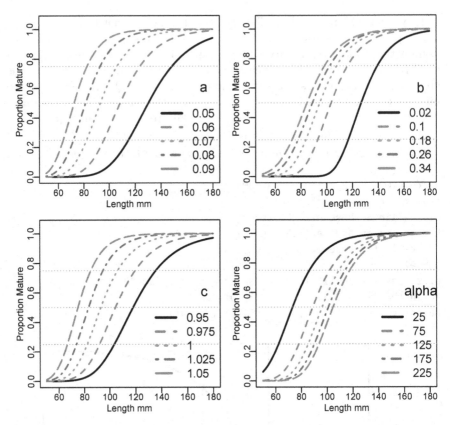

FIGURE 5.10 The proportion mature at length for hypothetical examples using the Schnute and Richards Unified Growth curve. The parameters when not varying were set at a=0.07, b=0.2, c=1.0, and alpha=100.

Schnute's (1981) model appears to have been used more in the literature than the more general Schnute and Richards (1990) model, but this is natural when the intent of the Schnute and Richard's model is to demonstrate that all these different curves can be unified and thus share a degree of commonality. This emphasizes that these curves provide primarily empirical descriptions of biological processes, which themselves are emergent properties of the populations being studied. Providing strict biological interpretations of their parameters and expecting nature to always provide for plausible or sensible parameter values is an extrapolation too far. An example is the situation described by Knight (1968), who was writing about growth but really discussing the reality or otherwise of giving explicit interpretations to the parameters of empirical models. He stated: "More important is the distorted point of view engendered by regarding L_∞ a fact of nature rather than as a mathematical artifact of the data analysis" (Knight, 1968, p 1306). Models that are descriptive rather than explanatory (hypothetical about a process) remain empirical descriptions and care is needed when trying to interpret their parameters.

5.7 Recruitment

5.7.1 Introduction

The main contributors to the production of biomass within a stock are the recruitment of new individuals to a population and the growth of individuals (where the definition of a stock means we need not consider immigration). Once again there is a huge literature on recruitment processes (Cushing, 1988; Myers and Barrowman, 1996; Myers, 2001), but here we will focus mainly on how it is described when attempting to include its dynamics within a model. As with other static models of growth and maturity such models are assumed to remain constant through time, although the option exists within stock assessment models of time-blocking different parameter sets within particular delineated time periods (Wayte, 2013). Here we will focus on the simple descriptions.

Recruitment to fish populations naturally tends to be highly variable, with the degree of variation varying among species. If you think I am using the term variation and related terms too often in a single sentence, perhaps that will reinforce the notion that recruitment really is highly variable. The main problem for fisheries scientists is whether recruitment is determined by the spawning stock size or environmental variation or some combination of both. This harks back to the debate in population dynamics from the 1950s about whether populations were controlled by density-dependent or density-independent effects (Andrewartha and Birch, 1954). This issue of variation has led to a commonly held but mistaken belief that the number of recruits to a fishery is effectively independent of the adult stock size over most of the observed range of stock sizes. This can be a dangerous assumption (Gulland, 1983). The notion of no recruitment relationship with spawning stock size suggests that scientists and managers can ignore stock recruitment relationships unless there is clear evidence that recruitment is not independent of stock size. The notion of no stock recruitment relationship existing derives from data on such relationships commonly appearing to be very scattered, with no obvious pattern.

Despite these arguments the modelling of recruitment processes has importance within fisheries models because when they are used to provide management advice forwards from the present, the model dynamics are often projected forwards while examining the effect of alternative harvest or effort levels. By including uncertainty in those projections it is possible to estimate the relative risk of alternative management arrangements failing to achieve desired management objectives (Francis, 1992). But to make such projections means having information about a stock's productivity. Assuming that the dynamics of individual growth have been covered, estimates of a time-series of recruitment levels or some defined stock recruitment relationship can be used to produce estimates of future recruitment levels for such projections.

In this chapter, we will consider only mathematical descriptions of stock recruitment relationships and mostly ignore the biology behind those relationships, although there are always exceptions. We will review just two of the most commonly used mathematical models of stock recruitment and will discuss their use in stock assessment models of varying complexity.

5.7.2 Properties of "Good" Stock Recruitment Relationships

A good introduction to the biological processes behind stock recruitment relationships is given by Cushing (1988), who provides an overview of the sources of egg and larval mortality along with good examples and a bibliography on the subject. As with the other fundamental production processes we have discussed, there is an enormous literature on the biology of stock recruitment relations and their modifiers.

A great variety of influences, both biological and physical, have been recorded as affecting the outcome of recruitment. We will not be considering the biological details of any real species except to point out that the relation between stock size and resulting recruitment is not deterministic, and there can be a number of forms of feedback affecting the outcome. Various mathematical descriptions of stock recruitment relationships have been suggested, but we will only consider those by Beverton and Holt and Ricker, but others such as that by Deriso-Schnute (Deriso, 1980; Schnute, 1985) are also important.

Ricker (1975) listed four properties of average stock recruitment relationships that he considered desirable:

- A stock recruitment curve should pass through the origin; that is, when stock size is zero, there should be no recruitment. This assumes the observations being considered relate to the total stock, and that there is no "recruitment" made up of immigrants.

- Recruitment should not fall to zero at high stock densities. This is not a necessary condition, but while declines in recruitment levels with increases in stock densities have been observed, declines to a zero recruitment index have not. Even if a population was at equilibrium at maximum stock biomass, recruitment should still match natural mortality levels.

- The rate of recruitment (recruits-per-spawner) should decrease continuously with increases in parental stock. This is only reasonable when positive density-dependent mechanisms (compensatory) are operating (for example, an increase in stock leads to an increase in larval mortality). But if negative density-dependent mechanisms (depensatory) are operating (for example, predator saturation and Allee effects; Begon and Mortimer, 1986), then this may not always hold.

- Recruitment must exceed parental stock over some part of the range of possible parental stocks. Strictly, this is only true for species spawning once before dying (e.g., salmon). For longer-lived, multi-spawning species, this should be interpreted as recruitment must be high enough over existing stock sizes to more than replace losses due to annual natural mortality.

Hilborn and Walters (1992) suggested two other general properties that they considered associated with good stock recruitment relationships:

- The average spawning stock recruitment curve should be continuous, with no sharp changes over small changes of stock size. They are referring to continuity, such that average recruitment should vary smoothly with stock size, related to condition 3 above.

- The average stock recruitment relationship is constant over time. This is stationarity, where the relationship does not change significantly through time. This assumption seems likely to fail in systems where the ecosystem, of which the exploited population is a part, changes markedly. But within models can be handled using time-blocks of parameters (for an example see Wayte, 2013).

5.7.3 Recruitment Overfishing

The term "overfishing" is commonly discussed in two contexts, growth overfishing and recruitment overfishing. Growth overfishing is where a stock is fished so hard that eventually most individuals are caught at a relatively small size. This is a problem in terms of yield-per-recruit (YPR). The analysis of YPR focusses upon balancing stock losses due to total mortality against the stock gains from individual growth with an aim to determining the optimum size and age at which to begin harvesting the species (where optimum can take a number of meanings). Growth overfishing is where the fish are being caught before they have time to reach this optimal size.

Recruitment overfishing is said to occur when a stock is fished so hard that the spawning stock size is reduced below the level at which it, as a population, can produce enough new recruits to replace those dying (either naturally or otherwise). Obviously, such a set of circumstances could not continue for long, and sadly, recruitment overfishing is usually a precursor to fishery collapse. Although keep in mind that fishery collapse usually means that a fishery is no longer economically viable, it does not mean literal extinction.

While growth overfishing is relatively simple to detect (is the stock at YPR optimum or not? Although, of course, there can be complications). Unfortunately, the same cannot be said about the detection of recruitment overfishing, which could require a determination of the relation between mature or spawning stock size and recruitment levels. This has proven to be a difficult task for very many fisheries. Instead, it has become common, with the advent of formal harvest strategies, to identify a spawning biomass level deemed to constitute an unacceptable risk to subsequent recruitment. A very common limit reference point is taken to be 20% of unfished spawning biomass ($0.2B_0$). The earliest reference to a Limit Reference Point depletion level of $20\%B_0$ appears to be Beddington & Cooke (1983). They explained a constraint they imposed on their analyses of potential yields from different stocks in the following manner:

"... an escapement level of 20% of the expected unexploited spawning stock biomass is used. This is not a conservative figure, but it represents a lower limit where recruitment declines might be expected to be observable" (Beddington & Cooke, 1983, p9-10).

The most influential document giving rise to the notion that $B_{20\%}$ (note the different ways of depicting $0.2B_0$) is a reasonable depletion level to use as an indicator of potential recruitment overfishing was a document prepared for the National Marine Fisheries Service in the USA (Restrepo *et al.*, 1998). In fact, they recommend $0.5B_{MSY}$ but consider $B20\%$ to be an acceptable proxy for that figure. However, it is important to note that this is only a 'rule of thumb' and there is no empirical basis that links the proxy $B_{LIM} = B_{20\%}$ or to $0.5B_{MSY}$. Indeed, selecting $0.5B_{MSY}$ for some species could result in B_{LIM} much lower than $B_{20\%}$.

5.7.4 Beverton and Holt Recruitment

Historically, Beverton and Holt's (1957) stock recruitment curve was useful because of its simple interpretation, which also meant it could be derived from first principles. Being mathematically tractable was also important to them as, at the time, only analytical methods were feasible. In fact, its continued use appears to stem a great deal from inertia and tradition. You should note that if we are to treat the Beverton–Holt curve simply as a mathematical description, then effectively any curve, with the good properties listed earlier, could be used. One talks of "the" Beverton-Holt stock recruitment model, however, it comes in more than one form. There are two common forms used:

$$R_y = \frac{B_{y-1}}{\alpha + \beta B_{y-1}} exp(N(0, \sigma_R^2)) \tag{5.15}$$

where R_y is the recruitment in year y, B_y is the spawning biomass in year y, α and β are the two Beverton-Holt parameters, and $exp(N(0, \sigma_R^2))$ represents the assumption that any relationship between predicted model values and observations from nature will be log-normally distributed (often represented as e^ε). The β value determines the asymptotic limit ($= 1/\beta$), while the differing values of α are inversely related to the rapidity with which each curve attains the asymptote, thus determining the relative steepness (a keyword, which we will hear much more about) near the origin (the smaller the value of α, the quicker the recruitment reaches a maximum). Of course, this is an average relationship and the scatter about the curve is as important as the curve itself. A common alternative re-parameterized formulation is:

$$R_y = \frac{aB_{y-1}}{b + B_{y-1}} e^\varepsilon \tag{5.16}$$

where a is the maximum recruitment ($a = 1/\beta$), and b is the spawning stock ($b = \alpha/\beta$) needed to produce, on average, half maximum recruitment ($a/2$). The use of $y-1$ for the biomass that gives rise to the recruits in year y has implications for the time of spawning. If spawning occurs in December and settlement of larval forms occurs in the following year then this would be correct, if they both occurred in the same year then obviously the subscript would need changing. The thing to note is that biological reality, in this case relating to timing, can even creep into very simple models. Mathematical models can provide wonderful representations of nature but one needs to be aware of the biology of the animals being modelled to avoid simple mistakes!

It is clear from **Figure** 5.11 that the initial steepness of the Beverton–Holt curve along with the asymptotic value captures the important aspects of the behaviour of the equation. The asymptote is given by the value of the parameter a, while the initial steepness is approximated by the value of ($a/b = 1/\alpha$), which happens when B_y is relatively small.

```
#plot the MQMF bh function for Beverton-Holt recruitment   Fig. 5.11
B <- 1:3000
bhb <- c(1000,500,250,150,50)
parset()
plot(B,bh(c(1000,bhb[1]),B),type="l",ylim=c(0,1050),
     xlab="Spawning Biomass",ylab="Recruitment")
for (i in 2:5) lines(B,bh(c(1000,bhb[i]),B),lwd=2,col=i,lty=i)
legend("bottomright",legend=bhb,col=c(1:5),lwd=3,bty="n",lty=c(1:5))
abline(h=c(500,1000),col=1,lty=2)
```

5.7.5   Ricker Recruitment

An alternative to the Beverton-Holt curve was suggested by Ricker (1954, 1958), but this also has more than one parameterization:

$$R_y = aS_y e^{-bS_y} e^\varepsilon \tag{5.17}$$

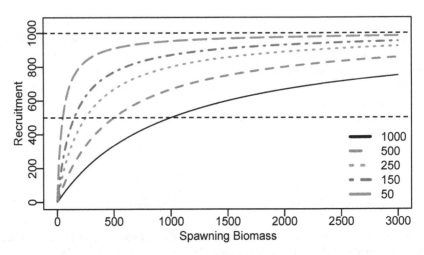

FIGURE 5.11 The Beverton-Holt stock recruitment curve with a constant a = 1000 and five decreasing b values leading to increasingly productive curves.

where R_y is the recruitment from S_y, the spawning stock, in year y, a is the recruits-per-spawner at low stock levels, and b relates to the rate of decrease of recruits-per-spawner as S increases. The e^ε indicates Log-Normal residual errors between the relationship and observed data. Note well that parameters a and b are very different from those in the Beverton-Holt equation. This stock recruitment curve does not attain an asymptote but instead exhibits a decline in recruitment levels at higher stock levels **Figure** 5.12.

```
#plot the MQMF ricker function for Ricker recruitment  Fig. 5.12
B <- 1:20000
rickb <- c(0.0002,0.0003,0.0004)
parset()
plot(B,ricker(c(10,rickb[1]),B),type="l",xlab="Spawning Biomass",
             ylab="Recruitment")
for (i in 2:3)
   lines(B,ricker(c(10,rickb[i]),B),lwd=2,col=i,lty=i)
legend("topright",legend=rickb,col=1:3,lty=1:3,bty="n",lwd=2)
```

The idea behind the decline in recruitment with spawning biomass is that it relates to either competitive or predatory effects (cannibalism) of some cohorts on others. Various mechanisms have been proposed including cannibalism of juveniles by adults, density-dependent transmission of disease, damage by spawning adults of each other's spawning sites (occurs primarily in rivers with fish like salmon), and finally, density-dependent growth combined with size-dependent predation. Each of these mechanisms can lead to different interpretations of the parameters of the Ricker curve.

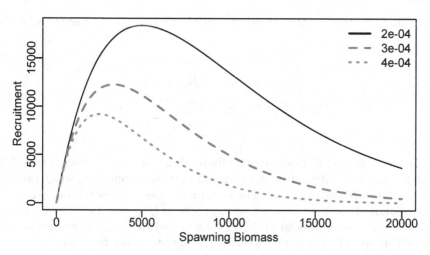

FIGURE 5.12 Two Ricker stock recruitment curves both with a constant $a = 10$ but with different b values. Note that the b value mainly influences the level of decline in recruitment with increasing biomass and has little effect on the initial steepness.

Once again, the distinction between whether the equation should be interpreted as an explanatory statement about the observable world instead of just a convenient empirical description of the average recruitment becomes important. In addition, while the parameters can certainly be given a real-world interpretation, the equations still tend to be overly simplistic and are best regarded as an empirical description rather than an explanation of events (Punt and Cope, 2019).

5.7.6 Deriso's Generalized Model

Deriso (1980) proposed a generalized equation for which the Beverton–Holt and the Ricker stock recruitment curves are special cases. Schnute (1985) restructured Deriso's equation to produce an even more flexible version with even greater flexibility:

$$R_y = \alpha S_y (1 - \beta \gamma S_y)^{1/\gamma} \tag{5.18}$$

In this Schnute (1985) generalized case there are three parameters α, β, and γ, the first two should always be positive but γ can be positive or negative.

By modifying the γ value, different special cases arise as in **Figure** 5.13:

$$
\begin{aligned}
\gamma = -\infty &\quad R_y = \alpha B_y \\
\gamma = -1 &\quad R_y = \alpha B_y / (1 + \beta B_y) \\
\gamma \to 0 &\quad R_y = \alpha B_y e^{-\beta B_y} \\
\gamma = 1 &\quad R_y = \alpha B_y (1 - \beta B_y)
\end{aligned}
\tag{5.19}
$$

The arrow in the third case means "approaches", as in γ approaches zero. The first case is a density-independent constant rate of recruitment, which can also be obtained by setting $\beta = 0$. The next three cases correspond respectively to the standard stock recruitment relationships of Beverton–Holt (1957), Ricker (1954, 1958), and Schaefer (1954), a form of logistic curve. Of course, in each case care would be required to select appropriate values for α and β.

There are some mathematically unstable properties to the Deriso–Schnute model as should be clear if we were to set $\gamma = 0$. This would lead to a mathematical singularity (divide by zero). The parameter limits should always be $\gamma \to 0$, from either the negative or the positive direction. With this equation there are many combinations of parameters that can give rise to implausible stock recruitment relationships. A major value of this equation is demonstrating the relationships between the different curves; one would not use the Deriso-Schnute in a fitted model, though it might be useful in a simulation model.

```
# plot of three special cases from Deriso-Schnute curve  Fig. 5.13
deriso <- function(p,B) return(p[1] * B *(1 - p[2]*p[3]*B)^(1/p[3]))
B <- 1:10000
plot1(B,deriso(c(10,0.001,-1),B),lwd=2,
      xlab="Spawning Biomass",ylab="Recruitment")        # BH
lines(B,deriso(c(10,0.0004,0.25),B),lwd=2,col=2,lty=2)  # DS
lines(B,deriso(c(10,0.0004,1e-06),B),lwd=2,col=3,lty=3) # Ricker
lines(B,deriso(c(10,0.0004,0.5),B),lwd=2,col=1,lty=3)   # odd line
legend(x=7000,y=8500,legend=c("BH","DS","Ricker","odd line"),
       col=c(1,2,3,1),lty=c(1,2,3,3),bty="n",lwd=3)
```

5.7.7   Re-Parameterized Beverton-Holt Equation

In the chapter *Model Parameter Estimation* the parameters of a Beverton-Holt stock-recruitment curve were estimated for the *tigers* data-set concerning tiger prawns. Estimates of relative abundance can equate to recruitment levels in short-lived prawn species, but with longer-lived species, assessed using age-structured models, estimating such parameters would be relatively abstract. An important development for more complicated stock assessment models was made by Francis (1992) who re-parameterized the Beverton-Holt stock

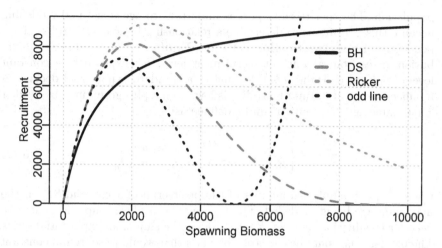

FIGURE 5.13 A comparison of the Beverton-Holt (BH), Ricker, and Deriso-Schnute (DS) stock recruitment curves, as implemented within the Deriso-Schnute generalized equation **Equ**(5.19). For the DS curve, a $\gamma = 0.5$ leads to unrealistic outcomes (odd line).

recruitment relationship in terms of the initial steepness of the curve, h, and the initial recruitment, R_0, which was derived from the unfished or virgin spawning biomass B_0. He first used these ideas in what was an age-structured surplus-production model for orange roughy (*Haplostethus atlanticus*), for which he had catch data, an index of relative abundance from surveys, and biological data relating to growth and weight at age. Using such data he was able to fit the model in a more defensible manner using his more meaningful parameterization of recruitment.

Francis (1992) defined steepness, designated h, as the deterministic number of recruits arising from the stock when the mature biomass is reduced to 20% of its unfished level. He started the description using:

$$R_0 = \frac{A_0}{\alpha + \beta A_0} \tag{5.20}$$

and thus, by defining steepness, $h = 0.2B_0$, we obtain:

$$hR_0 = \frac{0.2A_0}{\alpha + \beta 0.2A_0} \tag{5.21}$$

where α and β are the Beverton-Holt parameters, and A_0 is the total mature biomass per-recruit from the stable age distribution found in an unfished population. The "per-recruit" part is important to age-structured modelling because this permits us to determine a relationship between R_0 and B_0 independently of the Beverton-Holt equation, **Equ**(5.15). The stable age distribution arises from a constant recruitment level, R_0, exposed to a constant level of natural mortality (M), leading to a standard exponential decline on numbers-at-age. If natural mortality is low, then a plus group may be needed. That stable age distribution can be defined as:

$$n_{0,a} = \begin{cases} R_0 e^{-Ma} & a < t_{max} \\ R_0 e^{-Mt_{max}} / \left(1 - e^{-M}\right) & a = t_{max} \end{cases} \tag{5.22}$$

where $n_{0,a}$ is the unfished number of fish per-recruit of age a, and t_{max} is the maximum age being modelled. The t_{max} acts as a plus group and hence the need for the division by $(1-e^{-M})$, to provide the sum of an exponential series. The biomass A_0 would be the stock biomass that would arise from a constant recruitment level of one. Thus, with a biomass of A_0, distributed across a stable age distribution, the resulting recruitment level would be $R_0 = 1$ (and *vice versa*).

$$A_0 = \left(\sum_m n_{0,a} w_a \right) \tag{5.23}$$

where m is the age at maturity, $n_{0,a}$ is the unfished number of animals per-recruit of age a, and w_a is the weight of an animal of age a. One can also include part of natural mortality into this equation for it to become equivalent to recruitment part-way through a year. A_0 acts as a scaling factor because a stable age distribution will arise in the unfished population given any constant recruitment level. The magnitude of A_0 will be scaled by the estimated virgin biomass, B_0, but its value, relative to the constant recruitment needed to maintain the stable age distribution, R_0, will remain the same. In practice, one would derive A_0 from biological information and estimate R_0 in the stock assessment model:

$$B_0 = R_0 \left(\sum_m n_{0,a} w_a \right) = (R_0 A_0) \tag{5.24}$$

Francis (1992) then also used his definition of recruitment steepness to reparameterize the Beverton-Holt parameters a, b, α, and β (see appendix to this chapter):

$$a = \frac{4hR_0}{(5h - 1)} \qquad b = \frac{B_0(1 - h)}{(5h - 1)} \tag{5.25}$$

Substituting these into **Equ**(5.16) we obtain:

$$R_y = \frac{\frac{4hR_0}{(5h-1)}B_y}{\frac{B_0(1-h)}{(5h-1)} + B_y} \tag{5.26}$$

Dropping the upper level $(5h - 1)$ to the lower level:

$$R_y = \frac{4hR_0B_y}{\frac{(5h-1)B_0(1-h)}{(5h-1)} + (5h - 1)B_y} \tag{5.27}$$

then cancelling out the $(5h - 1)$ where possible leaves the Beverton-Holt redefined in terms of steepness h, B_0, and R_0, which all have more meaningful interpretations and using **Equ**(5.25), or **Equ**(5.28), would be more easily estimable in an assessment model:

$$R_y = \frac{4hR_0B_y}{B_0(1 - h) + B_y(5h - 1)} \tag{5.28}$$

5.7.8 Re-Parameterized Ricker Equation

In a similar fashion, The Ricker stock recruitment equation can be re-parameterized in terms of steepness h, B_0, and R_0:

$$R_y = \frac{R_0B_y}{B_0}exp\left(h\left(1 - \frac{B_y}{B_0} \right) \right) \tag{5.29}$$

Both **Equs**(5.25) and (5.28) are now the more commonly used parameterizations used in stock assessments and simulation models that attempt to include recruitment.

5.8 Selectivity

5.8.1 Introduction

Another class of static models used within stock assessment models (and simulations) relate to the selectivity of fishing gear for particular species. The idea being that if a population of an exploited species is available in an area then if a particular fishing gear is used (a trawl, a Danish seine, a gill-net, a lobster pot, etc.) then the construction of the gear and how it is used will influence which members of the available population will be vulnerable to the gear if they encounter it. The notion of selectivity within, say, an age-structured assessment model is complicated by the notion of availability. If, for example, the main fishing grounds for a species are in waters deeper than 250 m and predominantly only smaller juveniles are present in shallower waters then if one were to estimate the selectivity of the gear using data only from the shallower water one would likely get a different answer from an estimate made with data from deeper waters. In fact, selectivity curves estimated within assessment models should really be regarded as selectivity/availability curves.

The choice of the shape of a selectivity curve is an important decision. Some fishing gears typically have selectivity described by specific equations. Thus, the selectivity of trawl gears are often described using a logistic equation, whereas long-line gear (using hooks) is often described using a dome-shaped selectivity function.

One can only fit such models of selectivity within stock assessment models if there are age- or size-composition data available from the catches. As we saw in the *Model Parameter Estimation* chapter, one would usually fit such composition data using multinomial likelihoods. Within a stock assessment model one would normally model the stock dynamics in terms of expected numbers-at-age, which would imply certain numbers-at-size. So, when attempting to generate predicted catch composition data (of either ages or sizes) one needs to multiply the available numbers-at-age or size by the predicted selectivity. Thus, during the fitting process the composition data contribute to the estimation of the selectivity parameters.

Here we will merely illustrate alternative selectivity equations and show their different properties.

5.8.2 Logistic Selection

There are many different equations used to describe the selectivity characteristics of different fishing gears but an extremely common one is the standard logistic or S-shaped curve, which is typical for trawl gear selectivity. It implies that the vulnerability to the gear will gradually increase as age or size increases until 100% are vulnerable if they encounter the gear (this gradual increase to 100% is the same as we saw in maturity). Two equations are commonly used, the first is described by the **MQMF** function `logist()`:

$$s_a = \frac{1}{1 + e^{-\log(19)(a - a_{50})/\delta}} \tag{5.30}$$

where s_a is the selectivity-at-age (a proportion) for age a, a_{50} is the age at which 50% selectivity occurs, and δ is the gradient of the curve defined as the number of years from a selectivity of 50% to a selectivity of 95%. Where we talk of age we could equally be talking of size. The upper bound of δ is 95% (δ is actually $L95 - L50$). An alternative logistic curve, that we have already seen in the section *Selectivity in YPR* within the *Simple Population Models* chapter, and was also used to describe maturity-at-age, is defined in the function `mature()`:

$$s_a = \frac{1}{1 + \left(e^{(\alpha+\beta a)}\right)^{-1}} = \frac{e^{(\alpha+\beta a)}}{1 + e^{(\alpha+\beta a)}} \tag{5.31}$$

where α and β are the logistic parameters, and $-\alpha/\beta$ is the age at a selectivity of 0.5 (50%). The inter-quartile distance (literally quantile 25% to quantile 75%) is defined as $IQ = 2\log{(3)}/\beta$ (see the **MQMF** function `mature()` for an implementation of this function). Generally, in age-structured modelling one needs length- or age-composition data so that the gear selectivity combined with fishery availability can be estimated directly. When working with yield-per-recruit calculations the usual reason to include a form of selectivity is to determine the optimum age at which to begin applying fishing mortality. For this reason what is known as knife-edged selectivity is often used, which essentially identifies the specific age below which there is no selection and above which there is 100% selection. This is implemented in the **MQMF** function `logist()`; knife-edged selectivity no longer tends to be used in full age-structured stock assessment models, although it is still used in delay-difference models (Schnute, 1985; Hilborn and Walters, 1992).

```
#Selectivity curves from logist and mature functions  See Fig. 5.14
ages <- seq(0,50,1);   in50 <- 25.0
sel1 <- logist(in50,12,ages)        #-3.65/0.146=L50=25.0
sel2 <- mature(-3.650425,0.146017,sizeage=ages)
sel3 <- mature(-6,0.2,ages)
sel4 <- logist(22.0,14,ages,knifeedge = TRUE)
plot1(ages,sel1,xlab="Age Years",ylab="Selectivity",cex=0.75,lwd=2)
lines(ages,sel2,col=2,lwd=2,lty=2)
lines(ages,sel3,col=3,lwd=2,lty=3)
lines(ages,sel4,col=4,lwd=2,lty=4)
abline(v=in50,col=1,lty=2); abline(h=0.5,col=1,lty=2)
legend("topleft",c("25_eq5.30","25_eq5.31","30_eq5.31","22_eq5.30N"),
       col=c(1,2,3,4),lwd=3,cex=1.1,bty="n",lty=c(1:4))
```

5.8.3   Dome-Shaped Selection

A selectivity pattern that ascends to a peak, may possibly have a plateau, and then has a descending limb is termed dome-shaped and is typical of fishing gear such as gill-nets and hook-based gear such as long-lines. Having so many components such selectivity curves tend to be much more complex because the ascending, plateau, and descending parts all need to be joined together. Modern approaches to fitting such models tend to use auto-differentiation software such as AD-Model Builder or related software (Bull *et al*, 2012; Fournier

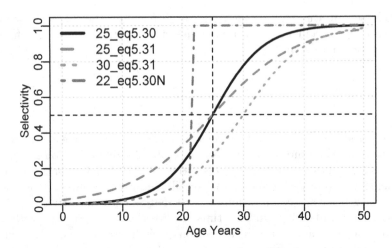

FIGURE 5.14 Examples of the logistic S-shaped curve from the `logist()` and `mature()` functions. The dashed red and solid black curves have the same *L50*, though different gradients. The dotted green illustrate the effect of varying the b parameter of the `mature()` function, while the hashed blue curve illustrates knife-edged selection. The legend depicts the *L50* and the equation used.

et al, 2012; Kristensen *et al*, 2016). This means component models within an assessment model need to be differentiable so the joints between the three components of a domed selectivity curve need to be continuous (Methot and Wetzell, 2013; Hurtado-Ferro *et al*, 2014). Such an equation would have at least five components the ascending limb, *asc*, the flat top where selectivity equals 1.0, the descending limb, *dsc*, and two joining functions, J_1 and J_2, between between the primary three, **Equ**(5.32):

$$s_L = asc\left(1 - J_{1,L}\right) + J_{1,L}\left(\left(1 - J_{2,L}\right) + J_{2,L}dsc\right) \tag{5.32}$$

where s_L is the selectivity of length L. The various component functions are defined as:

$$
\begin{aligned}
asc &= 1 - (1 - \lambda_5)\left(\frac{1 - \exp\left((m_L - \lambda_1)^2/\lambda_3\right)}{1 - \exp\left((m_{\min} - \lambda_1)^2/\lambda_3\right)}\right) \\
dsc &= 1 - \exp\left(-\left((m_L - \lambda_2)^2/\lambda_4\right)\right) \\
J_{1,L} &= 1/\left(1 + \exp\left(-(m_L - \lambda_1)/\left(1 + |m_L + \lambda_1|\right)\right)\right) \\
J_{2,L} &= 1/\left(1 + \exp\left(-(m_L - \lambda_2)/\left(1 + |m_L + \lambda_2|\right)\right)\right)
\end{aligned}
\tag{5.33}
$$

where m_L is the mean length of length class L, m_{min} is the mean size of the smallest length class, λ_1 is the size at which selectivity reaches 1.0, λ_2 is the size at which selectivity starts decreasing from 1.0 (if λ_1 and λ_2 are equal there would be no plateau), λ_3 affects the slope of the ascending limb, asc, λ_4 affects the slope of the descending limb, dsc, λ_5 is the log of the selectivity at m_{min}, and λ_6 is the log of the selectivity of the m_{max}, **Figure** 5.15.

There are thus six parameters, λ_1 to λ_6, required, along with the mean lengths of the length classes used, to define this domed selectivity curve.

```
#Examples of domed-shaped selectivity curves from domed. Fig. 5.15
L <- seq(1,30,1)
p <- c(10,11,16,33,-5,-2)
plot1(L,domed(p,L),type="l",lwd=2,ylab="Selectivity",xlab="Age Years")
p1 <- c(8,12,16,33,-5,-1)
lines(L,domed(p1,L),lwd=2,col=2,lty=2)
p2 <- c(9,10,16,33,-5,-4)
lines(L,domed(p2,L),lwd=2,col=4,lty=4)
```

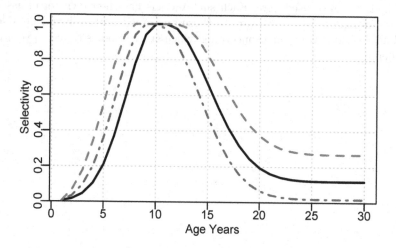

FIGURE 5.15 Three examples of domed-shaped selectivity curves produced by the function domed(), with changes to the initial age at achieving a selectivity of 1.0 (10, 8, 9), and the age (11, 12, 10) at which that stops, as well as the selectivity of the final age class.

5.9   Concluding Remarks for Static Models

Stock assessment models that are more complex than the surplus production models we have considered are mostly combinations of dynamic models of the progression of cohorts mixed up with a collection of static models of the kind illustrated in this chapter. Thus, a knowledge of models of growth, maturity, selectivity, and recruitment is essential for any understanding of the structure of more advanced models. Sometimes one would estimate their parameters outside of an assessment model but often the estimation would be part of fitting the whole stock assessment model. The advantage of fitting all the separate model components at once is that any interactions between the components can be automatically accounted for.

This chapter has provided an introductory foundation to quite an array of static models. The objective being that if and when you come across alternatives you will be able to use them and fit them in a manner similar to the methods we have used here. Each such models has their own set of assumptions. As long as you are aware of those assumptions you should be able to defend decisions you make concerning which static models to use in your own modelling.

5.10 Appendix: Derivation of Fabens Transformation

Fabens (1965) transformed the von Bertalanffy growth curve so that it matched the data made available from tagging programmes (the length and date at tagging, and the length and date at recapture). The length-at-age version of the von Bertalanffy curve is:

$$\hat{L}_t = L_\infty \left(1 - e^{(-K(t-t_0))}\right) + N(0,\sigma) \tag{5.34}$$

which, in a tagging context has a length after a given time increment (Δt) defined thus:

$$L_{t+\Delta t} = L_\infty - L_\infty e^{-K((t+\Delta t)-t_0)} \tag{5.35}$$

Being in the exponential term the contribution from the Δt can be extracted:

$$L_{t+\Delta t} = L_\infty - L_\infty e^{-K(t-t_0)} e^{-K\Delta t} \tag{5.36}$$

Now the growth increment ΔL expected to occur during Δt can be defined as L_t subtracted from $L_{t+\Delta t}$:

$$\begin{aligned} \Delta L = \left(L_{t+\Delta t} - L_t\right) = \\ \left(L_\infty - L_\infty e^{-K(t-t_0)} e^{-K\Delta t}\right) - \left(L_\infty - L_\infty e^{(-K(t-t_0))}\right) \end{aligned} \tag{5.37}$$

We can remove the brackets, which will alter the negation in the second term to a plus and then re-arrange the separated terms:

$$\Delta L = L_\infty - L_\infty e^{-K(t-t_0)} e^{-K\Delta t} - L_\infty + L_\infty e^{(-K(t-t_0))} \tag{5.38}$$

The two isolated L_∞ terms can be cancelled out to improve the clarity of the demonstration, but they will be returned a little later. In addition we can re-arrange the two remaining exponential terms:

$$\Delta L = L_\infty e^{(-K(t-t_0))} - L_\infty e^{-K(t-t_0)} e^{-K\Delta t} \tag{5.39}$$

the re-arrangement makes it possible to extract out a $L_\infty e^{(-K(t-t_0))}$ term to simplify the whole:

$$\Delta L = L_\infty e^{(-K(t-t_0))} \left(1 - e^{-K\Delta t}\right) \tag{5.40}$$

So now we can shift the Δt term to the left and return the $L_\infty - L_\infty$ term:

$$\frac{\Delta L}{\left(1 - e^{-K\Delta t}\right)} = L_\infty - L_\infty + L_\infty e^{(-K(t-t_0))} \tag{5.41}$$

If we add some brackets, which will reverse the plus to a minus, we can finally reach the point of recognizing the combination of terms that equals L_t:

$$\frac{\Delta \hat{L}}{\left(1 - e^{-K\Delta t}\right)} = L_\infty - \left(L_\infty - L_\infty e^{(-K(t-t_0))}\right) = (L_\infty - L_t) \tag{5.42}$$

which, by returning the Δt term to the right results in the classical Fabens growth increment equation:

$$\Delta L = (L_\infty - L_t)\left(1 - e^{-K\Delta t}\right) \tag{5.43}$$

It is important to note that this transformation also transforms the Normal random residuals from the length-at-age equation, which, of course, are used when estimating the parameters. This implies the parameters, even though they have the same names, are not strictly comparable.

5.11   Appendix: Reparameterization of Beverton-Holt

Francis (1992) provides definitions of the Beverton–Holt parameters in terms of the more biologically meaningful terms relating to steepness (h), virgin mature biomass (B_0), and virgin recruitment (R_0). Rather than use

$$R_y = \frac{B_{y-1}}{\alpha + \beta B_{y-1}} \tag{5.44}$$

the alternative formulation is used, although, at the end, we will revert to using the relationships $\alpha = b/a$ and $\beta = 1/a$.

Thus we will use:

$$R_y = \frac{aB_y}{b + B_y} \tag{5.45}$$

where R_y is the recruitment in year y, B_y is the spawning stock size in year y, and a and b are the Beverton–Holt parameters. In an unfished equilibrium state we can couch the equation as:

$$R_0 = \frac{aB_0}{b + B_0} \tag{5.46}$$

Steepness (h) is defined as the recruitment obtained at 20% of virgin biomass:

$$hR_0 = \frac{0.2aB_0}{b + 0.2B_0} \tag{5.47}$$

If the equilibrium equation for R_0 is substituted into this equation we obtain:

$$h\frac{aB_0}{b + B_0} = \frac{0.2aB_0}{b + 0.2B_0} \tag{5.48}$$

and hence:

$$h = \frac{(0.2aB_0)(b + B_0)}{(b + 0.2B_0)(aB_0)} = \frac{0.2(b + B_0)}{b + 0.2B_0} \tag{5.49}$$

Multiplying through leads to:

$$hb + 0.2hB_0 = 0.2b + 0.2B_0 \tag{5.50}$$

and exchanging terms to isolate b and B_0 and multiplying by 5 to remove the 0.2 values:

$$5hb - b = B_0 - hB_0 \tag{5.51}$$

and simplifying leads to:

$$b(5h - 1) = B_0(1 - h) \tag{5.52}$$

thus b can be reparameterized as:

$$b = \frac{B_0(1 - h)}{(5h - 1)} \tag{5.53}$$

This version of b can be used in our original equation and re-arranged to do something similar for the a parameter:

$$R_0 = \frac{aB_0}{\frac{B_0(1-h)}{(5h-1)} + B_0} \tag{5.54}$$

which converts to:

$$R_0 \frac{B_0 (1-h)}{(5h-1)} + R_0 B_0 = aB_0 \tag{5.55}$$

Dividing through by B_0 and then multiplying the second R_0 on the left hand of the = by $5h$–1 allows a simplification:

$$\frac{R_0 - hR_0 + 5hR_0 - R_0}{5h-1} = \frac{4hR_0}{5h-1} = a \tag{5.56}$$

Remember that $\alpha = b/a$ and $\beta = 1/a$, so we can finish with

$$\alpha = b \times 1/a = \frac{B_0 (1-h)}{(5h-1)} \times \frac{5h-1}{4hR_0} = \frac{B_0(1-h)}{4hR_0} \tag{5.57}$$

and

$$\beta = 1/a = \frac{5h-1}{4hR_0} \tag{5.58}$$

as defined by Francis (1992). This has redefined the Beverton-Holt model parameters in terms of h, B_0, and R_0. The relationship required between B_0 and R_0 is required but we cannot use the equilibrium equation for that so instead we assume that the unfished population is in equilibrium with a stable age distribution. The mature biomass generated per recruit from the stable age distribution (A_0) therefore defines the required relationship between B_0 and R_0.

6

On Uncertainty

6.1 Introduction

Fitting a model to a set of data involves searching for parameter estimates that optimize the relationship between the observed data and the predictions of the model. The model parameter estimates are taken to represent properties of the population in which we are interested. While it should be possible to find optimum parameter values in each situation, it remains the case that whatever data is used constitutes only one set of samples from the population. Assuming it were possible to take different, independent samples of the same kinds of data from the same population, if they were then analysed independently, they would very likely lead to at least slightly different parameter estimates. Thus, the exact value of the estimated parameters is not really the most important issue when fitting a model to data, rather, we want to know how repeatable such estimates are likely to be if we were to have the luxury of multiple independent samples. That is, the parameters are just estimates and we want to know how confident we can be in those estimates. For example, it is common that highly variable data usually leads to each model parameter potentially having a wide distribution of plausible estimates; typically they would be expected to have wide confidence intervals.

In this chapter we will explore alternative ways available for characterizing at least a proportion of the uncertainty inherent in any modelling situation. While there can be many sources of potential uncertainty that can act to hamper our capacity to manage natural resources, only some of these can be usefully investigated here. Some sources of uncertainty influence the variability of data collected, other sources can influence the type of data available.

6.1.1 Types of Uncertainty

There are various ways of describing the different types of uncertainty that can influence parameter estimates from fisheries models and, thus, how such estimates can be used. These are often called sources of error, usually in the sense of residual error rather than as a mistake having been made. Unfortunately, the term "error", as in residual error, has the potential to lead to confusion, so it is best to use the term "uncertainty". Although as long as you are aware of the issue it should not be a problem for you.

Francis and Shotton (1997) listed six different types of uncertainty relating to natural resource assessment and management, while Kell *et al* (2005) following Rosenberg and Restrepo (1994), contract these to five. Alternatively, they can all be summarized under four headings (Haddon, 2011), some with sub-headings, as follows:

- Process uncertainty: underlying natural random variation in demographic rates (such as growth, inter-annual mean recruitment, inter-annual natural mortality) and other biological properties and processes.

- Observation uncertainty: sampling error and measurement error reflect the fact that samples are meant to represent a population but remain only a sample. Inadequate or non-representative data collection would contribute to observational uncertainty as would any mistakes or the deliberate misreporting or lack of reporting of data, which is not unknown in the world of fisheries statistics.

- Model uncertainty: relates to the capacity of the selected model structure to describe the dynamics of the system under study:

  - Different structural models may provide different answers and uncertainty exists over which is the better representation of nature.

  - The selection of the residual error structure for a given process is a special case of model uncertainty that can have important implications for parameter estimates.

  - Estimation uncertainty is an outcome of the model structure having interactions or correlations between parameters such that slightly different parameter sets can lead to identical values of log-likelihood (to the limit of precision used in the numerical model fitting).

- Implementation uncertainty: where the effects or extent of management actions may differ from those intended. Poorly defined management options may lead to implementation uncertainty.

  - Institutional uncertainty: inadequately defined management objectives leading to unworkable management.

  - Time-lags between making decisions and implementing them can lead to greater variation. Assessments are often made a year or more after fisheries data is collected and management decisions often take another year or more to be implemented.

Here we are especially concerned with process, observational, and model uncertainty. Each can influence model parameter estimates and model outcomes and predictions. Implementation uncertainty is more about how the outcomes from a model or models are used when managing natural resources. This can have important effects on the efficiency of different management strategies based on the outcomes of stock assessment models but it does not influence those immediate outcomes (Dichmont *et al*, 2006).

Model uncertainty can be both quantitative and qualitative. Thus, different models that use exactly the same data and residual structures could be compared to one another and the best fitting selected. Such models may be considered to be related but different. However, where models are incommensurate, for example, when different residual error structures are used with the same structural model, they can each generate an optimum fit and model selection must be based on factors other than just quality of fit. Such models do not grade smoothly into each other but constitute qualitatively and quantitatively different descriptions of the system under study. Model uncertainty is one of the driving forces behind model selection (Burnham and Anderson, 2002). Even where there is only one model developed invariably this has been implicitly selected from many possible models. Working with more than one type of model in a given situation (perhaps a surplus production model along with a fully age-structured model) can often lead to insights that using one model alone would be missed. Sadly, with decreasing resources generally made available for stock assessment modelling, working with more than one model is now an increasingly rare option.

Model and implementation uncertainty are both important in the management of natural resources. We have already compared the outcomes from alternative models, however, here we are going to concentrate on methods that allow us to characterize the confidence with which we can accept the various parameter estimates and other model outputs obtained from given models that have significance for management advice.

There are a number of strategies available for characterizing the uncertainty around any parameter estimates or other model outputs. Some methods focus on data issues while other methods focus on the potential distributions of plausible parameter values given the available data. We will consider four different approaches:

1) *bootstrapping*, focusses on the uncertainty inherent in the data samples and operates by examining the implications for the parameter estimates had somewhat different samples been taken,

2) *asymptotic errors* use a variance-covariance matrix between parameter estimates to describe the uncertainty around those parameter values,

3) *likelihood profiles* are constructed on parameters of primary interest to obtain more specific distributions of each parameter, and finally,

4) *Bayesian marginal posteriors* characterize the uncertainty inherent
   in estimates of model parameters and outputs.

Each of these approaches permit the characterization of the uncertainty of
both parameters and model outputs and provide for the identification of
selected central percentile ranges around each parameter's mean or expectation (e.g., $\bar{x} \pm 90\% CI$, in some case may not be symmetric).

We will use a simple surplus production model to illustrate all of these methods, although they should be applicable more generally.

6.1.2   The Example Model

This is the same model as described in the section on fitting a dynamic model
using Log-Normal likelihoods in the chapter on *Model Parameter Estimation*.
We will use this in many of the examples in this chapter and it has relatively
simple population dynamics, **Equ(6.1)**:

$$B_{t=0} = B_{init}$$
$$B_{t+1} = B_t + rB_t \left(1 - \frac{B_t}{K}\right) - C_t \tag{6.1}$$

where B_t represents the available biomass in year t, with B_{init} being the
biomass in the first year of available data ($t = 0$), taking account of any
initial depletion when records begin. r is the intrinsic rate of natural increase
(a population growth rate term), and K is the carrying capacity or unfished
available biomass, commonly referred to elsewhere as B_0 (not to be confused
with B_{init}). Finally, C_t is the total catch taken in year t. To connect these
dynamics to observations from the fishery other than the catches we use an
index of relative abundance (I_t, often catch-per-unit-effort or cpue, but could
be a survey index);

$$I_t = \frac{C_t}{E_t} = qB_t \quad or \quad C_t = qE_tB_t \tag{6.2}$$

where I_t is the catch-rate (CPUE or cpue) in year t, E_t is the effort in year t,
and q is known as the catchability coefficient (which can also change through
time but we have assumed it to be constant). Because the q merely scales the
stock biomass against catches, we will use the closed form of estimating catchability to reduce the number of parameters requiring estimation (Polacheck
et al, 1993):

$$q = e^{\frac{1}{n}\sum \log\left(\frac{I_t}{\hat{B}_t}\right)} = \exp\left(\frac{1}{n}\sum \log\left(\frac{I_t}{\hat{B}_t}\right)\right) \tag{6.3}$$

We will use the **MQMF** *abdat* data-set and fit the model here, then we can examine the uncertainty in that model fit in the following sections. We will be fitting the model by minimizing the negative log-likelihood with residuals based upon the Log-Normal distribution. Simplified these become:

$$-LL(y|\theta, \hat{\sigma}, I) = \frac{n}{2}\left(\log(2\pi) + 2\log(\hat{\sigma}) + 1\right) + \sum_{i=1}^{n}\log(I_t) \qquad (6.4)$$

where θ is the vector of parameters (r, K, and B_{init}), the I_t are the n observed cpue values across the years t, and the maximum likelihood estimate of $\hat{\sigma}^2$ is defined as:

$$\hat{\sigma}^2 = \sum_{t=1}^{n} \frac{\left(\log(I_t) - \log(\hat{I}_t)\right)^2}{n} \qquad (6.5)$$

Note the division by n and not $n - 1$. Given the only non-constant values are the $\log(\hat{I}_t)$, the result obtained from using maximum likelihood will be the same as when using a least-squared residual approach (as long as the observed and predicted cpue are both log-transformed). Nevertheless, using maximum likelihood methods has advantages over sum-of-squares when we are examining uncertainty. Strictly, the $\hat{\sigma}$ value is also a model parameter but we are treating it specially here just to illustrate the equivalence with least-squared approaches.

```
#Fit a surplus production model to abdat fisheries data
data(abdat); logce <- log(abdat$cpue)
param <- log(c(0.42,9400,3400,0.05))
label=c("r","K","Binit","sigma") # simpspm returns
bestmod <- nlm(f=negLL,p=param,funk=simpspm,indat=abdat,logobs=logce)
outfit(bestmod,title="SP-Model",parnames=label) #backtransforms
```

```
# nlm solution:  SP-Model
# minimum      :  -41.37511
# iterations   :  20
# code         :  2 >1 iterates in tolerance, probably solution
#              par        gradient     transpar
# r       -0.9429555  6.707523e-06    0.38948
# K        9.1191569 -9.225209e-05 9128.50173
# Binit    8.1271026  1.059296e-04 3384.97779
# sigma   -3.1429030 -8.161433e-07    0.04316
```

The model fit can be visualized as in **Figure** 6.1. The simple dynamics of the Schaefer model appear to provide a plausible description of these observed abalone catch-rate data. In reality, there were changes to legal minimum sizes, the introduction of zonation within the fishery, and other important changes across the time-series, so this result can only be considered an approximation at very best and should only be considered as providing an example of the methodology. The Log-Normal residuals are observed/predicted, and if multiplied by the predicted would obviously return the observed (grey-line). This is illustrated here because we are about to use this simple relationship when bootstrapping the data.

The Maximum Sustainable Yield (MSY) can be estimated from the Schaefer model simply as MSY $= rK/4$, which means this optimal fit implies an MSY $=$ 888.842t. The two questions we will attempt to answer in each of the following sections will be: what is the plausible spread of the predicted cpue around the observed data in **Figure** 6.1? And, what would be the 90th percentile confidence bounds around the mean MSY estimate?

```
#plot the abdat data and the optimum sp-model fit  Fig. 6.1
predce <- exp(simpspm(bestmod$estimate,abdat))
optresid <- abdat[,"cpue"]/predce #multiply by predce for obsce
ymax <- getmax(c(predce,abdat$cpue))
plot1(abdat$year,(predce*optresid),type="l",maxy=ymax,cex=0.9,
      ylab="CPUE",xlab="Year",lwd=3,col="grey",lty=1)
points(abdat$year,abdat$cpue,pch=1,col=1,cex=1.1)
lines(abdat$year,predce,lwd=2,col=1)  # best fit line
```

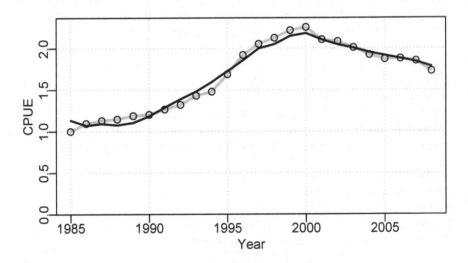

FIGURE 6.1 The optimum fit of the Schaefer surplus production model to the abdat data-set plotted in linear-space (solid red-line). The grey line passes through the data points to clarify the difference with the predicted line.

6.2 Bootstrapping

Data sampled from a population are treated as being (assumed to be) representative of that population and of the underlying probability density distribution of expected sample values. This is a very important assumption first suggested by Efron (1979). He asked the question: when a sample contains or is all of the available information about a population, why not proceed as if the sample really is the population for purposes of estimating the sampling distribution of the test statistic? Thus, given an original sample of n observations, bootstrap samples would be random samples of n observations taken from the original sample **with replacement**. Bootstrap samples (i.e., random re-sampling from the sample data values with replacement) are assumed to approximate the distribution of values that would have arisen from repeatedly sampling the original sampled population. Each of these bootstrapped samples is treated as an independent random sample from the original population. This resampling with replacement appears counter-intuitive to some but can be used to fit standard errors, percentile confidence intervals, and to test hypotheses. The name bootstrap is reported to derive from the story *The Adventures of Baron Munchausen*, in which the Baron escaped drowning by pulling himself up by his own bootstraps and thereby escaping from a water well (Efron and Tibshirani, 1993).

Efron (1979) first suggested bootstrapping as a practical procedure and Bickel and Freedman (1981) provided a demonstration of the asymptotic consistency of the bootstrap (convergent behaviour as the number of bootstrap samples increased). Given this demonstration, the bootstrap approach has been applied to numerous standard applications, such as multiple regression (Freedman, 1981; ter Braak, 1992) and stratified sampling (Bickel and Freedman, 1984, who found a limitation). Efron eventually converted the material he had been teaching to senior-level students at Stanford into a general summary of progress (Efron and Tibshirani, 1993).

6.2.1 Empirical Probability Density Distributions

The assumption is that given a sample from a population, the non-parametric, maximum likelihood estimate of the population's probability density distribution is the sample itself. That is, if the sample consists of n observations $(x_1, x_2, x_3, ..., x_n)$, the maximum likelihood, non parametric estimator of the probability density distribution is the function that places probability mass $1/n$ on each of the n observations x_i. It must be emphasized that this is not saying that all values have equal likelihood; instead, it implies that each observation has equal likelihood of occurring, although there may be multiple observations having the same value (make sure you are clear about this distinction before proceeding!). If the population variable being sampled has a modal value then one expects, some of the time, to obtain the same or similar values near that mode more often that values out at the extremes of the samples distribution.

Bootstrapping consists of applying Monte Carlo procedures, sampling with replacement but from the original sample itself, as if it were a theoretical statistical distribution (analogous to the Normal, Gamma, and Beta distributions). Sampling with replacement is consistent with a population that is essentially infinite. Therefore, we are treating the sample as representing the total population.

In summary, bootstrap methods are used to estimate the uncertainty around the value of a parameter or model output. This is done by summarizing many bootstrap parameter estimates from replicate samples derived from replacing the true population sample by one estimated from the original sample from the population.

6.3 A Simple Bootstrap Example

To gain an appreciation of how to implement a bootstrap in R it is sensible to start with a simple example. The Australian Northern prawn fishery, within the Gulf of Carpentaria and across to Joseph Bonaparte Bay, along the top right-hand side of the continent, is a mixed fishery that captures multiple species of prawn (Dichmont *et al*, 2006; Robins and Somers, 1994). We will use an example of the catches of tiger prawns (*Penaeus semisulcatus* and *P. esculentus*) and endeavour prawns (*Metapenaeus endeavouri* and *M. ensis*) taken between 1970 to 1992. There appears to be a correlation between those catches, **Figure** 6.2, but there is a good deal of scatter in the data. The endeavour prawns are invariably a bycatch in the much more valuable tiger prawn fishery, and this is reflected in their relative catch quantities.

The prawn catch data are relatively noisy, which is not unexpected with prawn catches. That endeavour and tiger prawn catches are correlated should also not be surprising. The endeavour prawns are generally taken as bycatch in the more valuable tiger prawn fishery, so one would expect the total tiger prawn catch to have some relationship with the total catch of endeavour prawns. If, on the other hand, you were to plot the banana prawn catch relative to the tiger prawn catch no such relationship would be expected because these are two almost independent fisheries (mostly in the same areas) but one fished during daytime the other at night. Exploring such relationships in mixed fisheries can often leads to hypotheses concerning the possibility of interactions between species or between individual fisheries.

```
#regression between catches of NPF prawn species Fig. 6.2
data(npf)
model <- lm(endeavour ~ tiger,data=npf)
plot1(npf$tiger,npf$endeavour,type="p",xlab="Tiger Prawn (t)",
      ylab="Endeavour Prawn (t)",cex=0.9)
abline(model,col=1,lwd=2)
correl <- sqrt(summary(model)$r.squared)
pval <- summary(model)$coefficients[2,4]
label <- paste0("Correlation ",round(correl,5)," P = ",round(pval,8))
text(2700,180,label,cex=1.0,font=7,pos=4)
```

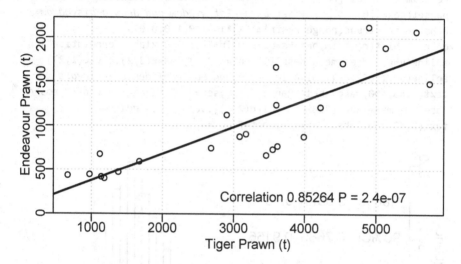

FIGURE 6.2 The positive correlation between the catches of endeavour and tiger prawns in the Australian Northern Prawn Fishery between 1970–1992 (data from Robins and Somers, 1994).

While the regression is highly significant ($P < 1e^{-06}$), the variability of the prawn catches means it is difficult to know with what confidence we can believe the correlation. Estimating confidence intervals around correlation coefficients is not usually straightforward; fortunately, using bootstrapping, this can be done easily. We might take 5000 bootstrap samples of the 23 pairs of data from the original data-set and each time calculate the correlation coefficient. Once finished we can calculate the mean and various quantiles. In this case we are not bootstrapping single values but pairs of values; we have to take pairs to maintain any intrinsic correlation. Re-sampling pairs can be done in R by first sampling the positions along each vector of data, with replacement, to identify which pairs to take for each bootstrap sample.

```
# 5000 bootstrap estimates of correlation coefficient Fig. 6.3
set.seed(12321)      # better to use a less obvious seed, if at all
N <- 5000                             # number of bootstrap samples
result <- numeric(N)          #a vector to store 5000 correlations
for (i in 1:N) {              #sample index from 1:23 with replacement
    pick <- sample(1:23,23,replace=TRUE)    #sample is an R function
    result[i] <- cor(npf$tiger[pick],npf$endeavour[pick])
}
rge <- range(result)                    # store the range of results
CI <- quants(result)       # calculate quantiles; 90%CI = 5% and 95%
restrim <- result[result > 0] #remove possible -ve values for plot
parset(cex=1.0)            # set up a plot window and draw a histogram
bins <- seq(trunc(range(restrim)[1]*10)/10,1.0,0.01)
outh <- hist(restrim,breaks=bins,main="",col=0,xlab="Correlation")
abline(v=c(correl,mean(result)),col=c(4,3),lwd=c(3,2),lty=c(1,2))
abline(v=CI[c(2,4)],col=4,lwd=2) # and 90% confidence intervals
text(0.48,400,makelabel("Range ",rge,sep=" ",sigdig=4),font=7,pos=4)
label <- makelabel("90%CI ",CI[c(2,4)],sep=" ",sigdig=4)
text(0.48,300,label,cex=1.0,font=7,pos=4)
```

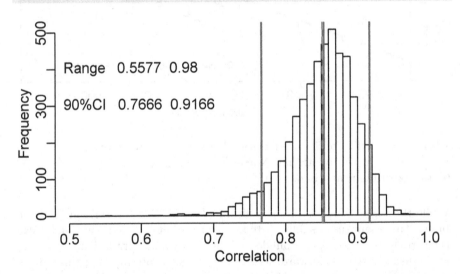

FIGURE 6.3 5000 bootstrap estimates of the correlation between endeavour and tiger prawn catches with the original mean in dashed green and bootstrap mean and 90% CI in solid blue. Possible negative correlations have been removed for plotting purposes (though none occurred).

While the resulting distribution of correlation values is certainly skewed to the left, we can be confident that the high correlation in the original data is a fair representation of the available data on the relationship.

6.4 Bootstrapping Time-Series Data

Even though the prawn catches in the previous example were taken across a series of years the particular year in which the data arrived was not relevant to the correlation between them. Thus, we could just bootstrap the data pairs and continue with the analysis. However, when bootstrapping information concerning the population dynamics of a species it is necessary to maintain the time-series nature of the data, where the values (of numbers or biomass) in one year depend in some way on the values in the previous year. For example, in a surplus production model the order in which the observed data enter the analysis is a vital aspect of the stock dynamics, so naively bootstrapping data pairs is not a valid or sensible option. The solution we adopt is that we obtain an optimum model fit with its predicted cpue time-series and we then bootstrap the individual residuals at each point. In each cycle the bootstrap-sampled residuals are applied to the optimum predicted values so as to generate a new bootstrapped 'observed' data series, which is then refitted. you will remember that the observed values can be derived from multiplying the optimum predicted cpue values by the Log-Normal residuals, **Figure** 6.1. Given the optimum solution from the original data the Log-Normal residual for a particular year t is:

$$resid_t = \frac{I_t}{\hat{I}_t} = \exp\left(\log(I_t) - \log(\hat{I}_t)\right) \qquad (6.6)$$

where I_t refers to the observed cpue in year t, and $\hat{I}_t$ is the optimum predicted cpue in year t. An optimum solution would imply a time-series of optimum predicted cpue values and a time-series of associated Log-Normal residuals. Given we are using multiplicative Log-Normal residuals, once we take a bootstrap sample of the residuals (random samples with replacement) we would need to multiply the optimum predicted cpue time-series by the bootstrap series of residuals. Log-Normal residuals would be expected to center around 1.0 with lower values constrained by zero and upper values not constrained; hence the skew that can occur with Log-Normal distributions:

$$I_t^* = \hat{I}_t \times \left(\frac{I_t}{\hat{I}_t}\right)^* = \exp\left(\log\left(\hat{I}_t\right) + \left(\log\left(I_t\right) - \log\left(\hat{I}_t\right)\right)^*\right) \qquad (6.7)$$

where the $*$ superscript denotes either a bootstrap value, as in I_t^* or a bootstrap sample, as in $(I_t/\hat{I}_t)^*$. Had we been using simple additive Normal random residuals then we would have used the equation on the right but without the log-transformation and exponentiation. For the surplus production model the Log-Normal version can be implemented thus (see also Table 6.1):

```
# fitting Schaefer model with log-normal residuals with 24 years
data(abdat); logce <- log(abdat$cpue) # of abalone fisheries data
param <- log(c(r= 0.42,K=9400,Binit=3400,sigma=0.05)) #log values
bestmod <- nlm(f=negLL,p=param,funk=simpspm,indat=abdat,logobs=logce)
optpar <- bestmod$estimate        # these are still log-transformed
predce <- exp(simpspm(optpar,abdat))        #linear-scale pred cpue
optres <- abdat[,"cpue"]/predce      # optimum log-normal residual
optmsy <- exp(optpar[1])*exp(optpar[2])/4
sampn <- length(optres)           # number of residuals and of years
```

TABLE 6.1: The *abdat* data-set with the associated optimum predicted cpue (predce), and the optimum residuals (optres).

year	catch	cpue	predce	optres	year	catch	cpue	predce	optres
1985	1020	1.000	1.135	0.881	1997	655	2.051	1.998	1.027
1986	743	1.096	1.071	1.023	1998	494	2.124	2.049	1.037
1987	867	1.130	1.093	1.034	1999	644	2.215	2.147	1.032
1988	724	1.147	1.076	1.066	2000	960	2.253	2.180	1.033
1989	586	1.187	1.105	1.075	2001	938	2.105	2.103	1.001
1990	532	1.202	1.183	1.016	2002	911	2.082	2.044	1.018
1991	567	1.265	1.288	0.983	2003	955	2.009	2.003	1.003
1992	609	1.320	1.388	0.951	2004	936	1.923	1.952	0.985
1993	548	1.428	1.479	0.966	2005	941	1.870	1.914	0.977
1994	498	1.477	1.593	0.927	2006	954	1.878	1.878	1.000
1995	480	1.685	1.724	0.978	2007	1027	1.850	1.840	1.005
1996	424	1.920	1.856	1.034	2008	980	1.727	1.782	0.969

Typically one would conduct 1000 bootstrap samples as a minimum. Note we have set *bootfish* to a matrix rather than a data.frame. If you remove the `as.matrix` so that bootfish becomes a data.frame instead compare the time it takes to do that and see the advantage of using a matrix in computer-intensive work.

```
# 1000 bootstrap Schaefer model fits; takes a few seconds
start <- Sys.time() # use of as.matrix faster than using data.frame
bootfish <- as.matrix(abdat)  # and avoid altering original data
N <- 1000;  years <- abdat[,"year"] # need N x years matrices
columns <- c("r","K","Binit","sigma")
results <- matrix(0,nrow=N,ncol=sampn,dimnames=list(1:N,years))
bootcpue <- matrix(0,nrow=N,ncol=sampn,dimnames=list(1:N,years))
parboot <- matrix(0,nrow=N,ncol=4,dimnames=list(1:N,columns))
for (i in 1:N) {  # fit the models and save solutions
  bootcpue[i,] <- predce * sample(optres, sampn, replace=TRUE)
  bootfish[,"cpue"] <- bootcpue[i,] #calc and save bootcpue
```

```
  bootmod <- nlm(f=negLL,p=optpar,funk=simpspm,indat=bootfish,
        logobs=log(bootfish[,"cpue"]))
  parboot[i,] <- exp(bootmod$estimate)    #now save parameters
  results[i,] <- exp(simpspm(bootmod$estimate,abdat))   #and predce
}
cat("total time = ",Sys.time()-start, "seconds    \n")
```

```
# total time =  3.831626 seconds
```

On the computer I used when writing this, the bootstrap took about 4 seconds to run. That was made up of taking the bootstrap sample of 24 years and fitting the surplus production model to the sample, with each iteration taking about 0.004 seconds. This is worth knowing with any computer-intensive method that requires fitting models or calculating likelihoods many times over. Knowing the time expected to run the analysis assists in designing the scale of those analyses. The surplus production model takes very little time to fit, but if a complex age-structured model takes say 1 minute to fit then 1000 replicates (a minimum by modern standards, and more would invariably provide for more precise results) would take over 16 hours! There is still very much a place for optimizing the speed of code where possible; We will discuss ways of optimizing for speed using the **Rcpp** package later in this chapter when we discuss approximating Bayesian posterior distributions.

The bootstrap samples can usually be plotted to provide a visual impression of the quality of a model fit, **Figure** 6.4. It would be possible to apply the quants() function to the *results* matrix containing the bootstrap estimates of the optimum predicted cpue and thereby plot percentile confidence intervals within the grey bounds of the plot, but here the outcome is so tight that it would mostly just make the plot untidy.

The values in years 1988 and 1989 have the largest residuals and so can never be exceeded.

```
# bootstrap replicates in grey behind main plot Fig 6.4
plot1(abdat[,"year"],abdat[,"cpue"],type="n",xlab="Year",
      ylab="CPUE") # type="n" just lays out an empty plot
for (i in 1:N)       # ready to add the separate components
  lines(abdat[,"year"],results[i,],lwd=1,col="grey")
points(abdat[,"year"],abdat[,"cpue"],pch=16,cex=1.0,col=1)
lines(abdat[,"year"],predce,lwd=2,col=1)
```

With 1000 replicates there remain some less well-defined parts in each plot, especially in the later years. For this reason one must be careful not just to plot an outline of the grey trajectories combined (perhaps defined using chull()), otherwise relatively thin areas of the predicted trajectory space may become un-noticed. Generally, doing 2000–5000 bootstraps may appear to be overkill, but to avoid actual gaps in the trajectory space then such numbers

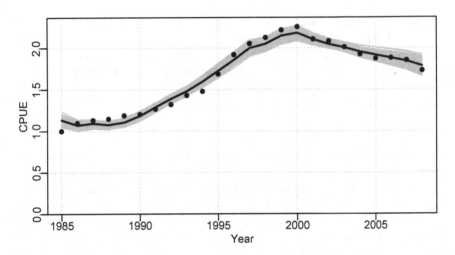

FIGURE 6.4 1000 bootstrap estimates of the optimum predicted cpue from the abdat data-set for an abalone fishery. Black points are the original data, the black line is the optimum predicted cpue from the original model fit, and the grey trajectories are the 1000 bootstrap estimates of the predicted cpue.

can be advantageous. Such diagrams can be helpful, but the principle findings relate to the model parameters and outputs, such as the MSY. We can use the bootstrap estimates of r and K to estimate the bootstrap estimates of MSY, and all can be plotted as histograms and, using quants(), we can identify whatever percentile confidence bounds we wish. quants() defaults to extracting the 0.025, 0.05, 0.5, 0.95, and 0.975 quantiles (other ranges can be entered), allowing for the identification of the central 95% and 90% confidence bounds as well as the median.

```
#histograms of bootstrap parameters and model outputs Fig. 6.5
dohist <- function(invect,nmvar,bins=30,bootres,avpar) { #adhoc
  hist(invect[,nmvar],breaks=bins,main="",xlab=nmvar,col=0)
  abline(v=c(exp(avpar),bootres[pick,nmvar]),lwd=c(3,2,3,2),
      col=c(3,4,4,4))
}
msy <- parboot[,"r"]*parboot[,"K"]/4 #calculate bootstrap MSY
msyB <- quants(msy)        #from optimum bootstrap parameters
parset(plots=c(2,2),cex=0.9)
bootres <- apply(parboot,2,quants); pick <- c(2,3,4) #quantiles
dohist(parboot,nmvar="r",bootres=bootres,avpar=optpar[1])
dohist(parboot,nmvar="K",bootres=bootres,avpar=optpar[2])
dohist(parboot,nmvar="Binit",bootres=bootres,avpar=optpar[3])
```

```
hist(msy,breaks=30,main="",xlab="MSY",col=0)
abline(v=c(optmsy,msyB[pick]),lwd=c(3,2,3,2),col=c(3,4,4,4))
```

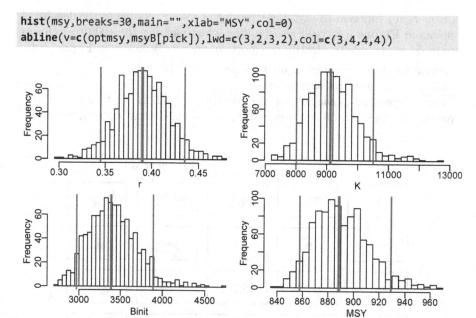

FIGURE 6.5 The 1000 bootstrap estimates of each of the first three model parameters and MSY as histograms. In each plot the two fine outer lines define the inner 90% confidence bounds around the median, the central vertical line denotes the optimum estimates, but these are generally immediately below the medians, except for the *Binit*.

Once again, with only 1000 bootstrap estimates, the histograms are not really smooth representations of the empirical distribution for all values of the parameters and outputs in question. More replicates would smooth the outputs and stabilize the quantile or percentile confidence bounds. Even so it is possible to generate an idea of the the precision with which such parameters and outputs are generated.

6.4.1 Parameter Correlation

We can also examine any correlations between the parameters and model outputs by plotting each against the others using the R function pairs(). The expected strong correlation between r, K, and B_{init} is immediately apparent. The lack of correlations with the value of sigma is typical of how variation within the residuals should behave. More interesting is the fact that the *msy*, which is a function of r and K, exhibits a reduced correlation with the other parameters. This reduction is a reflection of the negative correlation between r and K which acts to cancel out the effects of changes to each other. We can thus have similar values of *msy* for rather different values of r and K. The use of the `rgb()` function to vary the intensity of colour relative to the density of points in **Figure** 6.6 could also be used when plotting the 1000 trajectories to make it possible to identify the most common trajectories in **Figure** 6.4.

```
#relationships between parameters and MSY  Fig. 6.6
parboot1 <- cbind(parboot,msy)
 # note rgb use, alpha allows for shading, try 1/15 or 1/10
pairs(parboot1,pch=16,col=rgb(red=1,green=0,blue=0,alpha = 1/20))
```

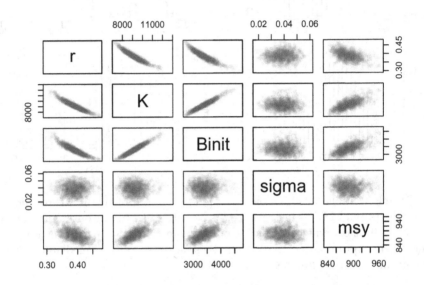

FIGURE 6.6 The relationships between the 1000 bootstrap estimates of the optimum parameter estimates and the derived MSY values for the Schaefer model fitted to the *abdat* data-set. Full-colour intensity derived from a minimum of 20 points. More bootstrap replicates would fill out these intensity plots.

6.5 Asymptotic Errors

The concept of confidence intervals (often 90% or 95% CI) is classically defined in Snedecor and Cochran (1967, 1989), and very many others, as:

$$\bar{x} \pm t_\nu \frac{\sigma}{\sqrt{n}} \tag{6.8}$$

where $\bar{x}$ is the mean of the sample of n observations (in fact, any parameter estimate), σ is the sample standard deviation, a measure of sample variation, and t_ν is the t-distribution with $\nu = (n-1)$ degrees of freedom (see also rt(), dt(), pt(), and qt()). If we had multiple independent samples it would be possible to estimate the standard deviation of the group of sample means. The $\sigma/\sqrt{n}$ is known as the standard error of the variable x, and is one analytical way of estimating the expected standard deviation of a set of sample means when there is only one sample (bootstrapping to generate multiple samples is another way, though then it would be better to use percentile CI directly). As we are dealing with normally distributed data, such classical confidence intervals are distributed symmetrically around the mean or expectation. This is fine when dealing with single samples but the problem we are attempting to solve is to determine with what confidence we can trust the parameter estimates and model outputs obtained when fitting a multi-parameter model to data.

In such circumstance asymptotic standard errors as in **Equ**(6.8) can be produced by estimating the variance-covariance matrix for the model parameters in the vicinity of the optimum parameter set. In practice, the gradient of the maximum likelihood or sum of squares surface at the optimum of a multi-parameter model is assumed to approximate multivariate Normal distribution and can be used to characterize any relationships between the various parameters. These relationships are the basis of generating the required variance-covariance matrix. This matrix is used to generate standard errors for each parameter, as in **Equ**(6.8), and these can then be used to estimate approximate confidence intervals for the parameters. They are approximate because this method assumes that the fitted surface near the optimum is regular and smooth, and that the surface is approximately multivariate Normal (= symmetrical) near the optimum. This means the resulting standard errors will be symmetric around the optimal solution, which may or may not be appropriate. Nevertheless, as a first approximation, asymptotic standard errors can provide an indication of the inherent variation around the parameter estimates.

The Hessian matrix describes the local curvature or gradient of the maximum likelihood surface near the optimum. More formally, this is made up of the second-order partial derivatives of the function describing the likelihood of different parameter sets. That is, the rate of change of the rate of change of each parameter relative to itself and all the other parameters. All Hessian matrices are square. If we only consider r, K, and $Binit$, three of the four parameters of the Schaefer model, **Equ(6.1)**, the Hessian describing these three would be:

$$H(f) = \left\{ \begin{array}{ccc} \frac{\partial^2 f}{\partial r^2} & \frac{\partial^2 f}{\partial r \partial K} & \frac{\partial^2 f}{\partial r \partial B_{init}} \\ \frac{\partial^2 f}{\partial r \partial K} & \frac{\partial^2 f}{\partial K^2} & \frac{\partial^2 f}{\partial K \partial B_{init}} \\ \frac{\partial^2 f}{\partial r \partial B_{init}} & \frac{\partial^2 f}{\partial K \partial B_{init}} & \frac{\partial^2 f}{\partial B_{init}^2} \end{array} \right\} \tag{6.9}$$

If the function f, for which we are calculating the second-order partial differentials, uses log-likelihoods to fit the model to the data then the variance-covariance matrix is the inverse of the Hessian matrix. However, and note this well, if the f function uses least-squares to conduct the analysis then formally the variance-covariance matrix is the product of the residual variance at the optimum fit and the inverse of the Hessian matrix. The residual variance reflects the number of parameters estimated:

$$Sx^2 = \frac{\sum (x - \hat{x})^2}{n - p} \tag{6.10}$$

which is the sum-of-squared deviations, between the observed and predicted values, divided by the number of observations (n) minus the number of parameters (p).

Thus, either the variance-covariance matrix $(\mathbf{A})$ is estimated by inverting the Hessian, $\mathbf{A} = \mathbf{H}^{-1}$ (when using maximum likelihood), or, when using least-squares, we multiply the elements of the inverse of the Hessian by the residual variance:

$$\mathbf{A} = Sx^2 \mathbf{H}^{-1} \tag{6.11}$$

Here we will focus on using maximum likelihood methods but it is worthwhile knowing the difference in procedure between using maximum likelihood and least-squares methodologies just in case (it is recommended to always use maximum likelihood when using asymptotic errors).

The estimate of the standard error for each parameter in the θ vector is obtained by taking the square root of the diagonal elements (the variances) of the variance-covariance matrix:

$$StErr(\theta) = \sqrt{diag(\mathbf{A})} \tag{6.12}$$

In Excel we used the method of finite differences to estimate the Hessian (Haddon, 2011) but this approach does not always perform well with strongly correlated parameters. Happily, in R many of the non-linear solvers available provide the option of automatically generating an estimate of the Hessian when fitting a model.

```
#Fit Schaefer model and generate the Hessian
data(abdat)
param <- log(c(r= 0.42,K=9400,Binit=3400,sigma=0.05))
 # Note inclusion of the option hessian=TRUE in nlm function
bestmod <- nlm(f=negLL,p=param,funk=simpspm,indat=abdat,
               logobs=log(abdat[,"cpue"]),hessian=TRUE)
outfit(bestmod,backtran = TRUE) #try typing bestmod in console
 # Now generate the confidence intervals
vcov <- solve(bestmod$hessian)       # solve inverts matrices
sterr <- sqrt(diag(vcov)) #diag extracts diagonal from a matrix
optpar <- bestmod$estimate      #use qt for t-distrib quantiles
U95 <- optpar + qt(0.975,20)*sterr # 4 parameters hence
L95 <- optpar - qt(0.975,20)*sterr # (24 - 4) df
cat("\n                 r      K     Binit     sigma \n")
cat("Upper 95% ",round(exp(U95),5),"\n") # backtransform
cat("Optimum   ",round(exp(optpar),5),"\n")#\n =linefeed in cat
cat("Lower 95% ",round(exp(L95),5),"\n")
```

```
# nlm solution:
# minimum       :  -41.37511
# iterations  :  20
# code          :  2 >1 iterates in tolerance, probably solution
#          par        gradient     transpar
# 1 -0.9429555   6.707523e-06      0.38948
# 2  9.1191569  -9.225209e-05   9128.50173
# 3  8.1271026   1.059296e-04   3384.97779
# 4 -3.1429030  -8.161433e-07      0.04316
# hessian     :
#                [,1]          [,2]         [,3]         [,4]
# [1,] 3542.8630987  2300.305473    447.63247  -0.3509669
# [2,] 2300.3054733  4654.008776  -2786.59928  -4.2155105
# [3,]  447.6324677 -2786.599276   3183.93947  -2.5662898
# [4,]   -0.3509669    -4.215511    -2.56629  47.9905538
#
#                r       K     Binit    sigma
# Upper 95%  0.45025 10948.12 4063.59 0.05838
# Optimum    0.38948 9128.502 3384.978 0.04316
# Lower 95%  0.33691 7611.311 2819.693 0.0319
```

6.5.1 Uncertainty about the Model Outputs

Asymptotic standard errors can also provide approximate confidence intervals around the model outputs, such as the *MSY* estimate, although this requires a somewhat different approach. This uses the assumption that the log-likelihood surface about the optimum solution is approximated by a multivariate Normal distribution. This is usually defined as:

$$\text{Multivariate Normal} = N(\mu, \Sigma) \qquad (6.13)$$

where, in the cases considered here, μ is the vector of optimal parameter estimates (a vector of means), and Σ is the variance-covariance matrix for the vector of optimum parameters.

Once these inputs are estimated, we can generate random parameter vectors by sampling from the estimated multivariate Normal distribution that has a mean of the optimum parameter estimates and a variance-covariance matrix estimated from the inverse Hessian. Such random parameter vectors can be used, in the same way as with bootstrap parameter vectors, to generate percentile confidence intervals around parameters and model outputs. Using the multivariate Normal (some write it as multivariate Normal) any parameter correlations between parameters are automatically accounted for.

6.5.2 Sampling from a Multivariate Normal Distribution

Base R does not have functions for working with the multivariate Normal distribution, but one can use a couple of R packages that include a suitable function. The **MASS** library (Venables and Ripley, 2002) includes a suitable random number generator while the **mvtnorm** library has an even wider array of multivariate probability density functions. Here we will use **mvtnorm**.

```
# Use multivariate normal to generate percentile CI     Fig. 6.7
library(mvtnorm) # use RStudio, or install.packages("mvtnorm")
N <- 1000 # number of multi-variate normal parameter vectors
years <- abdat[,"year"];  sampn <- length(years)  # 24 years
mvncpue <- matrix(0,nrow=N,ncol=sampn,dimnames=list(1:N,years))
columns <- c("r","K","Binit","sigma")
 # Fill parameter vectors with N vectors from rmvnorm
mvnpar <- matrix(exp(rmvnorm(N,mean=optpar,sigma=vcov)),
                nrow=N,ncol=4,dimnames=list(1:N,columns))
 # Calculate N cpue trajectories using simpspm
for (i in 1:N) mvncpue[i,] <- exp(simpspm(log(mvnpar[i,]),abdat))
msy <- mvnpar[,"r"]*mvnpar[,"K"]/4 #N MSY estimates
 # plot data and trajectories from the N parameter vectors
plot1(abdat[,"year"],abdat[,"cpue"],type="p",xlab="Year",
      ylab="CPUE",cex=0.9)
for (i in 1:N) lines(abdat[,"year"],mvncpue[i,],col="grey",lwd=1)
points(abdat[,"year"],abdat[,"cpue"],pch=16,cex=1.0)#orig data
lines(abdat[,"year"],exp(simpspm(optpar,abdat)),lwd=2,col=1)
```

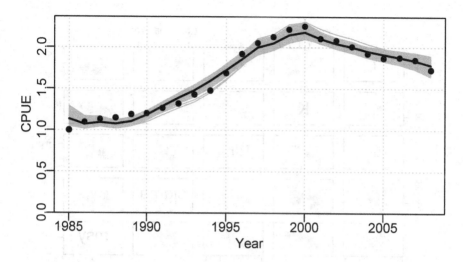

FIGURE 6.7 The 1000 predicted cpue trajectories derived from random parameter vectors sampled from the multivariate Normal distribution defined by the optimum parameters and their related variance-covariance matrix.

As with the bootstrap example, even a sample of 1000 trajectories were insufficient to fill in the trajectory space completely so that some regions were less dense than others. More replicates might fill in those spaces, as in **Figure** 6.7. The use of rgb() with the bootstrap cpue lines would also help identify less dense regions.

We can also plot out the implied parameter correlations (if any) using pairs(), as we did for the bootstrap samples as seen in **Figure** 6.8. Here, however, the results appear to be relatively evenly and cleanly distributed, relative to those from the bootstrapping process, see **Figure** 6.6. The smoother distributions reflect the fact that the values are all sampled from a well-defined probability density function rather than an empirical distribution. It would be reasonable to argue that the bootstrapping procedure would be expected to provide a more accurate representation of the properties of the data. However, which set of results better represents the population from which the original sample came from is less simple to answer with confidence. What really matters is whether their summary statistics differ and, considering **Table** 6.2, we can see that while the exact details of the distributions for each parameter and model output differ slightly, there does not appear to be any single pattern of one being wider or narrower than the other in a consistent fashion.

```
#correlations between parameters when using mvtnorm Fig. 6.8
pairs(cbind(mvnpar,msy),pch=16,col=rgb(red=1,0,0,alpha = 1/10))
```

Finally, we can illustrate the asymptotic confidence intervals by plotting out an array of histograms for parameters and *MSY* along with their predicted

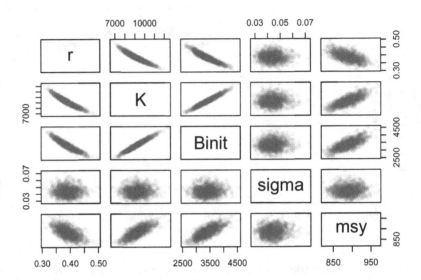

FIGURE 6.8 The relationships between the 1000 parameter estimates sampled from the estimated multivariate Normal distribution assumed to surround the optimum parameter estimates and the derived MSY values for the Schaefer model fitted to the *abdat* data-set.

confidence bounds, **Figure** 6.9. In this case we use the inner 90% bounds. The mean and median values are even closer than with the bootstrapping example, which is a further reflection of the symmetry of using the multivariate Normal distribution.

```
#N parameter vectors from the multivariate normal Fig. 6.9
mvnres <- apply(mvnpar,2,quants)  # table of quantiles
pick <- c(2,3,4)   # select rows for 5%, 50%, and 95%
meanmsy <- mean(msy)      # optimum bootstrap parameters
msymvn <- quants(msy)   # msy from multi-variate normal estimates

plothist <- function(x,optp,label,resmvn) {
  hist(x,breaks=30,main="",xlab=label,col=0)
  abline(v=c(exp(optp),resmvn),lwd=c(3,2,3,2),col=c(3,4,4,4))
} # repeated 4 times, so worthwhile writing a short function
par(mfrow=c(2,2),mai=c(0.45,0.45,0.05,0.05),oma=c(0.0,0,0.0,0.0))
par(cex=0.85, mgp=c(1.35,0.35,0), font.axis=7,font=7,font.lab=7)
plothist(mvnpar[,"r"],optpar[1],"r",mvnres[pick,"r"])
plothist(mvnpar[,"K"],optpar[2],"K",mvnres[pick,"K"])
```

```
plothist(mvnpar[,"Binit"],optpar[3],"Binit",mvnres[pick,"Binit"])
plothist(msy,meanmsy,"MSY",msymvn[pick])
```

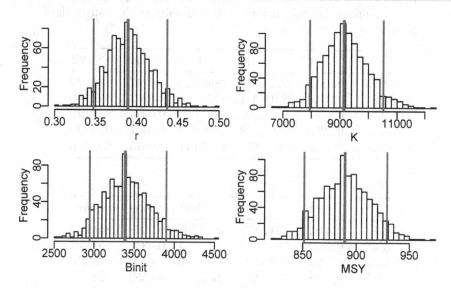

FIGURE 6.9 Histograms of the 1000 parameter estimates for r, K, Binit, and the derived MSY, from the multivariate Normal estimated at the optimum solution. In each plot, the vertical central green line denotes the arithmetic mean, the overlapping vertical thick blue line, the median, and the two fine outer blue lines the inner 90% confidence bounds around the median.

TABLE 6.2: A comparison of the bootstrap percentile confidence bounds on parameters with those derived from the Asymptotic estimate of the variance-covariance matrix. The top table relates to the bootstrapping and the bottom to the values from the multivariate Normal.

	r	K	Binit	sigma	msyB
2.5%	0.3353	7740.636	2893.714	0.0247	854.029
5%	0.3449	8010.341	2970.524	0.0263	857.660
50%	0.3903	9116.193	3387.077	0.0385	888.279
95%	0.4353	10507.708	3889.824	0.0501	929.148
97.5%	0.4452	11003.649	4055.071	0.0523	940.596

	r	K	Binit	sigma	msymvn
2.5%	0.3387	7717.875	2882.706	0.0328	844.626
5%	0.3474	7943.439	2945.264	0.0342	851.257
50%	0.3889	9145.146	3389.279	0.0432	889.216
95%	0.4368	10521.651	3900.034	0.0550	928.769
97.5%	0.4435	10879.571	4031.540	0.0581	934.881

6.6 Likelihood Profiles

The name 'Likelihood Profile' is suggestive of the process used to generate these analyses. In this chapter we have already optimally fitted a four-parameter model and examined ways of characterizing the uncertainty around those parameter estimates. One can imagine selecting a single parameter from the four and fixing its value a little way away from its optimum value. If one then refitted the model holding the selected parameter fixed but allowing the others to vary we can imagine that a new optimum solution would be found for the remaining parameters but that the negative log-likelihood would be larger than the optimum obtained when all four parameters were free to vary. This process is the essential idea behind the generation of likelihood profiles

The idea is to fit a model using maximum likelihood methods (minimization of the -ve log-likelihood) but only fitting specific parameters while keeping others constant at values surrounding the optimum value. In this way new "optimum" fits can be obtained while a given parameter has been given an array of fixed values. We can thus determine the influence of the fixed parameter(s) on the total likelihood of the model fit. An example may help make the process clear.

Once again using the Schaefer surplus production model on the *abdat* fishery data we obtain the optimum parameters $r = 0.3895$, $K = 9128.5$, $Binit = 3384.978$, and $sigma = 0.04316$, which imply an $MSY = 888.842t$.

```
#Fit the Schaefer surplus production model to abdat
data(abdat); logce <- log(abdat$cpue)    # using negLL
param <- log(c(r= 0.42,K=9400,Binit=3400,sigma=0.05))
optmod <- nlm(f=negLL,p=param,funk=simpspm,indat=abdat,logobs=logce)
outfit(optmod,parnames=c("r","K","Binit","sigma"))
```

```
# nlm solution:
# minimum    :  -41.37511
# iterations :  20
# code       :   2 >1 iterates in tolerance, probably solution
#              par       gradient     transpar
# r      -0.9429555  6.707523e-06     0.38948
# K       9.1191569 -9.225209e-05  9128.50173
# Binit   8.1271026  1.059296e-04  3384.97779
# sigma  -3.1429030 -8.161433e-07     0.04316
```

The problem of how to fit a restricted number of model parameters, while holding the remainder constant is solved by modifying the function used to minimize the negative log-likelihood. Instead of using the `negLL()` function to calculate the negative log-likelihood of the Log-Normally distributed cpue values we have used the **MQMF** function `negLLP()` (negative log-likelihood

profile). This adds the capacity to fix some of the parameters while solving for
an optimum solution by only varying the non-fixed parameters. If we look at
the R code for the negLLP() function and compare it to the negLL() function
we can see that the important differences, other than in the arguments, is in
the three lines of code before the *logpred* statement. See the help (?) page
for further details, though I hope you can see straightaway that if you ignore
the *initpar* and *notfixed* arguments, negLLP() should give the same answer as
negLL().

```
#the code for MQMF's negLLP function
negLLP <- function(pars, funk, indat, logobs, initpar=pars,
                   notfixed=c(1:length(pars)),...) {
  usepar <- initpar  #copy the original parameters into usepar
  usepar[notfixed] <- pars[notfixed] #change 'notfixed' values
  npar <- length(usepar)
  logpred <- funk(usepar,indat,...) #funk uses the usepar values
  pick <- which(is.na(logobs))  # proceed as in negLL
  if (length(pick) > 0) {
    LL <- -sum(dnorm(logobs[-pick],logpred[-pick],exp(pars[npar]),
                log=T))
  } else {
    LL <- -sum(dnorm(logobs,logpred,exp(pars[npar]),log=T))
  }
  return(LL)
} # end of negLLP
```

For example, to determine the precision with which the r parameter has been
estimated we can force it to take on constant values from about 0.3 to 0.45
while fitting the other parameters to obtain an optimum fit and then plot
up the total likelihood with respect to the given r value. As with all model
fitting in R we need two functions, one to generate the required predicted
values and the other to act as a wrapper to bring the observed and predicted
values together to be used by the minimizer, in this case nlm(). To proceed
we use negLLP() to allow for some parameters to be fixed in value. nlm()
works by iteratively modifying the input parameters in the direction that
ought to improve the model fit and then feeding those parameters back into
the input model function that generates the predicted values against which
the observed are compared. We therefore need to arrange the model function
to continually return the particular parameter values that we want fixed to
the initial set values. The code in negLLP() is one way in which to do this.
The changes required include having an independent set of the original *pars*
in *initpar*, which must contain the specified fixed parameters, and a *notfixed*
argument identifying which values from *initpar* to over-write with values from
pars, which will become modified by nlm.

It is best practice to check that negLLP() generates the same solution as the use of negLL(), even though inspection of the code assures us that it will (I am no longer surprised to find out that I can make mistakes), by allowing all parameters to vary (the default setting).

```
#does negLLP give same answers as negLL when no parameters fixed?
param <- log(c(r= 0.42,K=9400,Binit=3400,sigma=0.05))
bestmod <- nlm(f=negLLP,p=param,funk=simpspm,indat=abdat,logobs=logce)
outfit(bestmod,parnames=c("r","K","Binit","sigma"))

# nlm solution:
# minimum    :  -41.37511
# iterations :  20
# code       :  2 >1 iterates in tolerance, probably solution
#            par          gradient     transpar
# r      -0.9429555   6.707523e-06     0.38948
# K       9.1191569  -9.225209e-05  9128.50173
# Binit   8.1271026   1.059296e-04  3384.97779
# sigma  -3.1429030  -8.161433e-07     0.04316
```

Happily, as expected, we obtain the same solution and so now we can proceed to examine the effect of fixing the value of r and re-fitting the model. Using hindsight (so much better than reality), we have selected r values between 0.325 and 0.45. We can set up a loop to sequentially apply these values and fit the respective models, saving the solutions as the loop proceeds. Below we have tabulated the first few results, **Table** 6.3, to illustrate exactly how the negative log-likelihood increases as the fixed value of r is set further and further away from its optimum value.

```
#Likelihood profile for r values 0.325 to 0.45
rval <- seq(0.325,0.45,0.001)  # set up the test sequence
ntrial <- length(rval)         # create storage for the results
columns <- c("r","K","Binit","sigma","-veLL")
result <- matrix(0,nrow=ntrial,ncol=length(columns),
                dimnames=list(rval,columns))# close to optimum
bestest <- c(r= 0.32,K=11000,Binit=4000,sigma=0.05)
for (i in 1:ntrial) {  #i <- 1
  param <- log(c(rval[i],bestest[2:4])) #recycle bestest values
  parinit <- param #to improve the stability of nlm as r changes
  bestmodP <- nlm(f=negLLP,p=param,funk=simpspm,initpar=parinit,
              indat=abdat,logobs=log(abdat$cpue),notfixed=c(2:4),
              typsize=magnitude(param),iterlim=1000)
  bestest <- exp(bestmodP$estimate)
  result[i,] <- c(bestest,bestmodP$minimum)  # store each result
}
minLL <- min(result[,"-veLL"]) #minimum across r values used.
```

TABLE 6.3: The first 12 records from the 126 rows of the nlm solutions that are used to make the likelihood profile on r. The strong correlation between r, K, and Binit is, once again, apparent.

	r	K	Binit	sigma	-veLL
0.325	0.325	11449.17	4240.797	0.0484	-38.61835
0.326	0.326	11403.51	4223.866	0.0483	-38.69554
0.327	0.327	11358.24	4207.082	0.0481	-38.77196
0.328	0.328	11313.35	4190.442	0.0480	-38.84759
0.329	0.329	11268.83	4173.945	0.0478	-38.92242
0.33	0.330	11224.69	4157.589	0.0477	-38.99643
0.331	0.331	11180.91	4141.373	0.0475	-39.06961
0.332	0.332	11137.49	4125.293	0.0474	-39.14194
0.333	0.333	11094.43	4109.350	0.0472	-39.21339
0.334	0.334	11051.72	4093.540	0.0471	-39.28397
0.335	0.335	11009.11	4077.752	0.0469	-39.35364
0.336	0.336	10967.34	4062.316	0.0468	-39.42239

6.6.1 Likelihood Ratio-Based Confidence Intervals

Venzon and Moolgavkar (1988) described a method of obtaining what they call "approximate $1 - \alpha$ profile-likelihood-based confidence intervals", which were based upon a re-arrangement of the usual approach to the use of a likelihood ratio test. The method relies on the fact that likelihood ratio tests asymptotically approach the χ^2 distribution as the sample size gets larger, so this method is only approximate. Not surprisingly, likelihood ratio tests are based upon a ratio of two likelihoods, or if dealing with log-likelihoods, the subtraction of one from another:

$$\frac{L(\theta)_{max}}{L(\theta)} = e^{LL(\theta)_{max} - LL(\theta)} \tag{6.14}$$

where $L(\theta)$ is the likelihood of the θ parameters, the max subscript denotes the maximum likelihood (assuming all other parameters are also optimally fitted), and $LL(\theta)$ is the equivalent log-likelihood. The expected log-likelihoods for the actual confidence intervals for a single parameter, assuming all others remain at the optimum (as in a likelihood profile), are given by the following (Venzon and Moolgavkar, 1988):

$$2 \times \left[LL(\theta)_{max} - LL(\theta) \right] \leq \chi^2_{1,1-\alpha}$$
$$LL(\theta) = LL(\theta)_{max} - \frac{\chi^2_{1,1-\alpha}}{2} \tag{6.15}$$

where $\chi^2_{1,1-\alpha}$ is the $1 - \alpha$th quantile of the χ^2 distribution with 1 degree of freedom (e.g., for 95% confidence intervals, $\alpha = 0.95$ and $1-\alpha = 0.05$, and $\chi^2_{1,1-\alpha} = 3.84$.

For a single parameter θ_i, the approximate 95% confidence intervals are therefore those values of θ_i for which two times the difference between the corresponding log-likelihood and the overall optimal log-likelihood is less than or equal to 3.84 ($\chi^2_{1,1-\alpha} = 3.84$). Alternatively (bottom line in **Equ**(6.15)), one can search for the θ_i that generates a log-likelihood equal to the maximum log-likelihood minus half the required χ^2 value (i.e., $LL(\theta)_{max}-1.92$, when with one degree of freedom). If conducting a two-parameter likelihood surface, then one would search for the θ_i with χ^2 set at 5.99 (2 degrees of freedom), that is one would subtract 2.995 from the maximum likelihood (and so on for higher degrees of freedom).

We can plot up the likelihood profile for the r parameter and include these approximate 95% confidence intervals. Examine the code within the function plotprofile() to see the steps involved in calculating the confidence intervals from the results of the analysis.

```
#Likelihood profile on r from the Schaefer model Fig. 6.10
plotprofile(result,var="r",lwd=2)  # review the code
```

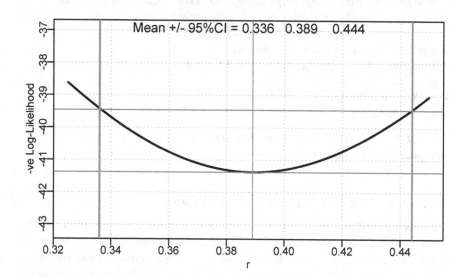

FIGURE 6.10 A likelihood profile for the r parameter from the Schaefer surplus production model fitted to the *abdat* data-set. The horizontal solid lines are the minimum and the minimum minus 1.92 (95% level for χ^2 with 1 degree of freedom, see text). The outer vertical lines are the approximate 95% confidence bounds around the central mean of 0.389.

We can repeat this analysis for the other parameters in the Schaefer model. For example, we can look at the K parameter in a similar way using almost identical code. Note the stronger asymmetry in the profile plot for K. There is slight asymmetry in the likelihood profile for r (subtract the 95% CI from the mean estimate), but visually it is obvious with the K parameter. The optimum parameter estimate for K is 9128.5, but the maximum likelihood in the profile data, points to a value of 9130. This is merely a reflection of the step length of the likelihood profile. It is currently set to 10; if it is set to 1 then we can obtain a closer approximation, although, of course, the analysis would take 10 times as long to run.

```
#Likelihood profile for K values 7200 to 12000
Kval <- seq(7200,12000,10)
ntrial <- length(Kval)
columns <- c("r","K","Binit","sigma","-veLL")
resultK <- matrix(0,nrow=ntrial,ncol=length(columns),
                  dimnames=list(Kval,columns))
bestest <- c(r= 0.45,K=7500,Binit=2800,sigma=0.05)
for (i in 1:ntrial) {
  param <- log(c(bestest[1],Kval[i],bestest[c(3,4)]))
  parinit <- param
  bestmodP <- nlm(f=negLLP,p=param,funk=simpspm,initpar=parinit,
              indat=abdat,logobs=log(abdat$cpue),
              notfixed=c(1,3,4),iterlim=1000)
  bestest <- exp(bestmodP$estimate)
  resultK[i,] <- c(bestest,bestmodP$minimum)
}
minLLK <- min(resultK[,"-veLL"])
#kable(head(result,12),digits=c(4,3,3,4,5))  # if wanted.
```

```
#Likelihood profile on K from the Schaefer model Fig 6.11
plotprofile(resultK,var="K",lwd=2)
```

6.6.2 -ve Log-Likelihoods or Likelihoods

We have plotted the -ve Log-Likelihoods, e.g., **Figure** 6.11, and these are fine when operating in log-space, but few people have clear intuitions about log-space. An alternative, with which it may be simpler to understand the principle behind the percentile confidence intervals, would be to back-transform the -ve log-likelihoods into simple likelihoods. Of course, a log value of -41 denotes a very small number when back-transformed, but we can re-scale those values in the process of ensuring that the values of all the likelihoods sum to 1.0.

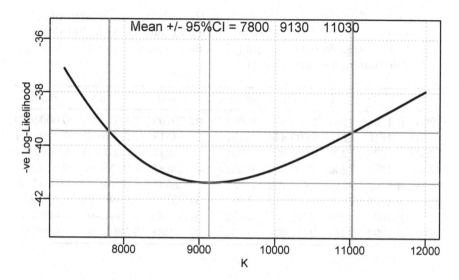

FIGURE 6.11 A likelihood profile for the K parameter from the Schaefer surplus production model fitted to the *abdat* data-set, conducted in the same manner as the r parameter. The red lines are the minimum and the minimum plus 1.92 (95% level for Chi2 with 1 degree of freedom, see text). The vertical thick lines are the approximate 95% confidence bounds around the mean of 9128.5.

To back-transform the negative log-likelihoods we need to negate them and then exponent them:

$$L = exp(-(-veLL))$$ (6.16)

So, for the likelihood profile applied to the K parameter we can see that the linear-space likelihoods begin to increase as the value of K approaches the optimum (Table 6.4), and if we plot those likelihoods the shape of the distribution can be seen to take on a shape more intuitively understandable. The tails stay above zero but are well away from the 95% intervals:

```
#translate -veLog-Likelihoods into Likelihoods
likes <- exp(-resultK[,"-veLL"])/sum(exp(-resultK[,"-veLL"]),
                                 na.rm=TRUE)
resK <- cbind(resultK,likes,cumlike=cumsum(likes))
```

TABLE 6.4: The first 8 records from the 481 rows of the nlm solutions that are used to make the likelihood profile on K. Included is the back-transformed -ve log-likelihoods, scaled to sum to 1.0, and their running cumulative sum.

	r	K	Binit	sigma	-veLL	likes	cumlike
7200	0.4731	7200	2689.875	0.0516	-37.09799	6.8863e-05	0.0000689
7210	0.4726	7210	2693.444	0.0515	-37.14518	7.2191e-05	0.0001411
7220	0.4720	7220	2697.023	0.0514	-37.19213	7.5660e-05	0.0002167
7230	0.4714	7230	2700.602	0.0513	-37.23881	7.9276e-05	0.0002960
7240	0.4709	7240	2704.182	0.0512	-37.28524	8.3044e-05	0.0003790
7250	0.4703	7250	2707.762	0.0511	-37.33141	8.6968e-05	0.0004660
7260	0.4698	7260	2711.341	0.0510	-37.37732	9.1054e-05	0.0005571
7270	0.4692	7270	2714.921	0.0509	-37.42298	9.5307e-05	0.0006524

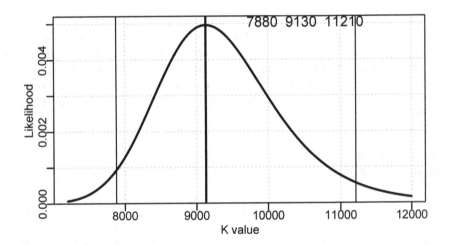

FIGURE 6.12 A likelihood profile for the K parameter from the Schaefer surplus production model fitted to the *abdat* data-set. In this case the -ve log-likelihoods have been back-transformed to likelihoods and scaled to sum to 1.0. The vertical lines are the approximate 95% confidence bounds around the mean. The top three numbers are the bounds and estimated optimum.

6.6.3 Percentile Likelihood Profiles for Model Outputs

Generally, the interest in stock assessment models relates to model outputs that are not directly estimated as parameters. As we have seen, generating confidence bounds around parameter estimates is relatively straightforward, but how do we provide similar estimates of uncertainty about model outputs such as the MSY (for the Schaefer model $MSY = rK/4$)? For example, an assessment model might estimate stock biomass, or the maximum sustainable yield, or some other performance measure that can be considered an indirect output from a model. Likelihood profiles for such model outputs can be produced by adding a penalty term to the negative log-likelihood that attempts to constrain the likelihood to the optimal target (see **Equ**(6.17)). In this way, the impact on the log-likelihood of moving away from the optimum can be characterized.

$$-veLL = -veLL - w \left(\frac{output - target}{target} \right)^2 \tag{6.17}$$

where *-veLL* is the negative log-likelihood, *output* is the variable of interest (in the example to follow this is the MSY), target is the optimum for that variable (the MSY from the overall optimal solution), and w is a weighting factor. The weighting factor should be as large as possible to generate the narrowest likelihood profile while still being able to converge on a solution. Below, we describe a specialized function negLLO() to deal with likelihood profiles around model outputs (this is not in **MQMF**). It is simply a modified version of the negLL() function that allows for the introduction of the weighting factor described in **Equ**(6.17). We can illustrate this by examining the likelihood profile around the optimum MSY value of 887.729t, which can be done across a range of 740–1050 tonnes, with a weighting of 900 (see Table 6.5).

```
#examine effect on -veLL of MSY values from 740 - 1050t
#need a different negLLP() function, negLLO(): O for output.
#now optvar=888.831 (rK/4), the optimum MSY, varval ranges 740-1050
#and wght is the weighting to give to the penalty
negLLO <- function(pars,funk,indat,logobs,wght,optvar,varval) {
  logpred <- funk(pars,indat)
  LL <- -sum(dnorm(logobs,logpred,exp(tail(pars,1)),log=T)) +
            wght*((varval - optvar)/optvar)^2  #compare with negLL
  return(LL)
} # end of negLLO
msyP <- seq(740,1020,2.5);
optmsy <- exp(optmod$estimate[1])*exp(optmod$estimate[2])/4
ntrial <- length(msyP)
wait <- 400
columns <- c("r","K","Binit","sigma","-veLL","MSY","pen",
             "TrialMSY")
resultO <- matrix(0,nrow=ntrial,ncol=length(columns),
                  dimnames=list(msyP,columns))
bestest <- c(r= 0.47,K=7300,Binit=2700,sigma=0.05)
for (i in 1:ntrial) {  # i <- 1
  param <- log(bestest)
```

```
bestmodO <- nlm(f=negLLO,p=param,funk=simpspm,indat=abdat,
                logobs=log(abdat$cpue),wght=wait,
                optvar=optmsy,varval=msyP[i],iterlim=1000)
  bestest <- exp(bestmodO$estimate)
  ans <- c(bestest,bestmodO$minimum,bestest[1]*bestest[2]/4,
           wait *((msyP[i] - optmsy)/optmsy)^2,msyP[i])
  resultO[i,] <- ans
}
minLLO <- min(resultO[,"-veLL"])
```

TABLE 6.5: The first and last 7 records from the 113 rows of the nlm solutions that are used to make the likelihood profile on MSY (one might do more). The row names are the trial MSY values and pen is the penalty value.

	r	K	Binit	sigma	-veLL	MSY	pen
740	0.389	9130.914	3385.871	0.0432	-30.16	888.883	11.22
742.5	0.389	9130.911	3385.872	0.0432	-30.53	888.883	10.84
745	0.389	9130.911	3385.872	0.0432	-30.90	888.883	10.47
747.5	0.389	9130.911	3385.872	0.0432	-31.26	888.883	10.11

	r	K	Binit	sigma	-veLL	MSY	pen
1012.5	0.389	9130.911	3385.872	0.0432	-33.63	888.883	7.74
1015	0.389	9130.911	3385.872	0.0432	-33.32	888.883	8.06
1017.5	0.389	9130.911	3385.872	0.0432	-32.99	888.883	8.38
1020	0.389	9130.911	3385.872	0.0432	-32.66	888.883	8.71

Unfortunately, determining the optimum weighting to apply to the penalty term inside what we have termed negLLO(), can only be done empirically (trial and error). I recommended using a weight of 900 because I had already found that at that level the 95% CI stabilized. But it took my trying from 100 up to 700 in steps of 100, and then using steps of 50 to discover this. You should try using weight values of 500, 700, 800, 900, and 950, to see the convergence to stable values. The suggestion of the largest value that will still permit stable solutions remains vague guidance. This makes this approach perhaps the trickiest to apply in a repeatable manner. If you were to try the analysis above using a *wght* = 400, then the 95% CI become 825–950, rather than 847.5–927.5. In such cases there is no deterministic solution so it becomes a case of trying different weights and searching for a solution that is repeatable and consistent.

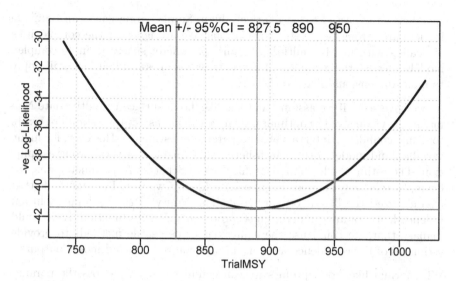

FIGURE 6.13 A likelihood profile for the MSY implied by the Schaefer surplus production model fitted to the *abdat* data-set. The horizontal red lines are the minimum and the minimum plus 1.92 (95% level for Chi2 with 1 degree of freedom, see text). The vertical lines are the approximate 95% confidence bounds around the mean of 887.729t.

The 95% CI obtained using this likelihood profile approach (Figure 6.13) are comparable to those from using a bootstrap (856.7–927.5) and from using asymptotic errors (849.7–927.4). Each of these methods could generate somewhat different answers as a function of the sample size selected. Nevertheless, the consistency between methods suggests they each generate a reasonable characterization of the variation inherent in the combination of this model and this data.

6.7 Bayesian Posterior Distributions

Fitting model parameters to observed data can be likened to searching for the location of the maximum likelihood on a multi-dimensional likelihood surface implied by a model and a given data-set. If the likelihood surface is very steep in the vicinity of the optimum, then the uncertainty associated with those parameters would be relatively small. Conversely, if the likelihood surface was comparatively flat near the optimum, this would imply that similar likelihoods could be obtained from rather different parameter values, and the uncertainty around those parameters and any related model outputs would therefore be expected to be relatively high. Similar arguments can be made if there are strong parameter correlations. We are ignoring the complexities of dealing with highly multi-dimensional parameter spaces in more complex models because the geometrical concepts of steepness and surface still apply to multi-dimensional likelihoods.

As we have seen, if we assume that the likelihood surface is multivariate Normal in the vicinity of the optimum, then we can use asymptotic standard errors to define confidence intervals around parameter estimates. However, for many variables and model outputs in fisheries that might be a very strong assumption. The estimated likelihood profile for the Schaefer K parameter was, for example, relatively skewed, **Figure** 6.12. Ideally, we would use methods that characterized the likelihood surface in the vicinity of the optimum solution independently of any predetermined probability density function. If we could manage that, we could use the equivalent of percentile methods to provide estimates of the confidence intervals around parameters and model outputs.

With formal likelihood profiles we can conduct a search across the parameter space to produce a two-dimensional likelihood surface. However, formal likelihood profiles for more than two parameters, perhaps using such a grid search, would be very clumsy and more and more impractical as the number of parameters increased further. What is needed is a method of integrating across many dimensions at once to generate something akin to a multi-dimensional likelihood profile. In fact, there are a number of ways of doing this, including Sampling Importance Resampling (SIR; see McAllister and Ianelli, 1997; and McAllister *et al*, 1994), and Markov Chain Monte Carlo (MCMC; see Gelman *et al* 2013). In the examples to follow we will implement the MCMC methodology. There are numerous alternative algorithms for conducting an MCMC, but we will focus on a relatively flexible approach called the Gibbs-within-Metropolis sampler (or sometimes Metropolis-Hastings-within-Gibbs). The Metropolis algorithm (Metropolis *et al*, 1953), which started as two-dimensional integration, was generalized by Hastings (1970), hence Metropolis-Hastings. In the literature you will find reference to the Markov Chain Monte Carlo but also the Monte Carlo Markov Chain. The former is used by the standard reference used in fisheries (Gelman *et al*, 2013) although personally I find the idea of a Monte Carlo approach to generating a Markov Chain more intuitively obvious. Despite this potential confusion I would suggest sticking with MCMC when writing and if necessary ignore my intuitions and use Markov Chain Monte Carlo. Such things are not sufficiently important to spend time worrying about, except sometimes the differences can actually mean something (such confusions remain a problem, but as long as you are aware of such pitfalls you can avoid them!).

An MCMC, obviously, uses a Markov chain to trace over the multi-dimensional likelihood surface. A Markov chain describes a process whereby each state is determined probabilistically from the previous state. A random walk would constitute one form of Markov chain, and its final state would be a description of a random distribution. However, the aim here is to produce a Markov chain whose final equilibrium state, the so-called *stationary distribution*, provides a description of the target or *posterior* distribution of Bayesian statistics.

The Markov chain starts with some combination of parameter values, the available observed data, and the model being used, and these together define a location in the likelihood space. Depending on the set of parameter values, the likelihood can obviously be small or large (with an upper limit of the model's maximum likelihood). The MCMC process entails iteratively stepping through the parameter space following a set of rules based on the likelihood of each new candidate parameter set relative to the previous set, to determine which steps will become part of the Markov chain. The decision being made each iteration is which parameter vectors will be accepted and which rejected? Each step of the process entails the production of a new candidate set of parameter values, which is done stochastically (hence Markov Chain Monte Carlo) and, in the Gibbs-within-Metropolis, one parameter at a time (Casella and George, 1992; Millar and Meyer, 2000). Each of these new candidate sets of parameters, combined with the available data and the model, defines a new likelihood. Whether this new parameter combination is accepted as the next step in the Markov chain depends on how much the likelihood has been changed. In all cases where the likelihood increases the step is accepted, which seems reasonable. Now comes the important part, where the likelihood decreases it can still be accepted if the ratio of the new likelihood relative to the old likelihood is larger than some uniform random number (between 0 and 1; see the equations that follow).

There is a further constraint that relates to parameter correlations and the fact that the Monte Carlo process is likely to lead to auto-correlation between sequential parameter sets. If there is significant serial correlation among parameter sets this can generate biased conclusions about the full extent of variation across the parameter space. The solution used is to thin out the resulting chain so that the final Markov chain will contain only every n^{th} point. The trick is to select this thinning rate so that the serial correlation between the nth parameter vectors is sufficiently small that it no longer influences the overall variance. Among other things we will explore this thinning rate in the practical examples.

6.7.1 Generating the Markov Chain

A Markov chain can be generated if the likelihood of an initial set of a specific model's parameters, θ_t, can be defined given a set of data x, and the Bayesian prior probability of the parameter set, $L(\theta_t)$:

$$L_t = L(\theta_t|x) \times L(\theta_t) \tag{6.18}$$

We can then generate a new trial or candidate (C) parameter set θ^C by randomly incrementing at least one of the parameters in $\theta^C = \theta_t + \Delta\theta_t$, which will alter the implied likelihood:

$$L^C = L(\theta^C|x) \times L(\theta_t) \tag{6.19}$$

If the ratio of L^C and L_t is greater than 1, then the jump from θ_t to θ^C is accepted into the Markov chain $(\theta_{t+1} = \theta^C)$:

$$r = \frac{L^C}{L_t} = \frac{L\left(\theta^C|x\right) \times L(\theta_t)}{L\left(\theta_t|x\right) \times L(\theta_t)} > 1.0 \tag{6.20}$$

Alternatively, if the ratio (r) is less than 1, then the jump is only accepted if the ratio r is greater than a newly selected uniform random number. If it is not accepted then θ_{t+1} reverts to θ_t and a new candidate cycle is begun:

$$\theta_{t+1} = \begin{cases} \theta^C & if\ [L^C/L_t > U\,(0,1)] \\ \theta_t & otherwise \end{cases} \tag{6.21}$$

In fact, because the maximum random uniform number between 0 and 1 is 1 then **Equ**(6.20) is not strictly necessary. We could just estimate the ratio and use **Equ**(6.21) (this is how it is implemented in the function do_MCMC()). If the candidate parameter vector is rejected, it reverts to the original and the cycle starts again with a new candidate parameter set. As the Markov chain develops it should trace out a multi-dimensional volume in parameter space. After sufficient iterations it should converge on the stationary distribution. After some expansion on some of these theoretical details we will go through some worked examples to illustrate all of these ideas.

6.7.2   The Starting Point

Starting off the Markov Chain simply entails selecting a vector of parameters. One solution is to select a vector close to but not identical to the maximum likelihood optimum solution. Gelman and Rubin (1992) recommend using a sample from an over-dispersed distribution akin to the expected distribution, but that assumes one has an idea of what over-dispersed distribution one should use to start with, and choosing this remains something of an art form (Racine-Poon, 1992). As further stated by Racine-Poon (1992): "However, the target distributions are often multi-modal, especially for high-dimensional problems, and I am not too confident about automating the process." However, the starting points used can be influential on the outcome so Gelman and Rubin (1992) recommend starting multiple chains from very different starting points (rather than just one very long chain; Geyer, 1992) and then ensure they all converge on the same final distribution. All these recommendations arose in

the early 1990s when computers were much slower than now, so the discussion
that occurred over multiple chains or very long single chains (see, for example,
much of Volume 7(4) of *Statistical Science*) is no longer such an issue and there
is no real reason not to run multiple very long chains (though it would still
be sensible to use the most efficient software as MCMCs invariably require a
large investment of time).

6.7.3 The Burn-In Period

An important point that relates to starting off the Markov chain(s) at points
possibly some way from the multi-dimensional mode of the posterior distribution, is that the first series of points generated are expected to make their
stochastic way towards the region of higher likelihoods. However, these early
points may add a clumsy and inappropriate tail to what ought to be the final
cloud of acceptable parameter vectors. Thus, the standard practice is simply
to delete the first m points where m is known as the *burn-in* period. Gelman
et al (2013) recommend deleting up to the first half (50%) of each chain as a
burn-in period, but at that point in their book they are discussing chains that
are only 100's of steps long, so, in fisheries, when dealing with much higher
dimensional problems, we need not be so drastic. A few hundred early points
or even less may often be sufficient, especially if one starts not too far from the
maximum likelihood estimate. But initial exploration of the types of chains
one might generate should enable a reasonable burn-in period to be found for
each problem.

6.7.4 Convergence to the Stationary Distribution

The discussion in the 1990s concerning how many chains and how long they
should be was driven by the need to demonstrate that the generated Markov
Chain had converged to a stable solution (the *stationary distribution* or *posterior distribution*, when dealing with Bayesian statistics). Both Gelman and
Rubin (1992) and Geyer (1992) presented empirical methods for determining
the evidence for convergence. These generally revolved around comparing the
variance within chains with the variance between chains, or the variance in
one part of a Markov Chain with another part of the same chain. A visual
approach was simply to plot the marginal distributions of important parameters from multiple chains, or portions of a chain, on top of each other. If
they matched sufficiently well one could assume that convergence had been
achieved, though nobody seems to define exactly what would be classed as
sufficient. Being empirical, all of these methods cannot guarantee that convergence is complete, but, as yet, more analytical methods remain undiscov-
ered. Here we will focus on multiple chains with simple comparative statistics
(mean, median, quantiles, and variance) and marginal distribution plots, but
there are various such diagnostics (Geweke, 1989; Gelman and Rubin, 1992;
Geyer, 1992). The R package **coda** also has implementations of many of the

diagnostic tools and can import simple matrix outputs, such as we will generate, using a simple statement such as post = coda::mcmc(posterior). Here, however, we will continue to tabulate and plot in explicit R so the reader can easily see the mechanics. The problem of deciding whether to use an already excellent package, such as **coda**, or write custom plotting and tabulating functions oneself is something to decide for yourself, but only after you know how to write your own functions.

6.7.5   The Jumping Distribution

How each new candidate parameter vector is generated depends upon what is known as the jumping distribution. There are many options available to act as this jumping distribution, but commonly, for each parameter in turn in the set θ, a standard Normal random deviate $N(0,1)$ is generated, and this is scaled, α_i, to suit the scale of the parameter, i, being incremented, **Equ**(6.22). If there was information about parameter correlations, it would be possible to use a multivariate Normal distribution to generate a full new candidate parameter vector at each step; using a standard multivariate Normal remains a possibility so that parameter correlations are ignored. However, incrementing each parameter in turn before applying the Metropolis-Hastings approach (Gibbs-within-Metropolis) seems intuitively simpler to understand (though not always the most efficient).

$$\theta_i^C = \theta_{t,i} + N(0,1) \times \alpha_i \qquad (6.22)$$

The scaling by α_i as the procedure cycles through each of the parameters in θ is important because if the jumps in parameter space are too large, then the success rate of the jumps may be very low, but if they are too small, the success rate may be too high, and it may take an enormous number of iterations, and time, to fully explore the multi-dimensional likelihood surface and converge on a stationary distribution. There is an element of trial and error in this procedure with no fixed or simple rule as to what scaling factor to use. The acceptance rate of the new candidate values is an indicator of the efficiency of the performance. A simple rule-of-thumb might be to scale the Normal random deviate to approximate between 0.5% and 1.0% of the original parameter value. An acceptance rate of 0.2 to 0.4 (20–40 %) might be a reasonable target to aim for when adjusting the increments to the parameter set. The scaling values α_i will generally differ between parameters and when developing a new analysis may require some detailed searching for some acceptable values. It is possible to build such adaptive searching into the code used to run an MCMC. Generally, when conducting an MCMC analysis one would use software customized for executing such analyses. For example, Gelman *et al* (2013) have an appendix entitled *Example of Computation in R and Bugs*, but now the recommendation is to use software named *Stan* rather than *Bugs* (see https://mc-stan.org/, where there is ample documentation).

However, here, so as to ensure the exposition remains transparent with no black boxes, we will keep things purely within R while emphasizing that such analyses require a degree of customization to suit each case.

If one uses a multivariate Normal to increment all parameters at once then the scaling factors will become smaller than when incrementing single parameters, and the reduction is related to the number of parameters being varied at once.

6.7.6   Application of MCMC to the Example

Once again we will use the *abdat* data with the Schaefer surplus production model fit as the example explored to illustrate the methods. With more complex models (more parameters) the time to conduct the analyses can expand greatly, but the same underlying principles apply.

```
#activate and plot the fisheries data in abdat  Fig. 6.14
data(abdat)   # type abdat in the console to see contents
plotspmdat(abdat) #use helper function to plot fishery stats vs year
```

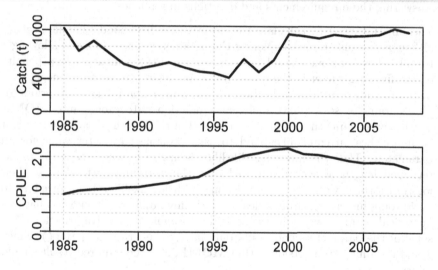

FIGURE 6.14 The time-series of cpue and catch from the *abdat* data-set.

6.7.7   Markov Chain Monte Carlo

The equations describing the Gibbs-within-Metropolis appear relatively straightforward and even simple. However, their implementation involves quite a few more details. As we saw in the chapter on *Model Parameter Estimation* when calculating Bayesian statistics we need a method for calculating the likelihood of a set of parameters, but we also need a method for calculating their prior probability (even where we attribute a non-informative prior to the analysis). There are a number of other pre-requisites needed to implement the MCMC :

- the functions required to calculate the negative log-likelihood and the prior probability of each candidate parameter set.

- with what parameter set are we going to start the MCMC process, and how long should we run the MCMC process before we start storing the accepted parameter vectors (how big a burn-in)?

- how often in the process should we consider accepting a result (thinning rate)?

- how many independent Markov Chains are we going to generate, and how long should the chains we intend to generate be before we stop the MCMC process?

- how does one select an appropriate set of weightings (the α_i) in a given situation?

Answering these requirements and questions in sequence:

The function used to calculate the negative log-likelihood remains negLL() although, because we are exploring the parameter space, to avoid allowing r to fall below zero one should perhaps use negLL1(), which places a penalty on the first parameter in the vector if it approaches zero (examine negLL1()'s help and code). Here we will always be assuming non-informative priors for each parameter so we have included a function calcprior() into **MQMF** which merely sums the log of the reciprocal of the length of each chain. That is, it ascribes an equal likelihood to each possible parameter value for all parameters. Thus the individual likelihoods are $1/N$, where N is the predetermined length of each chain, including the burn-in length. We use their log-transformed value as we are dealing with log-likelihoods in an effort to avoid rounding errors through dealing with tiny numbers. Should you have an analytical problem where it was desired to use an informative prior for all or some of the parameters one would only have to re-write calcprior(), which would then mask the version within **MQMF**, and the required result should follow.

It is standard practice to have what is termed a *burn-in* period where the MCMC process is run for a number of iterations from its starting point without storing the results. This rejection of early results is done to ensure that the Markov Chain has begun exploring the model's likelihood surface rather than wandering around in very low likelihood space. This will, of course, depend upon what starting values one would be using to begin the Markov Chain. One could start the MCMC with the maximum likelihood solution but this is generally discouraged as potentially biasing the outcome. However, if one included a burn-in period of a few hundred steps the accepted starting point would have moved away from the optimum solution. Nevertheless, it is good practice to begin the Markov Chain candidate parameter vectors somewhat away from the optimum solution and, with multiple chains, to start each chain at a different point.

Looking at the equations for the Gibbs-within-Metropolis one might gain the impression that each step through the process of selecting a candidate parameter vector and testing it can lead to the Markov Chain being incremented. This might be the case if there were no parameter correlations or auto-correlation between successive draws. However, in an effort to avoid such auto-correlation between successive steps in the Markov Chain it is common to only consider a step for inclusion in the chain after a given number of iterations have passed. Such chain thinning is designed to break any levels of auto-correlation. We already know that the Schaefer parameters are strongly correlated and will use that knowledge as the basis for exploring what degree of chain thinning is required to remove such auto-correlation in the Markov Chain. A potential confusion is that with the Gibbs-within-Metropolis we need a thinning step for each parameter. Thus a desired thinning step of 4 would require do_MCMC() receiving 4 x 4 = 16 as the *thinstep* argument.

The objective when generating a Markov Chain is that it should converge on the stationary distribution (the posterior distribution), which would provide a comprehensive characterization of the uncertainty of the subject model with its data. But how does one determine whether such convergence has been achieved. One approach is to generate more than one Markov Chain, starting from different origins. If they all converge so that the marginal distributions for each parameter are very similar across replicate chains this would be evidence of convergence. This could be visualized graphically. However, rather than only using graphical indicators, diagnostic statistics should be used to indicate whether convergence to a stationary distribution has occurred. There are many such statistics available, but here we will only mention some simple strategies. As with any nonlinear solver, it is a good idea to start the MCMC process from a wide range of initial values. Any diagnostic tests for identifying whether the MCMC has achieved the target stationary distribution would literally consider the convergence of the different Markov sequences.

Of course, it would be necessary to discard the so-called burn-in phase before making any comparisons. The burn-in phase is where the Markov chain may only be traversing sparse likelihood space before it starts to characterize the posterior distribution. Gelman *et al* (2013) recommend the conservative option of discarding the first half of each sequence, but the actual fraction selected should be determined by inspection. The simplest diagnostic statistics involve comparing the means and variances of either different sequences, or different portions of the same sequence. If the mean values from different sequences (or subsets of a sequence) are not significantly different, then the sequence(s) can be said to have converged. Similarly, if the within-sequence (or subset of a sequence) variance is not significantly different from the between-sequence, then convergence can be identified. Using a single sequence may appear convenient, but if convergence is relatively slow and this was unknown, then relying on a single sequence may provide incorrect answers. The title of Gelman and Rubin (1992a) states the problem clearly: "A single sequence from the Gibbs sampler gives a false sense of security." Suffice to say that it is better to use multiple starting points to give multiple sequences along with an array of diagnostic statistics and graphics to ensure that the conclusions that one draws from any MCMC simulation are not spurious (Gelman *et al*, 2013). Such approaches are termed computer-intensive for a good reason. Determining how long a sequence to generate for each chain is also a question that only has an empirical answer. One needs to run the chains until convergence has occurred. This may happen quickly, or it may take a very long time. The more parameters in a model the longer this generally takes. Highly uncertain or unbalanced models may even fail to converge, which would be an inefficient method of identifying such a problem.

The weightings given to the increments generated for each parameter are another thing that can only really be determined empirically. By developing criteria this process can be automated and is often developed during a long burn-in phase, but here we will utilize trial and error and aim for an acceptance rate between 20–40%.

6.7.8 A First Example of an MCMC

We will illustrate some of the ideas discussed above in our first example. To do that we will generate a Markov Chain of 10000 steps starting from close to the maximum likelihood solution but with a burn-in phase of 50 iterations (to move into the informative likelihood space), and a chain thinning rate of 4 steps (this will not avoid auto-correlation between successive points because a thinning each 4 steps with 4 parameters means no thinning, but more on that later). Using hindsight (because I ran this multiple times) I have set the α_i to values that lead to acceptance rates between 20–40% of trials. Finally, we will also run three short chains of 100 iterations with no burn-in phase and different starting points to illustrate the effect of the burn-in and the use of different starting points. By using the set.seed() function we will also ensure that we obtain repeatable results (which would not usually be a sensible option).

Most of the work is done by the **MQMF** function do_MCMC(). You should examine the help and the code for this function and trace back where the equations describing the Gibbs-within-Metropolis are operating and how the prior probability is included; you should also be able to see how the acceptance rate is calculated.

```
# Conduct MCMC analysis to illustrate burn-in. Fig. 6.15
data(abdat); logce <- log(abdat$cpue)
```

```
fish <- as.matrix(abdat) # faster to use a matrix than a data.frame!
begin <- Sys.time()        # enable time taken to be calculated
chains <- 1                # 1 chain per run; normally do more
burnin <- 0                # no burn-in for first three chains
N <- 100                   # Number of MCMC steps to keep
step <- 4         # equals one step per parameter so no thinning
priorcalc <- calcprior # define the prior probability function
scales <- c(0.065,0.055,0.065,0.425) #found by trial and error
set.seed(128900) #gives repeatable results in book; usually omitted
inpar <- log(c(r= 0.4,K=11000,Binit=3600,sigma=0.05))
result1 <- do_MCMC(chains,burnin,N,step,inpar,negLL,calcpred=simpspm,
             calcdat=fish,obsdat=logce,priorcalc,scales)
inpar <- log(c(r= 0.35,K=8500,Binit=3400,sigma=0.05))
result2 <- do_MCMC(chains,burnin,N,step,inpar,negLL,calcpred=simpspm,
             calcdat=fish,obsdat=logce,priorcalc,scales)
inpar <- log(c(r= 0.45,K=9500,Binit=3200,sigma=0.05))
result3 <- do_MCMC(chains,burnin,N,step,inpar,negLL,calcpred=simpspm,
             calcdat=fish,obsdat=logce,priorcalc,scales)
burnin <- 50 # strictly a low thinning rate of 4; not enough
step <- 16   # 16 thinstep rate = 4 parameters x 4 = 16
N <- 10000   # 16 x 10000 = 160,000 steps + 50 burnin
inpar <- log(c(r= 0.4,K=9400,Binit=3400,sigma=0.05))
result4 <- do_MCMC(chains,burnin,N,step,inpar,negLL,calcpred=simpspm,
             calcdat=fish,obsdat=logce,priorcalc,scales)
post1 <- result1[[1]][[1]]
post2 <- result2[[1]][[1]]
post3 <- result3[[1]][[1]]
postY <- result4[[1]][[1]]
cat("time  = ",Sys.time() - begin,"\n")
cat("Accept = ",result4[[2]],"\n")
```

```
# time   =  10.83922
# Accept =  0.3471241 0.3437158 0.354289 0.3826251
```

Now we can plot out the 10000-long chain as a set of points and on top of those impose the three shorter chains for which no burn-in phase is used. This highlights the importance of the burn-in phase but also illustrates how different chains begin their exploration of the likelihood space independently (here only represented in two dimensions). If these three chains had been longer we would expect them to traverse a broader area within the space occupied by the grey dots.

```
#first example and start of 3 initial chains for MCMC Fig. 6.15
parset(cex=0.85)
P <- 75  # the first 75 steps only start to explore parameter space
plot(postY[,"K"],postY[,"r"],type="p",cex=0.2,xlim=c(7000,13000),
```

```
    ylim=c(0.28,0.47),col=8,xlab="K",ylab="r",panel.first=grid())
lines(post2[1:P,"K"],post2[1:P,"r"],lwd=1,col=1)
points(post2[1:P,"K"],post2[1:P,"r"],pch=15,cex=1.0)
lines(post1[1:P,"K"],post1[1:P,"r"],lwd=1,col=1)
points(post1[1:P,"K"],post1[1:P,"r"],pch=1,cex=1.2,col=1)
lines(post3[1:P,"K"],post3[1:P,"r"],lwd=1,col=1)
points(post3[1:P,"K"],post3[1:P,"r"],pch=2,cex=1.2,col=1)
```

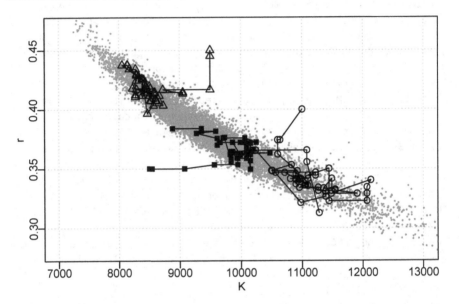

FIGURE 6.15 The first 75 points in three separate MCMC chains starting from different origins (triangles, squares, circles). No burn-in was set for these short chains so records begin from the starting points. The grey dots are 10000 points from a single fourth chain, with a 50-point burn-in and a thinning rate of four, giving an approximate idea of the stationary distribution towards which all chains should converge.

We can also examine details of the parameter correlation through plotting each parameter against the other using the `pairs()` function and the `rgb()` function to shade the colouring; this enables us to visualize the softer tail of points towards the larger values of K and smaller values of r. To see the effect change the *alpha* argument (the 1/50) to 1/1, using a divisor of 50 appears to be a reasonable compromise for illustrating the gradient of density in this case (of 10000 points).

```
#pairs plot of parameters from the first MCMC Fig. 6.16
posterior <- result4[[1]][[1]]
```

```
msy <-posterior[,1]*posterior[,2]/4
pairs(cbind(posterior[,1:4],msy),pch=16,col=rgb(1,0,0,1/50),font=7)
```

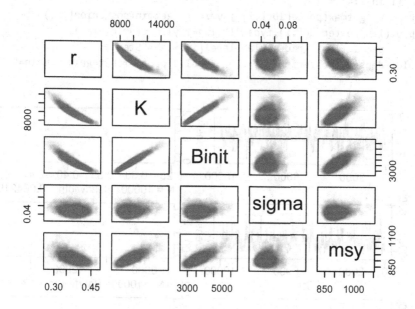

FIGURE 6.16 The relationship between the 10000 samples from the posterior distributions of the Schaefer model parameters and MSY. Normally one would use a much longer thinning step than 4 to characterize the posterior. Full colour in the plot comes from at least 50 points.

The accepted parameter vectors that make up the posterior distribution can each be plotted individually against the replicate number to provide a trace of each parameter. The ideal is to obtain what is commonly called a "hairy caterpillar", which is especially evident in the trace for the σ parameter in **Figure** 6.17. The other traces, with their spikes up and down are suggestive of a degree of auto-correlation within each trace. This is also suggested by the complementary patterns of variation apparent in the r and K parameter traces. The marginal distributions provide an initial examination of the shape of the empirical distribution of each parameter. It is quite possible to have more than one mode or a flatter top depending on the data one has. However, an irregular shape suggests a lack of convergence.

```
#plot the traces from the first MCMC example Fig. 6.17
posterior <- result4[[1]][[1]]
par(mfrow=c(4,2),mai=c(0.4,0.4,0.05,0.05),oma=c(0.0,0,0.0,0.0))
```

```
par(cex=0.8, mgp=c(1.35,0.35,0), font.axis=7,font=7,font.lab=7)
label <- colnames(posterior)
N <- dim(posterior)[1]
for (i in 1:4) {
  ymax <- getmax(posterior[,i]); ymin <- getmin(posterior[,i])
  plot(1:N,posterior[,i],type="l",lwd=1,ylim=c(ymin,ymax),
       panel.first=grid(),ylab=label[i],xlab="Step")
  plot(density(posterior[,i]),lwd=2,col=2,panel.first=grid(),main="")
}
```

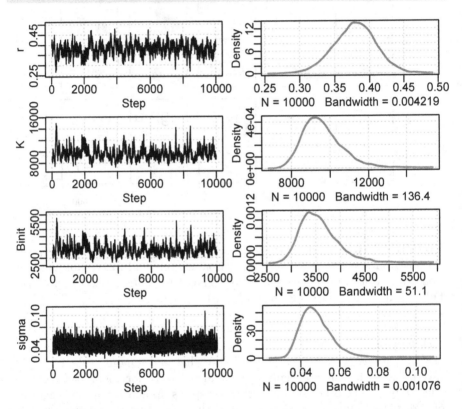

FIGURE 6.17 The traces for the four Schaefer model parameters along with the implied marginal distribution of each parameter. The obvious auto-correlation within traces should be improved if the thinning step were increased to 128, 256, or, as we shall see, much longer.

We can examine the degree of any auto-correlation among sequential steps in the Markov Chain by using the R function acf(). This correlates the values in a vector with itself, then with a lag of 1, then a lag of 2, and so on. One expects a correlation of 1 with a lag of 0, but, ideally, the correlation between sequential terms in the Markov Chain should drop to insignificance very quickly if we are to avoid under-representing the total variation. Compare the auto-correlogram for each parameter with its trace and marginal distribution. The visual difference between *sigma*, with relatively low serial correlation, and the other parameters, should be apparent.

```
#Use acf to examine auto-correlation with thinstep = 16    Fig. 6.18
posterior <- result4[[1]][[1]]
label <- colnames(posterior)[1:4]
parset(plots=c(2,2),cex=0.85)
for (i in 1:4) auto <- acf(posterior[,i],type="correlation",lwd=2,
                   plot=TRUE,ylab=label[i],lag.max=20)
```

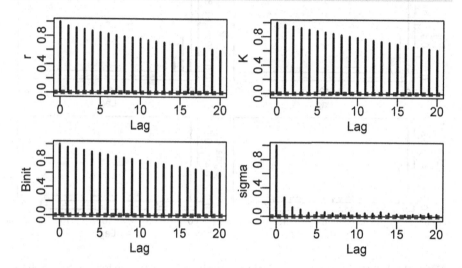

FIGURE 6.18 The auto-correlation exhibited in the traces for the four Schaefer model parameters. This is with a thinning step of 16 = 4 for each of four parameters. Clearly a large increase is needed to remove the strong correlations that occur.

If we run an MCMC with a much larger thinning rate, hopefully we will be able to see the reduction in serial correlation. Here we increase the thinning rate 128 times ($4 \times 4 = 16$ to $4 \times 128 = 512$) and plot the outcome. Even though we have reduced the length of the chain to 1000, the total number of steps is (512 x 1000) + (512 x 100) = 563200, so we can expect this to take a little longer than the initial MCMC run. It is always a good idea to have some idea about how long a run will take. These examples all run in at most a few minutes; most MCMC runs for serious models take many hours and sometimes days.

```
#setup MCMC with thinstep of 128 per parameter  Fig 6.19
begin=gettime()
scales <- c(0.06,0.05,0.06,0.4)
inpar <- log(c(r= 0.4,K=9400,Binit=3400,sigma=0.05))
```

```
result <- do_MCMC(chains=1,burnin=100,N=1000,thinstep=512,inpar,
                  negLL,calcpred=simpspm,calcdat=fish,
                  obsdat=logce,calcprior,scales,schaefer=TRUE)
posterior <- result[[1]][[1]]
label <- colnames(posterior)[1:4]
parset(plots=c(2,2),cex=0.85)
for (i in 1:4) auto <- acf(posterior[,i],type="correlation",lwd=2,
                           plot=TRUE,ylab=label[i],lag.max=20)
```

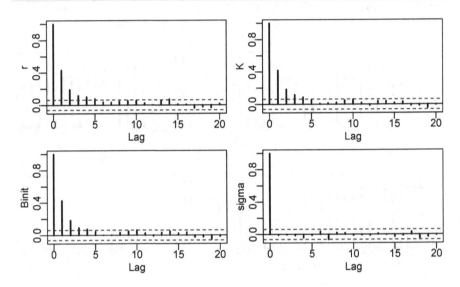

FIGURE 6.19 The auto-correlation exhibited in the traces for the four Schae-fer model parameters when the thinning step is 512 = 4 x 128, which is 128 times that used in the first auto-correlation diagram.

```
cat(gettime() - begin)
```

37.39104

There is clearly a marked reduction in the auto-correlation exhibited in the four parameter traces when the thinning step is increased from 4 to 128 (4 × 128 = 512), **Figure** 6.19. But even that 32x increase is not sufficient to reduce the correlations within lags of 2, 3, and 4 to insignificance. Despite that an examination of the traces generated with that thinning step clearly shows a difference (improvement) relative to the traces in **Figure** 6.17. However, we note that the 563200 steps (1100-length chain) took about 40 seconds on the desktop computer I am using, which is starting to get a little slow for interactive work. Had we used 10000 iterations at a thinning rate of 512 that would take nearly 7 minutes, which is beginning to get tediously long. More complex models might be another order of magnitude or more in duration, sometimes days! Thus, before we explore the effectiveness of even larger thinning steps, we will first find ways to speed each cycle significantly. Through continued trials it was found that doubling *thinstep* to 1024 (4 x 256) still had significant auto-correlation at lags of 1 and sometimes 2, and it took a four-fold increase to a *thinstep* = 2048 (4 × 3 × 128) to remove the auto-correlation from all variables.

Before looking for speedier ways of doing things, we will finish the examination of the standard illustrative plots that one might use when conducting an MCMC examination of the uncertainty or variation within a modelling framework.

6.7.9 Marginal Distributions

One approach to visualizing the posterior distribution is to examine the frequency distribution of the accepted values of each parameter. Plotting the outline from the density() function on top of the histogram can also improve our view of the distribution found by the MCMC. In the case below it appears that 1000 replicates is not sufficient to smooth each distribution, even with a thinning rate per parameter of 128. However, determining how many replicates are sufficient is really asking whether the posterior has converged on the stationary distribution. Rather than going by intuition and the appearance of various plots it is better to use various standard diagnostics. Most of the figures plotted relating to MCMC outputs using a thinning rate of 16 appear to suggest plausibly smooth solutions. Nevertheless, the auto-correlation is so large the outputs could be biased. It is always better to use quantifiable diagnostics.

```
# plot marginal distributions from the MCMC  Fig. 6.20
dohist <- function(x,xlab) { # to save a little space
  return(hist(x,main="",breaks=50,col=0,xlab=xlab,ylab="",
              panel.first=grid()))
}
# ensure we have the optimum solution available
param <- log(c(r= 0.42,K=9400,Binit=3400,sigma=0.05))
bestmod <- nlm(f=negLL,p=param,funk=simpspm,indat=abdat,
               logobs=log(abdat$cpue))
optval <- exp(bestmod$estimate)
posterior <- result[[1]][[1]] #example above N=1000, thin=512
par(mfrow=c(5,1),mai=c(0.4,0.3,0.025,0.05),oma=c(0,1,0,0))
par(cex=0.85, mgp=c(1.35,0.35,0), font.axis=7,font=7,font.lab=7)
np <- length(param)
for (i in 1:np) { #store invisible output from hist for later use
  outH <- dohist(posterior[,i],xlab=colnames(posterior)[i])
  abline(v=optval[i],lwd=3,col=4)
  tmp <- density(posterior[,i])
```

```
  scaler <- sum(outH$counts)*(outH$mids[2]-outH$mids[1])
  tmp$y <- tmp$y * scaler
  lines(tmp,lwd=2,col=2)
}
msy <- posterior[,"r"]*posterior[,"K"]/4
mout <- dohist(msy,xlab="MSY")
tmp <- density(msy)
tmp$y <- tmp$y * (sum(mout$counts)*(mout$mids[2]-mout$mids[1]))
lines(tmp,lwd=2,col=2)
abline(v=(optval[1]*optval[2]/4),lwd=3,col=4)
mtext("Frequency",side=2,outer=T,line=0.0,font=7,cex=1.0)
```

We could try to run multiple, independent chains and then determine whether they generate the same results (perhaps the mean, variance, and distribution shape). However, to ensure a stable posterior distribution, if we use a large thinning rate then it is likely that we will need a large number of iterations in each chain. Before we explore those options we will examine how we might speed up the MCMC process.

6.8   The Use of Rcpp

With simple models such as the examples we have used so far in this chapter, the time taken to run even 200,000 replicates (a thinning step length of 8 means this would involve 1.6 million likelihood calculations) is not overly onerous. However, with more complex models, or when trying to remove high levels of serial or auto-correlation, the time involved when running an MCMC can become tiresome. If you have a procedure that takes what feels like a long time to run then it will likely be worthwhile to profile that part of your code. This is easily done within R by using the Rprof() function. Once initiated, this function interrupts execution every now and then (defaults to 0.02 of a second) and determines what function is running at the moment of interruption. These occurrences are all logged and once the software has completed running one can apply the function summaryRprof() to discover which functions are active (Table 6.6) for the most time (self.time). One can then attempt to speed up those slow parts. The total.time includes the time spent in a function and any functions that it calls.

```
#profile the running of do_MCMC  using the now well known abdat
data(abdat); logce <- log(abdat$cpue); fish <- as.matrix(abdat)
param <- log(c(r=0.39,K=9200,Binit=3400,sigma=0.05))
Rprof(append=TRUE)  # note the use of negLL1()
result <- do_MCMC(chains=1,burnin=100,N=20000,thinstep=16,inpar=param,
```

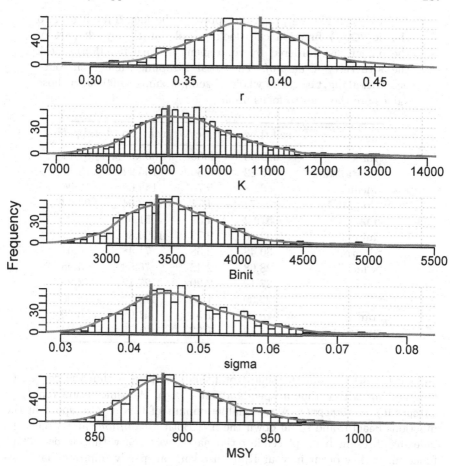

FIGURE 6.20 The marginal posterior distributions for the 1000 points at a thinning rate of 128. The vertical blue line, in each case, is the maximum likelihood optimum estimate. More replicates may be needed to smooth the distributions. The posterior mode is not necessarily the same as the maximum likelihood estimate.

```
                    infunk=negLL1,calcpred=simpspm,calcdat=fish,
                    obsdat=logce,priorcalc=calcprior,
                    scales=c(0.07,0.06,0.07,0.45))
Rprof(NULL)
outprof <- summaryRprof()
```

TABLE 6.6: The output from applying the Rprof() function to the do_MCMC() function call, looking only at the by.self part of the output list (ordered by time spent in each function), check the structure of *outprof*. The total sampling time is in out-prof$sampling.time. The whole of self.pct sums to 99.99 so those values are the ones to focus upon.

	self.time	self.pct	total.time	total.pct
"funk"	619.60	52.35	959.94	81.11
"[.data.frame"	75.22	6.36	155.92	13.17
"mean"	68.08	5.75	142.68	12.06
"do_MCMC"	55.74	4.71	1182.64	99.93
"max"	51.92	4.39	51.92	4.39
"infunk"	32.20	2.72	1094.86	92.51
"dnorm"	30.48	2.58	71.48	6.04
"which"	28.68	2.42	70.58	5.96
"names"	27.12	2.29	27.12	2.29
"["	22.30	1.88	178.22	15.06
"tail"	18.20	1.54	41.00	3.46
"priorcalc"	18.12	1.53	21.72	1.84

Obviously, virtually all the time (total.time) is spent within the do_MCMC() function but within that it spends about 80% of the time inside funk() = calcpred() = simpspm(). A reasonable amount of time is also spent in the functions mean(), and so on down the list. Both *[.data.frame* and *[* relate to indexing of results being placed into the matrix *posterior* within the do_MCMC() function. if they occur in your Rprof list you can partly improve that slow-down by converting the data.frame of input data into a matrix. You should compare the speed of comutation using each form of data. The R software is truly excellent but, even with the increases in speed in recent versions and with modern computers, nobody would claim it was rapid with regard to computer-intensive methods such as MCMC. It seems clear that if we could speed up the simpspm() function (i.e., *funk*) we might be able to obtain some significant speed gains when running an MCMC. Perhaps the best way would be to use *Stan* (see https://mc-stan.org/), but in the interests of keeping as much as possible within base R, we will examine an alternative.

One extremely useful approach to increasing the speed of execution is to com-bine the R-code with another computer language, which can be compiled into executable code rather than the interpreted code of R. Perhaps the simplest method for including C++ code into one's R-code is to use the **Rcpp** package (Eddelbuettel and Francois, 2011; Eddelbuettel, 2013; Eddelbuettel and Bal-muta, 2017). Obviously to use this one needs both a C++ compiler and the **Rcpp** package, both of which can be downloaded from the CRAN repository (most easily done within RStudio). If the reader is working on a Linux or Mac computer then they already have the GNU C++ compiler (used by **Rcpp**). On Windows the easiest way to install the GNU C++ compiler is to go to the CRAN home page, click on the Download R for Windows link, and then the Rtools link. Be sure to include the installation directory into your path. This also provides many tools useful for writing R packages. **Rcpp** provides a number of approaches for including C++ code with perhaps the simplest way being to use the `cppFunction()` to compile code at the start of each session. However, an even better way is to use the function `Rcpp::sourceCpp()` to load a C++ file from disk in the same way you might use `source()` to load a file of R code. So there are options, which are well worth exploring if you are going to pursue this strategy of speeding up your code.

6.8.1 Addressing Vectors and Matrices

In the following code block you can see that the `cppFunction()` requires the C++ code to be input as one long text string. One complication with using C++ is that in R, vectors, matrices, and arrays run from 1:N,... whereas in C++ an equivalent number of cells would run from 0:(N-1),.... The power of habit can make fools of us when we switch back and forth between R and C++ (or perhaps that is just me). For example, when developing the C++ code below I initially set *biom[0] = ep[3]*. So, I remembered to use 0 instead of 1 in the *biom[0]* part but promptly forgot the indexing issue again by setting the initial biomass level in the time-series of biomass to the value of *sigma* in the parameter vector rather than *Binit*. In R the *pars* variable contains *r* in index 1, *K* in index 2, B_{init} in 3, and *sigma* in 4, but in C++ the indices are 0, 1, 2, and 3. If these last few sentences are confusing to you then consider that to be a good thing because hopefully, if you go down this route of speeding your code, you will remember to be very careful over indexing variables within vectors and matrices. As you will quite possibly discover, if you use an index which points outside of an array in C++ (index 4 in a vector of 0, 1, 2, and 3, does not exist but does point to somewhere in memory!), this generally leads to R bombing out and the need for a restart. One quickly learns to save everything before running anything when developing C++ code within R, I recommend that you do the same.

Of course, programming in C++ and the use of **Rcpp** both have their complications, but reviewing these subjects is not the intent of this chapter or book. Nevertheless, hopefully this simple example will illustrate the rather remarkable advantages of using such methods when applying computer-intensive methods and succeeds in encouraging you to learn and use such methods in the right circumstances. Eddelbuettel (2013) and Wickham (2019) provide excellent introductions to the advantages of Rcpp.

6.8.2 Replacement for simpspm()

The code in the following chunk would need to be run before running any code that wanted to use the simpspmC() function instead of simpspm(), or one could place it in a separate R file and source it into your own code.

```r
library(Rcpp)
 #Send a text string containing the C++ code to cppFunction this will
 #take a few seconds to compile, then the function simpspmC will
 #continue to be available during the rest of your R session. The
 #code in this chunk could be included into its own R file, and then
 #the R source() function can be used to include the C++ into a
 #session. indat must have catch in col2 (col1 in C++), and cpue in
 #col3 (col2 in C++). Note the use of ; at the end of each line.
 #Like simpspm(), this returns only the log(predicted cpue).
cppFunction('NumericVector simpspmC(NumericVector pars,
             NumericMatrix indat, LogicalVector schaefer) {
    int nyrs = indat.nrow();
    NumericVector predce(nyrs);
    NumericVector biom(nyrs+1);
    double Bt, qval;
    double sumq = 0.0;
    double p = 0.00000001;
    if (schaefer(0) == TRUE) {
      p = 1.0;
    }
    NumericVector ep = exp(pars);
    biom[0] = ep[2];
    for (int i = 0; i < nyrs; i++) {
      Bt = biom[i];
      biom[(i+1)]=Bt+(ep[0]/p)*Bt*(1-pow((Bt/ep[1]),p))-
                      indat(i,1);
      if (biom[(i+1)] < 40.0) biom[(i+1)] = 40.0;
      sumq += log(indat(i,2)/biom[i]);
    }
    qval = exp(sumq/nyrs);
    for (int i = 0; i < nyrs; i++) {
      predce[i] = log(biom[i] * qval);
    }
    return predce;
}')
```

Once the cppFunction() code has been run then we can use the simpspmC() function anywhere the simpspm() function was used (see Table 6.7). One small complication is that simpspmC() expects the input data, *abdat*, to be a matrix, whereas *abdat* starts as a data.frame (which is actually a list, try class(abdat)). Inputting a data.frame instead of a matrix would cause the C++ function to fail, so, to fix this, in the code below, you will see that we have used the function as.matrix() to ensure the correct class of object is sent to simpspmC(), fortunately, a matrix is faster to use than a data.frame so we give that to simpspm() as well. We have also included the package **microbenchmark** so as to enable accurate comparisons of the speed of operation of the two different functions. Obviously this would need to be installed for it to be used (it can be omitted if you do not wish to install it). In the comparisons, on my Windows 10, 2018 XPS 13, simpspmC() typically took only a 20th the time taken by simpspm(), **Table** 6.8, going by the median times. The first use of the simpspmC() function sometimes has some very slow starts, which can influence the mean but the medians are less disturbed.

```
#Ensure results obtained from simpspm and simpspmC are same
library(microbenchmark)
data(abdat)
fishC <- as.matrix(abdat) # Use a matrix rather than a data.frame
inpar <- log(c(r= 0.389,K=9200,Binit=3300,sigma=0.05))
spmR <- exp(simpspm(inpar,fishC)) # demonstrate equivalence
 #need to declare all arguments in simpspmC, no default values
spmC <- exp(simpspmC(inpar,fishC,schaefer=TRUE))
out <- microbenchmark( # verything identical calling function
  simpspm(inpar,fishC,schaefer=TRUE),
  simpspmC(inpar,fishC,schaefer=TRUE),
  times=1000
)
out2 <- summary(out)[,2:8]
out2 <- rbind(out2,out2[2,]/out2[1,])
rownames(out2) <- c("simpspm","simpspmC","TimeRatio")
```

TABLE 6.7: The predictions from simpspm() and simpspmC() side-by-side to demonstrate that the code generates identical answers from the parameters $c(r= 0.389, K=9200, Binit=3300, sigma=0.05)$.

	spmR	spmC	spmR	spmC	spmR	spmC
1	1.1251	1.1251	1.4538	1.4538	2.1314	2.1314
2	1.0580	1.0580	1.5703	1.5703	2.0773	2.0773
3	1.0774	1.0774	1.7056	1.7056	2.0396	2.0396
4	1.0570	1.0570	1.8446	1.8446	1.9915	1.9915
5	1.0827	1.0827	1.9956	1.9956	1.9552	1.9552
6	1.1587	1.1587	2.0547	2.0547	1.9208	1.9208
7	1.2616	1.2616	2.1619	2.1619	1.8852	1.8852
8	1.3616	1.3616	2.2037	2.2037	1.8276	1.8276

We can now tabulate the output from the microbenchmarking

> TABLE 6.8: The output from the microbenchmark comparison of
> the simpspm and simpspmC functions. The values in microseconds
> are the minimum, the 25th quantile, the mean and median, the
> 75th quantile, the maximum, and the number of evaluations in the
> comparison. The TimeRatio is the second row divided by the first,
> so that in terms of the mean simpspmC takes ~ 7% of the time
> that simpspm takes. The actual value would vary between runs
> but not by much, although expect differences between computers
> and between versions of R (best to use the latest version, which
> tends to be fastest).

	min	lq	mean	median	uq	max	neval
simpspm	52.100	53.600	55.404	54.350	56.00	96.600	1000
simpspmC	2.800	3.100	4.384	3.700	3.90	782.700	1000
TimeRatio	0.054	0.058	0.079	0.068	0.07	8.102	1

The maximum for the simpspmC will sometimes be larger than for the simpspm, which is because the first time it is called sometimes takes longer. If
this happens to you then try running the comparison again and notice the
change in the maximum. The real comparison of interest, however, is how
much quicker on average it is to run an MCMC using simpspmC() rather than
simpspm(). We can do this without using **microbenchmark** as the time taken
is now appreciable for each run. Instead we use the **MQMF** function get-
time(), which provides time as seconds since the start of each day.

```
#How much does using simpspmC in do_MCMC speed the run time?
#Assumes Rcpp code has run, eg source("Rcpp_functions.R")
set.seed(167423) #Can use getseed() to generate a suitable seed
beginR <- gettime()  #to enable estimate of time taken
setscale <- c(0.07,0.06,0.07,0.45)
reps <- 2000  #Not enough but sufficient for demonstration
param <- log(c(r=0.39,K=9200,Binit=3400,sigma=0.05))
resultR <- do_MCMC(chains=1,burnin=100,N=reps,thinstep=128,
                   inpar=param,infunk=negLL1,calcpred=simpspm,
                   calcdat=fishC,obsdat=log(abdat$cpue),schaefer=TRUE,
                   priorcalc=calcprior,scales=setscale)
timeR <- gettime() - beginR
cat("time = ",timeR,"\n")
cat("acceptance rate = ",resultR$arate," \n")
postR <- resultR[[1]][[1]]
set.seed(167423)     # Use the same pseudo-random numbers and the
beginC <- gettime()  # same starting point to make the comparsion
```

```
param <- log(c(r=0.39,K=9200,Binit=3400,sigma=0.05))
resultC <- do_MCMC(chains=1,burnin=100,N=reps,thinstep=128,
                inpar=param,infunk=negLL1,calcpred=simpspmC,
                calcdat=fishC,obsdat=log(abdat$cpue),schaefer=TRUE,
                priorcalc=calcprior,scales=setscale)
timeC <- gettime() - beginC
cat("time = ",timeC,"\n")  # note the same acceptance rates
cat("acceptance rate = ",resultC$arate," \n")
postC <- resultC[[1]][[1]]
cat("Time Ratio = ",timeC/timeR)
```

```
# time =  18.37542
# acceptance rate =  0.319021 0.3187083 0.3282664 0.368747
# time =  3.603301
# acceptance rate =  0.319021 0.3187083 0.3282664 0.368747
# Time Ratio =  0.1960935
```

While the exact times taken will vary in each run, because of other processes your computer will be running, generally the use of simpspmC() takes only 1/5 or 20% of the time that using simpspm() takes. When dealing with minutes this may not seem that important, but once an MCMC run take 20–40 hours then a saving of 16–32 hours might be seen as more valuable. Of course there are potentially other ways to speed the process and this is generally worthwhile for the truly computer-intensive analyses.

The outcomes obtained from using either of the two functions are equivalent to running different chains, although only if the set.seed() function uses different values or, better, is not used at all. If you used the same seed then the resulting marginal distribution would be identical. With different seeds, but with only 2000 iterations, some deviations might be expected. Now that we have a faster method we can explore larger numbers of iterations while keeping larger thinning rates at the same time.

```
#compare marginal distributions of the 2 chains  Fig. 6.21
par(mfrow=c(1,1),mai=c(0.45,0.45,0.05,0.05),oma=c(0.0,0,0.0,0.0))
par(cex=0.85, mgp=c(1.35,0.35,0), font.axis=7,font=7,font.lab=7)
maxy <- getmax(c(density(postR[,"K"])$y,density(postC[,"K"])$y))
plot(density(postR[,"K"]),lwd=2,col=1,xlab="K",ylab="Density",
     main="",ylim=c(0,maxy),panel.first=grid())
lines(density(postC[,"K"]),lwd=3,col=5,lty=2)
```

6.8.3 Multiple Independent Chains

Best practice when conducting an MCMC analysis is to run multiple chains, but in practice the total number of chains generated tends to be a trade-off between the time available and having at least three or more chains. What

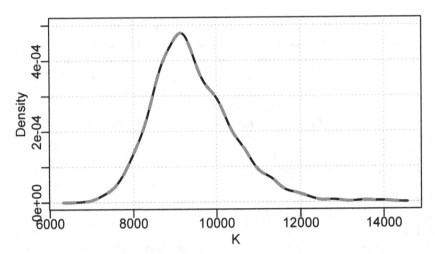

FIGURE 6.21 A comparison of the K parameter density distribution for the chains produced by the simpspm function (solid black line) and the simpspmC function (dashed blue line), with each chain having identical starting positions and the same random seed they lie on top of each other. Repeat these examples with different seeds, and or different starting positions to see the effect.

is important is to provide sufficient evidence to support the analyst's claim about the analysis having reached convergence. Here we will use just three chains, although in reality, for such a simple model, running more would be more convincing. For better speed we will now only use simpspmC() because each chain before thinning will be 10100 x 256 = 2585600 long (2585600 x 3 = 7756800, 7.7 million iterations).

```
#run multiple = 3 chains
setscale <- c(0.07,0.06,0.07,0.45)  # I only use a seed for
set.seed(9393074) # reproducibility within this book
reps <- 10000   # reset the timer
beginC <- gettime()  # remember a thinstep=256 is insufficient
resultC <- do_MCMC(chains=3,burnin=100,N=reps,thinstep=256,
                inpar=param,infunk=negLL1,calcpred=simpspmC,
                calcdat=fishC,obsdat=log(fishC[,"cpue"]),
                priorcalc=calcprior,scales=setscale,schaefer=TRUE)
cat("time = ",gettime() - beginC," secs  \n")
```

```
# time =  105.4873  secs
```

```
#3 chain run using simpspmC, 10000 reps, thinstep=256 Fig. 6.22
par(mfrow=c(2,2),mai=c(0.4,0.45,0.05,0.05),oma=c(0.0,0,0.0,0.0))
par(cex=0.85, mgp=c(1.35,0.35,0), font.axis=7,font=7,font.lab=7)
label <- c("r","K","Binit","sigma")
```

```
for (i in 1:4) {
    plot(density(resultC$result[[2]][,i]),lwd=2,col=1,
         xlab=label[i],ylab="Density",main="",panel.first=grid())
    lines(density(resultC$result[[1]][,i]),lwd=2,col=2)
    lines(density(resultC$result[[3]][,i]),lwd=2,col=3)
}
```

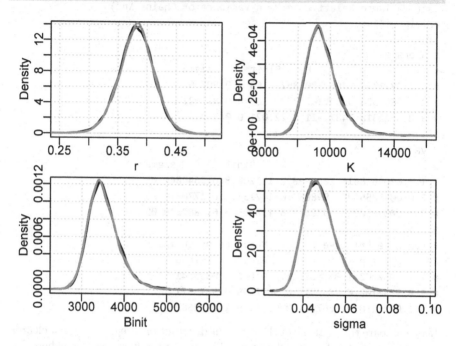

FIGURE 6.22 The variation between three chains in the marginal density distributions for the four Schaefer parameters using 10000 replicates at a thinning rate of 64 ($4 \times 64 = 256$), and the simpspmC function. Slight differences are apparent where the line is wider than average.

We can also generate summary statistics for the different chains. Indeed, there are many different diagnostic statistics and plots that can be used.

```
#generate summary stats from the 3 MCMC chains
av <- matrix(0,nrow=3,ncol=4,dimnames=list(1:3,label))
sig2 <- av  # do the variance
relsig <- av # relative to mean of all chains
for (i in 1:3) {
  tmp <- resultC$result[[i]]
  av[i,] <- apply(tmp[,1:4],2,mean)
  sig2[i,] <- apply(tmp[,1:4],2,var)
}
```

```
cat("Average \n")
av
cat("\nVariance per chain \n")
sig2
cat("\n")
for (i in 1:4) relsig[,i] <- sig2[,i]/mean(sig2[,i])
cat("Variance Relative to Mean Variance of Chains \n")
relsig
```

```
# Average
#            r        K      Binit      sigma
# 1 0.3821707 9495.580 3522.163 0.04805695
# 2 0.3809524 9530.307 3537.186 0.04811021
# 3 0.3822318 9487.911 3522.021 0.04810015
#
# Variance per chain
#                r         K      Binit         sigma
# 1 0.0009018616 1060498.2 151208.8 6.264484e-05
# 2 0.0008855405  998083.0 142153.1 6.177037e-05
# 3 0.0009080043  978855.6 138585.3 6.288734e-05
#
# Variance Relative to Mean Variance of Chains
#           r         K      Binit      sigma
# 1 1.0037762 1.0474275 1.0501896 1.0033741
# 2 0.9856108 0.9857815 0.9872949 0.9893677
# 3 1.0106130 0.9667911 0.9625155 1.0072582
```

If we compare the quantile of the different distributions we can obtain a clearer idea of the scale of any differences. The percent differences are where we directly compare the values in the 2.5% and 97.5% quantiles (the central 95% of the distribution) and the medians of the second and third marginal distributions, and only one comparison (the upper 97.5% point of Binit) is larger than 1%.

```
#compare quantile from the 2 most widely separate MCMC chains
tmp <- resultC$result[[2]] # the 10000 values of each parameter
cat("Chain 2 \n")
msy1 <- tmp[,"r"]*tmp[,"K"]/4
ch1 <- apply(cbind(tmp[,1:4],msy1),2,quants)
round(ch1,4)
tmp <- resultC$result[[3]]
cat("Chain 3 \n")
msy2 <- tmp[,"r"]*tmp[,"K"]/4
ch2 <- apply(cbind(tmp[,1:4],msy2),2,quants)
round(ch2,4)
cat("Percent difference ")
```

```
cat("\n2.5%  ",round(100*(ch1[1,] - ch2[1,])/ch1[1,],4),"\n")
cat("50%   ",round(100*(ch1[3,] - ch2[3,])/ch1[3,],4),"\n")
cat("97.5% ",round(100*(ch1[5,] - ch2[5,])/ch1[5,],4),"\n")

# Chain 2
#              r        K     Binit  sigma      msy1
# 2.5%   0.3206  7926.328 2942.254 0.0356  853.1769
# 5%     0.3317  8140.361 3016.340 0.0371  859.6908
# 50%    0.3812  9401.467 3489.550 0.0472  896.5765
# 95%    0.4287 11338.736 4214.664 0.0624  955.1773
# 97.5%  0.4386 11864.430 4425.248 0.0662  970.7137
# Chain 3
#              r        K     Binit  sigma      msy2
# 2.5%   0.3225  7855.611 2920.531 0.0355  853.0855
# 5%     0.3324  8090.493 3001.489 0.0371  859.3665
# 50%    0.3826  9370.715 3475.401 0.0471  895.8488
# 95%    0.4316 11248.955 4188.052 0.0626  952.1486
# 97.5%  0.4416 11750.426 4376.639 0.0665  966.2832
# Percent difference
# 2.5%   -0.6006 0.8922 0.7383  0.4636 0.0107
# 50%    -0.3871 0.3271 0.4055  0.2278 0.0812
# 97.5%  -0.6817 0.9609 1.0985 -0.5278 0.4564
```

6.8.4 Replicates Required to Avoid Serial Correlation

We saw earlier that if the thinning rate is too low there can be high levels of auto-correlation within the trace or sequence of each parameter. It was clear that increasing the number of thinning steps reduces the auto-correlation. What was less clear was how large the thinning rate needed to be to make such correlations insignificant.

Now we have a faster way of exploring such a question we can search for the scale of thinning rate required. We know from our earlier trials that a thinning rate of 128 per parameter still leaves significant correlations between lags of 2 to 4 steps, so, we should examine the effect of thinning rates of 1024 (4 x 256) and 2048 (4 x 512). To make these more strictly comparable we balance the thinning rates with the iterations so we use 2000 x 1024 and 1000 x 2048 and both = 2048000. The question is whether removing the auto-correlation permits a better grasp of the full variation in the different parameters. But to make the trials comparable the un-thinned chains need to be the same length to enable the same amount of exploration of the likelihood surface. Hence, with the smaller thinning rate we need, more iterations, we also need to account for the burn-in period (100 for one, 50 for the other). The thinning rate of 1024 still leaves a significant correlation at a lag of 1, which disappears with the thinning level of 2028, **Figure** 6.23. Even with `simpspmC()` this routine takes about 60 seconds.

The importance of removing within-sequence correlation is that if it is high it can interfere with the convergence onto the stationary distribution (because sequential points are correlated rather than tracing the full range of the likelihood surface) and the outcome can fail to capture the full range of the variation inherent in the model and available data being investigated.

```
#compare two higher thinning rates per parameter in MCMC
param <- log(c(r=0.39,K=9200,Binit=3400,sigma=0.05))
setscale <- c(0.07,0.06,0.07,0.45)
result1 <- do_MCMC(chains=1,burnin=100,N=2000,thinstep=1024,
                   inpar=param,infunk=negLL1,calcpred=simpspmC,
                   calcdat=fishC,obsdat=log(abdat$cpue),
                   priorcalc=calcprior,scales=setscale,schaefer=TRUE)
result2 <- do_MCMC(chains=1,burnin=50,N=1000,thinstep=2048,
                   inpar=param,infunk=negLL1,calcpred=simpspmC,
                   calcdat=fishC,obsdat=log(abdat$cpue),
                   priorcalc=calcprior,scales=setscale,schaefer=TRUE)
```

```
#autocorrelation of 2 different thinning rate chains Fig. 6.23
posterior1 <- result1$result[[1]]
posterior2 <- result2$result[[1]]
label <- colnames(posterior1)[1:4]
par(mfrow=c(4,2),mai=c(0.25,0.45,0.05,0.05),oma=c(1.0,0,1.0,0.0))
par(cex=0.85, mgp=c(1.35,0.35,0), font.axis=7,font=7,font.lab=7)
for (i in 1:4) {
  auto <- acf(posterior1[,i],type="correlation",plot=TRUE,
              ylab=label[i],lag.max=20,xlab="",ylim=c(0,0.3),lwd=2)
  if (i == 1) mtext(1024,side=3,line=-0.1,outer=FALSE,cex=1.2)
  auto <- acf(posterior2[,i],type="correlation",plot=TRUE,
              ylab=label[i],lag.max=20,xlab="",ylim=c(0,0.3),lwd=2)
  if (i == 1) mtext(2048,side=3,line=-0.1,outer=FALSE,cex=1.2)
}
mtext("Lag",side=1,line=-0.1,outer=TRUE,cex=1.2)
```

We can compare the two chains with different thinning rates in the same way that we compared the three replicate chains above. That is, we can plot their marginal distributions and compare their quantile distributions. Given the limited number of final replicates after thinning the marginal distributions are surprisingly similar, **Figure** 6.23. However, when their quantile distributions are compared, the differences observed between the central 95% of the distributions tended to be rather larger than the three chains compared in the *Multiple Independent Chains* section above. These differences do not appear to follow any particular direction, although there appears to be a suggestion

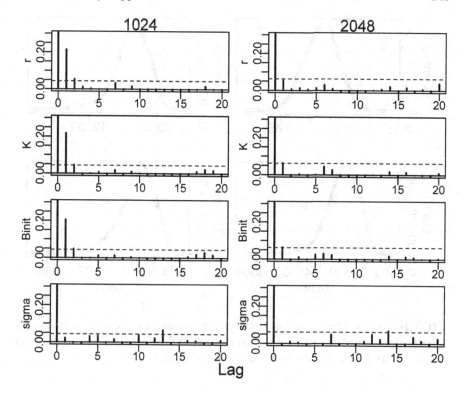

FIGURE 6.23 The auto-correlation of two chains across the four parameters of the Schaefer model with combined thinning rates of 1024 and 2048. Note the reduced maximum on the y-axis to make the differences between the two more apparent.

of some bias with a shift of the lower and upper bounds in the same direction, but more replicates would be needed to clarify this.

```
#visual comparison of 2 chains marginal densities  Fig. 6.24
parset(plots=c(2,2),cex=0.85)
label <- c("r","K","Binit","sigma")
for (i in 1:4) {
   plot(density(result1$result[[1]][,i]),lwd=4,col=1,xlab=label[i],
        ylab="Density",main="",panel.first=grid())
   lines(density(result2$result[[1]][,i]),lwd=2,col=5,lty=2)
}
```

Separate MCMC chains invariably differ to some extent, which is where the notion of a criterion of similarity becomes important. In the case of these two chains there are visible differences but the actual differences were all less

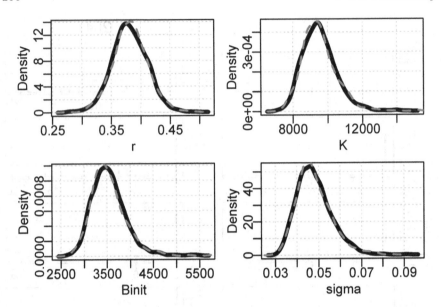

FIGURE 6.24 The variation between two chains in the marginal density distributions for the K parameter using 1000 and 2000 replicates at thinning rates of 2048 (dashed line) and 1024 (solid black line).

than 1% for the medians and in 8 out of 10 of the outer quantiles. One would generally trust the chain with the longer thinning rate.

```
#tabulate a summary of the two different thinning rates.
cat("1024 thinning rate \n")
posterior <- result1$result[[1]]
msy <-posterior[,1]*posterior[,2]/4
tmp1 <- apply(cbind(posterior[,1:4],msy),2,quants)
rge <- apply(cbind(posterior[,1:4],msy),2,range)
tmp1 <- rbind(tmp1,rge[2,] - rge[1,])
rownames(tmp1)[6] <- "Range"
print(round(tmp1,4))
posterior2 <- result2$result[[1]]
msy2 <-posterior2[,1]*posterior2[,2]/4
cat("2048 thinning rate \n")
tmp2 <- apply(cbind(posterior2[,1:4],msy2),2,quants)
rge2 <- apply(cbind(posterior2[,1:4],msy2),2,range)
tmp2 <- rbind(tmp2,rge2[2,] - rge2[1,])
rownames(tmp2)[6] <- "Range"
print(round(tmp2,4))
cat("Inner 95% ranges and Differences between total ranges \n")
cat("95% 1 ",round((tmp1[5,] - tmp1[1,]),4),"\n")
```

```
cat("95% 2 ",round((tmp2[5,] - tmp2[1,]),4),"\n")
cat("Diff  ",round((tmp2[6,] - tmp1[6,]),4),"\n")

# 1024 thinning rate
#             r        K    Binit  sigma      msy
# 2.5%   0.3221  7918.242 2943.076 0.0352 853.5243
# 5%     0.3329  8139.645 3016.189 0.0367 858.8872
# 50%    0.3801  9429.118 3499.826 0.0470 895.7376
# 95%    0.4289 11235.643 4172.932 0.0627 953.9948
# 97.5%  0.4392 11807.732 4380.758 0.0663 973.2185
# Range  0.2213  7621.901 2859.858 0.0612 238.5436
# 2048 thinning rate
#             r        K    Binit  sigma     msy2
# 2.5%   0.3216  7852.002 2920.198 0.0351 855.8295
# 5%     0.3329  8063.878 3000.767 0.0368 859.8039
# 50%    0.3820  9400.708 3482.155 0.0468 896.6774
# 95%    0.4313 11235.368 4184.577 0.0628 959.2919
# 97.5%  0.4434 11638.489 4456.164 0.0676 975.4358
# Range  0.2189  8156.444 3161.232 0.0546 257.1803
# Inner 95% ranges and Differences between total ranges
# 95% 1  0.1172 3889.49 1437.682 0.0311 119.6942
# 95% 2  0.1218 3786.487 1535.966 0.0325 119.6064
# Diff  -0.0024 534.5429 301.3746 -0.0066 18.6367
```

6.9 Concluding Remarks

The objective of this chapter was not to encourage people to write their own bootstrap, asymptotic error, likelihood profile, or MCMC functions but to allow them to explore these methods and gain the intuitions to enable them to use them with a critical awareness of their strengths and, equally importantly, their limitations. This is especially the case when implementing an MCMC analysis where one would preferably be using something like Stan, or Template Model Builder (Kristensen *et al*, 2016), or AD Model Builder (Fournier *et al*, 1998; Fournier *et al*, 2012). Nevertheless, we have gone into some detail with the application of Bayesian statistics through the use of MCMC because this really is the best way of capturing all of the uncertainty inherent in any particular modelling analysis. Having said that, there are many fisheries models where numerous parameters, such as natural mortality, recruitment steepness, and some selectivity parameters, are set as constants, which, in a Bayesian context, implies extremely informative priors. That is a somewhat artificial statement because if these parameters are not estimated then we need not account for any prior probability, but in principle this is what it implies. The usual solution to such issues is to examine likelihood profiles on such parameters or conduct sensitivity analyses where the implications of using different constants are examined. It would even be possible to use a standardized likelihood profile as a prior probability.

While it is true that the use of an MCMC can provide a more complete description of the variation within a modelling scenario, such analyses fail to capture model uncertainty, where a structurally different model may provide a somewhat different view of the population dynamics being assessed. This is where discussion usually occurs over the notion of model averaging. Although that raises the issue of which model is deemed most realistic and what weight to give to each when they may be completely incommensurate. Nevertheless, model uncertainty needs to be borne in mind when examining the outcome of any modelling. As stated by Punt and Hilborn (1997) "The most common approach is to select a single structural model and to consider the uncertainty in its parameters only. A more defensible alternative is to consider a series of truly different structural models. However, apart from being computationally more intensive, it is difficult to 'bound' the range of models considered. A related issue is how to determine how many model parameters should be considered uncertain."

The characterization of uncertainty is important because it provides some idea of how confident one can be when providing management advice. It is possible to get lost in the computational details and forget that the primary objective is to provide defensible management advice for some natural resource. There is no single way of doing that, and so if circumstances are such that an MCMC never converges, it is still possible to apply other methods and generate a narrative about an assessment and a stock's status through time. It obviously helps if one is aware of the methods and knows how to use and interpret what they can discover about a model and its data. In the end, however, it also helps to understand the history of a fishery and what other influences there may have been in addition to the catches removed.

7

Surplus Production Models

7.1 Introduction

In the previous chapters we have used and fitted what we have called static models that are stable through time (e.g., growth models using vB(), Gz(), or mm()). In addition, in the chapters *Model Parameter Estimation* and *On Uncertainty*, we have already introduced surplus production models (spm) that can be used to provide a stock assessment (e.g., the Schaefer model) and provide an example of dynamic models that use time-series of data. However, the development of the details of such models has been limited while we have focussed on particular modelling methods. Here we will examine surplus production models in more detail.

Surplus production models (alternatively Biomass Dynamic models; Hilborn and Walters, 1992) pool the overall effects of recruitment, growth, and mortality (all aspects of production) into a single production function dealing with undifferentiated biomass (or numbers). The term "undifferentiated" implies that all aspects of age and size composition, along with gender and other differences, are effectively ignored.

To conduct a formal stock assessment it is necessary, somehow, to model the dynamic behaviour and productivity of an exploited stock. A major component of these dynamics is the manner in which a stock responds to varied fishing pressure through time, that is, by how much does it increase or decrease in size. By studying the effects of different levels of fishing intensity it is often possible to assess a stock's productivity. Surplus production models provide for the simplest stock assessments available that attempt to generate a description of such stock dynamics based on fitting the model to data from the fishery.

In the 1950s, Schaefer (1954, 1957) described how to use surplus production models to generate a fishery's stock assessment. They have been developed in many ways since (Hilborn and Walters, 1992; Prager, 1994; Haddon, 2011; Winker *et al*, 2018), and it is a rapid survey of these more recent dynamic models that we will consider here.

7.1.1   Data Needs

The minimum data needed to estimate parameters for modern discrete versions of such models are at least one time-series of an index of relative abundance and an associated time-series of catch data. The catch data can cover more years than the index data. The index of relative stock abundance used in such simple models is often catch-per-unit-effort (cpue) but could be some fishery-independent abundance index (e.g., from trawl surveys, acoustic surveys), or both could be used.

7.1.2   The Need for Contrast

Despite occasional more recent use (Elder, 1979; Saila *et al*, 1979; Punt, 1994; Haddon, 1998), the use of surplus production models appears to have gone out of fashion in the 1980s. This was possibly because early on in their development it was necessary to assume the stocks being assessed were in equilibrium (Elder, 1979; Saila *et al*, 1979), and this often led to overly optimistic conclusions that were not supportable in the longer term. Hilborn (1979) analyzed many such situations and demonstrated that the data used were often too homogeneous; they lacked contrast in their effort levels and hence were uninformative about the dynamics of the populations concerned. For the data to lack contrast means that fishing catch and effort information is only available for a limited range of stock abundance levels and limited fishing intensity levels. A limited effort range means the range of responses to different fishing intensity levels will also be limited. A lack of such contrast can also arise when the stock dynamics are driven more by environmental factors than by the catches so that the stock can appear to respond to the fishery in unexpected ways (e.g., large changes to the stock despite no changes in catch or effort).

A strong assumption made with surplus production models is that the measure of relative abundance used provides an informative index of the relative stock abundance through time. Generally, the assumption is that there is a linear relationship between stock abundance and cpue, or other indices (though this is not necessarily a 1:1 relationship). The obvious risk is that this assumption is either false or can be modified depending on conditions. It is possible that cpue, for example, may become hyper-stable, which means it does not change even when a stock declines or increases. Or the variation of the index may increase so much as a result of external influences that detection of an abundance trend becomes impossible. For example, very large changes in cpue between years may be observed but not be biologically possible given the productivity of a stock (Haddon, 2018).

A different but related assumption is that the quality of the effort and subsequent catch-rates remains the same through time. Unfortunately, the notion of "effort creep" as a result of technological changes in fishing gear, changes in fishing behaviour or methods, or other changes in fishing efficiency, is invariably a challenge or problem for assessments that rely on cpue as an index of relative abundance. The notion of effort-creep implies that the effectiveness of effort increases so that any observed nominal cpue, based on nominal effort, will over-estimate the relative stock abundance (it is biased high). Statistical standardization of cpue (Kimura, 1981; Haddon, 2018) can address some of these concerns but obviously can only account for factors for which data is available. For example, if one introduces GPS plotters or colour echo-sounders into a fishery, which would tend to increase the effectiveness of effort, but have no record of which vessels introduced them and when, their positive influence on catch rates would not be able to be accounted for by standardization.

7.1.3 When Are Catch-Rates Informative?

A possible test of whether the assumed relationship between abundance and any index of relative abundance is real and informative can be derived from the implication that, *in a developed fishery*, if catches are allowed to increase then it is expected that cpue will begin to decline some time after. Similarly, if catches are reduced to be less than surplus production, perhaps through management or marketing changes, then cpue would be expected to increase some time soon after as stock size increases. The idea being that if catches are less than the current productivity of a stock then eventually the stock size and thereby cpue should increase and *vice-versa*. If catches declined through a lack of availability but stayed at or above the current productivity then, of course, the cpue could not increase and may even decline further despite, perhaps minor, reductions in catch. Emphasis was placed upon *developed* fisheries because when a fishery begins, any initial depletion of the biomass will lead to "windfall" catches (MacCall, 2009) as the stock is fished down, which in turn will lead to cpue levels that would not be able to be maintained once a stock has been reduced away from unfished levels.

The expectation is, therefore, that in a developed fishery cpue would be, in many cases, negatively correlated with catches, possibly with a time-lag between cpue changing in response to changes in catches. If we can find such a relationship this usually means there is some degree of contrast in the data, if we cannot find such a negative relationship this generally means the information content of the data, with regard to how the stock responds to the fishery, is too low to inform an assessment based only on catches and the index of relative abundance. That is, the cpue adds little more information that that available in the catches.

We will use the **MQMF** data-set *schaef* to illustrate these ideas. *schaef* contains the catches and cpue of the original yellowfin tuna data (Table 7.1) from Schaefer (1957), which was an early example of the use of surplus production models to conduct stock assessments.

```
#Yellowfin-tuna data from Schaefer 12957
data(schaef)
```

TABLE 7.1: The Schaefer (1957) yellowfin tuna fishery data from 1934–1955. Catches are '000s of lbs, effort is '000s of standard class4 clipper days, and cpue is '000s lbs/day.

year	catch	effort	cpue	year	catch	effort	cpue
1934	60913	5879	10.3611	1945	89194	9377	9.5120
1935	72294	6295	11.4844	1946	129701	13958	9.2922
1936	78353	6771	11.5719	1947	160134	20381	7.8570
1937	91522	8233	11.1165	1948	200340	23984	8.3531
1938	78288	6830	11.4624	1949	192458	23013	8.3630
1939	110417	10488	10.5279	1950	224810	31856	7.0571
1940	114590	10801	10.6092	1951	183685	18726	9.8091
1941	76841	9584	8.0176	1952	192234	31529	6.0971
1942	41965	5961	7.0399	1953	138918	36423	3.8140
1943	50058	5930	8.4415	1954	138623	24995	5.5460
1944	64094	6397	10.0194	1955	140581	17806	7.8951

The plot of catch, cpue, and their relationship, **Figure** 7.1, only exhibits a weak negative relationship between cpue and catch. If we examine the outcome of the regression conducted using lm() by using summary(model), we find this regression is only just significant at $P = 0.04575$. However, this reflects a correlation with no time-lags, i.e., a lag = 0. We do not know how many years may need to pass before we can detect a potential effect of a change in catch on cpue, so hence, we need to do a time-lagged correlation analysis between cpue and catch; for that we can use the base R cross-correlation function ccf().

```
#schaef fishery data and regress cpue and catch       Fig. 7.1
parset(plots=c(3,1),margin=c(0.35,0.4,0.05,0.05))
plot1(schaef[,"year"],schaef[,"catch"],ylab="Catch",xlab="Year",
      defpar=FALSE,lwd=2)
plot1(schaef[,"year"],schaef[,"cpue"],ylab="CPUE",xlab="Year",
      defpar=FALSE,lwd=2)
plot1(schaef[,"catch"],schaef[,"cpue"],type="p",ylab="CPUE",
      xlab="Catch",defpar=FALSE,pch=16,cex=1.0)
model <- lm(schaef[,"cpue"] ~ schaef[,"catch"])
abline(model,lwd=2,col=2)    # summary(model)
```

As before there is an indication that a time-lag = 0 is only just significant. However, the significant negative correlation of cpue on catch at a lag of 2 years, **Figure** 7.2, suggests that there would be sufficient contrast in this yellowfin tuna data to inform a surplus production model (there are also significant

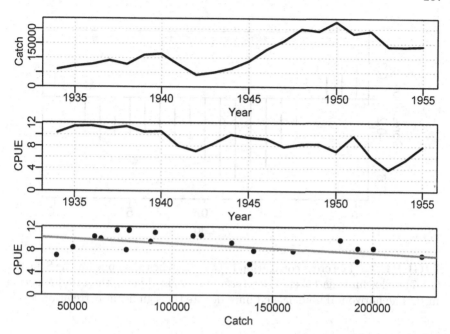

FIGURE 7.1 The catches and cpue by year, and their relationship, described by a regression, for the Schaefer (1957) yellowfin tuna fishery data.

effects at 1, 3, and 4 years). This correlation should become more apparent if we physically lag the CPUE data by two year, **Figure** 7.3.

```
#cross correlation between cpue and catch in schaef Fig. 7.2
parset(cex=0.85) #sets par parameters for a tidy base graphic
ccf(x=schaef[,"catch"],y=schaef[,"cpue"],type="correlation",
    ylab="Correlation",plot=TRUE)
```

```
#now plot schaef data with timelag of 2 years on cpue    Fig 7.3
parset(plots=c(3,1),margin=c(0.35,0.4,0.05,0.05))
plot1(schaef[1:20,"year"],schaef[1:20,"catch"],ylab="Catch",
      xlab="Year",defpar=FALSE,lwd=2)
plot1(schaef[3:22,"year"],schaef[3:22,"cpue"],ylab="CPUE",
      xlab="Year",defpar=FALSE,lwd=2)
plot1(schaef[1:20,"catch"],schaef[3:22,"cpue"],type="p",
      ylab="CPUE",xlab="Catch",defpar=FALSE,cex=1.0,pch=16)
model2 <- lm(schaef[3:22,"cpue"] ~ schaef[1:20,"catch"])
abline(model2,lwd=2,col=2)
```

The relationship between cpue and catches from the Schaefer (1957) yellowfin tuna fishery data with a negative lag of 2 years imposed on the cpue time-series

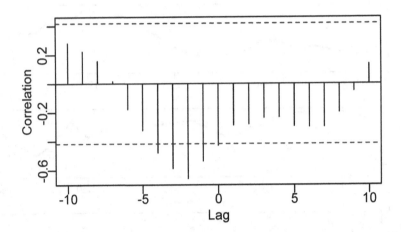

FIGURE 7.2 The cross correlation between catches and cpue for the Schaefer (1957) yellowfin tuna fishery data (*schaef*) obtained using the ccf() function in R. The horizontal blue dashed lines denote the significant correlation levels.

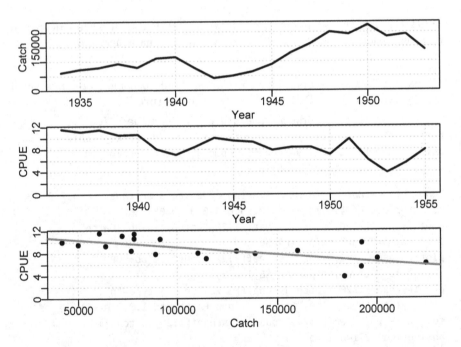

FIGURE 7.3 The catches and cpue, and their relationship, for the Schaefer (1957) yellowfin tuna fishery data. With a negative lag of 2 years on the cpue time-series the negative or inverse correlation becomes more apparent.

(rows 3:22 against row 1:20). The very small gradient is a reflection of the catches being reported in '000s of pounds.

```
#write out a summary of he regression model2
summary(model2)
```

```
#
# Call:
# lm(formula = schaef[3:22, "cpue"] ~ schaef[1:20, "catch"])
#
# Residuals:
#     Min      1Q   Median      3Q      Max
# -3.10208 -0.92239 -0.06399  1.04280  3.11900
#
# Coefficients:
#                        Estimate Std. Error t value Pr(>|t|)
# (Intercept)           1.165e+01  7.863e-01  14.814 1.59e-11
# schaef[1:20, "catch"] -2.576e-05  6.055e-06  -4.255 0.000477
#
# Residual standard error: 1.495 on 18 degrees of freedom
# Multiple R-squared:  0.5014,  Adjusted R-squared:  0.4737
# F-statistic:  18.1 on 1 and 18 DF,  p-value: 0.0004765
```

7.2 Some Equations

The dynamics of the stock being assessed are described using an index of relative abundance. The index, however it is made, is assumed to reflect the biomass available to the method used to estimate it (fishery-dependent cpue or an independent survey), and this biomass is assumed to be affected by the catches removed by the fishery. This means, if we are using commercial catch-per-unit-effort (cpue), that strictly we are dealing with the exploitable biomass and not the spawning biomass (which is the more usual objective of stock assessments). However, generally, the assumption is made that the selectivity of the fishery is close to the maturity ogive so that the index used is still indexing spawning biomass, at least approximately. Even so, exactly what is being indexed should be considered explicitly before drawing such conclusions.

In general terms, the dynamics are designated as a function of the biomass at the start of a given year t, though it could, by definition, refer to a different date within a year. Remember that the end of a year is effectively the same as the start of the next year although precisely which is used will influence how to start and end the analysis (e.g., from which biomass-year to remove the catches taken in a given year):

$$B_0 = B_{init}$$
$$B_{t+1} = B_t + rB_t\left(1 - \frac{B_t}{K}\right) - C_t \tag{7.1}$$

where B_{init} is the initial biomass at whatever time one's data begins. If data is available from the beginning of fishery then $B_{init} = K$, where K is the carrying capacity or unfished biomass. B_t represents the stock biomass at the start of year t, r represents the population's intrinsic growth rate, and $rB_t\left(1 - \frac{B_t}{K}\right)$ represents a production function in terms of stock biomass that accounts for recruitment of new individuals, any growth in the biomass of current individuals, natural mortality, and assumes linear density-dependent effects on population growth rate. Finally, C_t is the catch during the year t, which represents fishing mortality. The emphasis placed upon when in the year each term refers to is important, as it determines how the dynamics are modelled in the equations and the consequent R code.

In order to compare and then fit the dynamics of such an assessment model with the real world the model dynamics are also used to generate predicted values for the index of relative abundance in each year:

$$\hat{I}_t = \frac{C_t}{E_t} = qB_t \tag{7.2}$$

where $\hat{I}_t$ is the predicted or estimated mean value of the index of relative abundance in year t, which is compared to the observed indices to fit the model to the data, E_t is the fishing effort in year t, and q is the catchability coefficient (defined as the amount of biomass/catch taken with one unit of effort). This relationship also makes the strong assumption that the stock biomass is what is known as a dynamic-pool. This means that irrespective of geographical distance, any influences of the fishery, or the environment, upon the stock dynamics have an effect across the whole stock within each time-step used (often one year but could be less). This is a strong assumption, especially if any consistent spatial structure occurs within a stock or the geographical scale of the fishery is such that large amounts of time would be needed for fish in one region to travel to a different region. Again, awareness of these assumptions is required to appreciate the limitations and appropriately interpret any such analyses.

7.2.1 Production Functions

A number of functional forms have been put forward to describe the productivity of a stock and how it responds to stock size. We will consider two, the Schaefer model and a modified form of the Fox (1970) model, plus a generalization that encompasses both:

The production function of the Schaefer (1954, 1957) model is:

$$f(B_t) = rB_t \left(1 - \frac{B_t}{K}\right) \qquad (7.3)$$

while a modified version of the Fox (1970) model uses:

$$f(B_t) = \log(K)rB_t \left(1 - \frac{\log(B_t)}{\log(K)}\right) \qquad (7.4)$$

The modification entails the inclusion of the $\log(K)$ as the first term, which only acts to maintain the maximum productivity at approximately the equivalent of a similar set of parameters in the Schaefer model.

Pella and Tomlinson (1969) produced a generalized production function that included the equivalent to both the Schaefer and the Fox model as special cases. Here, we will use an alternative to their formulation that was produced by Polacheck et al (1993). This provides a general equation for the population dynamics that can be used for both the Schaefer and the Fox model, and gradations between, depending on the value of a single parameter p:

$$B_{t+1} = B_t + rB_t \frac{1}{p} \left(1 - \left(\frac{B_t}{K}\right)^p\right) - C_t \qquad (7.5)$$

where the first term, B_t, is the stock biomass at time t, and the last term is C_t, which is the catch taken in time t. The middle term is a more complex component that defines the production curve. This is made up of r, the intrinsic rate of population growth, B_t, the current biomass at time t, K, the carrying capacity or maximum population size, and p is a term that controls any asymmetry of the production curve. If p is set to 1.0 (the default in the **MQMF** discretelogistic() function), this equation simplifies to the classical Schaefer model (Schaefer, 1954, 1957). Polacheck et al (1993) introduced the equation above, but it tends to be called the Pella-Tomlinson (1969) surplus production model (though their formulation was different it has very similar properties).

The sub-term rB_t represents unconstrained exponential population growth (because in this difference equation, it is added to B_t), which, as long as $r > 0.0$, would lead to unending positive population growth in the absence of catches (such positive exponential growth is still exemplified by the world's human population; although plague in the 14th century reversed that trend for a brief but particularly unhappy time). The sub-term $(1/p)(1-(Bt/K)^p)$ provides a constraint on the exponential growth term because its value tends to zero as the population size increases. This acts as, and is known as, a density-dependent effect.

When p is set to 1.0 this equation becomes the same as the Schaefer model (linear density-dependence). But when p is set to a very small number, say 1e-08, then the formulation becomes equivalent to the Fox model's dynamics. Values of $p > 1.0$ lead to a production curve skewed to the left with the mode to the right of center. With p either > 1 or < 1, the density-dependence would no longer be linear in character. Generally, one would fix the p value and not attempt to fit it using data. Catch and an index of relative abundance alone would generally be insufficient to estimate the detailed effect on productivity of relative population size.

```
#plot productivity and density-dependence functions Fig. 7.4
prodfun <- function(r,Bt,K,p) return((r*Bt/p)*(1-(Bt/K)^p))
densdep <- function(Bt,K,p) return((1/p)*(1-(Bt/K)^p))
r <- 0.75; K <- 1000.0; Bt <- 1:1000
sp <- prodfun(r,Bt,K,1.0)  # Schaefer equivalent
sp0 <- prodfun(r,Bt,K,p=1e-08)  # Fox equivalent
sp3 <- prodfun(r,Bt,K,3) #left skewed production, marine mammal?
parset(plots=c(2,1),margin=c(0.35,0.4,0.1,0.05))
plot1(Bt,sp,type="l",lwd=2,xlab="Stock Size",
      ylab="Surplus Production",maxy=200,defpar=FALSE)
lines(Bt,sp0 * (max(sp)/max(sp0)),lwd=2,col=2,lty=2) # rescale
lines(Bt,sp3*(max(sp)/max(sp3)),lwd=3,col=3,lty=3)   # production
legend(275,100,cex=1.1,lty=1:3,c("p = 1.0 Schaefer","p = 1e-08 Fox",
                "p = 3 LeftSkewed"),col=c(1,2,3),lwd=3,bty="n")
plot1(Bt,densdep(Bt,K,p=1),xlab="Stock Size",defpar=FALSE,
      ylab="Density-Dependence",maxy=2.5,lwd=2)
lines(Bt,densdep(Bt,K,1e-08),lwd=2,col=2,lty=2)
lines(Bt,densdep(Bt,K,3),lwd=3,col=3,lty=3)
```

The Schaefer model assumes a symmetrical production curve with maximum surplus production or maximum sustainable yield (MSY) at $0.5K$ and the density-dependent term trends linearly from 1.0 at very low population sizes to zero as B_t tends towards K. The Fox model is approximated when p has a small value, say $p = 1e-08$, which generates an asymmetrical production curve with the maximum production at some lower level of depletion (in this case at $0.368K$, found using Bt[which.max(sp0)]). The density-dependent term is non-linear and the maximum productivity (MSY) occurs where the density-dependent term $= 1.0$. Without the re-scaling used in **Figure** 7.4 the Fox model is generally more productive than the Schaefer as a result of the density-dependent term becoming greater than 1.0 at stock sizes less than B_{MSY}, the

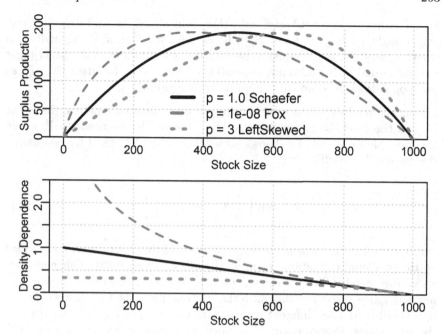

FIGURE 7.4 The effect of the p parameter on the Polacheck *et al*, 1993, production function (upper plot) and on the density-dependent term (lower plot). Note the rescaling of the productivity to match that produced by the Schaefer curve. Stock size could be biomass or numbers.

stock biomass that generates MSY. With values of $p > 1.0$, the maximum productivity occurs at higher stock sizes and with population growth rates increasing only almost linearly at lower stock sizes and density-dependent declines only occurring at rather higher stock levels. Such dynamics would be more typical of marine mammals than of fish.

The Schaefer model can be regarded as more conservative than the Fox in that it requires the stock size to be higher for maximum production and generally leads to somewhat lower levels of catch, though exceptions could occur because of the generally higher productivity from Fox-type models.

7.2.2 The Schaefer Model

For the Schaefer model we would set $p = 1.0$ leading to:

$$B_{t+1} = B_t + rB_t \left(1 - \frac{B_t}{K}\right) - C_t \tag{7.6}$$

Given a time-series of fisheries data there will always be an initial biomass, which might be $B_{init} = K$, or B_{init} being some fraction of K, depending on whether the stock was deemed to be depleted when data from the fishery first became available. It is also not impossible that B_{init} can be larger than K, as real populations tend not to exhibit a stable equilibrium.

Fitting the model to data would require at least three parameters, the r, the K, and the catchability coefficient q (B_{init} might also be needed). However, it is possible to use what is known as a "closed-form" method for estimating the catchability coefficient q:

$$\hat{q} = \exp\left(\frac{1}{n}\sum \log\left(\frac{I_t}{\hat{B}_t}\right)\right) \tag{7.7}$$

which is the back-transformed geometric mean of the observed CPUE divided by the predicted exploitable biomass (Polacheck et al, 1993). This generates an average catchability for the time-series. In circumstances where a fishery has undergone a major change such that the quality of the CPUE has changed, it is possible to have different estimates of catchability for different parts of the time-series. However, care is needed to have a strong defense for such suggested model specifications, especially as the shorter the time-series used to estimate q the more uncertainty will be associated with it.

7.2.3   Sum of Squared Residuals

Such a model can be fitted using least squares or, more properly, the sum of squared residual errors:

$$ssq = \sum\left(\log(I_t) - \log(\hat{I}_t)\right)^2 \tag{7.8}$$

the log-transformations are required as generally CPUE tends to be distributed log-normally and the least-squares method implies Normal random errors. The least squares approach tends to be relatively robust when first searching for a set of parameters that enable a model to fit to available data. However, once close to a solution more modelling options become available if one then uses maximum likelihood methods. The full Log-Normal log likelihood is:

$$L\left(data|B_{init}, r, K, q\right) = \prod_t \frac{1}{I_t\sqrt{2\pi\hat{\sigma}}} e^{\frac{-(\log I_t - \log \hat{I}_t)}{2\hat{\sigma}^2}} \tag{7.9}$$

Apart from the log transformations this differs from Normal PDF likelihoods by the variable concerned (here I_t) being inserted before the $\sqrt{2\pi\hat{\sigma}}$ term. Fortunately, as shown in *Model Parameter Estimation*, the negative log-likelihood can be simplified (Haddon, 2011), to become:

$$-veLL = \frac{n}{2}\left(\log(2\pi) + 2\log(\hat{\sigma}) + 1\right) \tag{7.10}$$

where the maximum likelihood estimate of the standard deviation, $\hat{\sigma}$ is given by:

$$\hat{\sigma} = \sqrt{\frac{\sum\left(\log(I_t) - \log(\hat{I}_t)\right)^2}{n}} \tag{7.11}$$

Note the division by n rather than by $n\text{-}1$. Strictly, for the Log-Normal (in **Equ**(7.10)), the *-veLL* should be followed by an additional term:

$$-\sum\log(I_t) \tag{7.12}$$

the sum of the log-transformed observed catch rates. But as this will be constant it is usually omitted. Of course, when using R we can always use the built-in probability density function implementations (see `negLL()` and `negLL1()`) so such simplifications are not strictly necessary, but they can remain useful when one wishes to speed up the analyses using **Rcpp**, although *Rcpp-syntactic-sugar*, which leads to C++ code looking remarkably like R code, now includes versions of `dnorm()` and related distribution functions.

7.2.4   Estimating Management Statistics

The Maximum Sustainable Yield can be calculated for the Schaefer model simply by using:

$$MSY = \frac{rK}{4} \tag{7.13}$$

However, for the more general equation using the p parameter from Polacheck *et al* (1993) one needs to use:

$$MSY = \frac{rK}{(p+1)^{\frac{(p+1)}{p}}} \tag{7.14}$$

which simplifies to **Equ**(7.13) when $p = 1.0$. We can use the **MQMF** function `getMSY()` to calculate **Equ**(7.14), which is one illustration of how the Fox model can imply greater productivity than the Schaefer.

```
#compare Schaefer and Fox MSY estimates for same parameters
param <- c(r=1.1,K=1000.0,Binit=800.0,sigma=0.075)
cat("MSY Schaefer = ",getMSY(param,p=1.0),"\n") # p=1 is default
cat("MSY Fox      = ",getMSY(param,p=1e-08),"\n")
```

```
# MSY Schaefer =  275
# MSY Fox       =  404.6674
```

Of course, if you fit the two models to real data this will generally produce different parameters for each and so the derived MSY's may be closer in value.

It is also possible to produce effort-based management statistics. The effort level that if maintained should lead the stock to achieve MSY at equilibrium is known as E_{MSY}:

$$E_{MSY} = \frac{r}{q(1+p)} \tag{7.15}$$

which collapses to $E_{MSY} = r/2q$ for the Schaefer model but remains general for other values of the p parameter. It is also possible to estimate the equilibrium harvest rate (proportion of stock taken each year) that should lead to B_{MSY}, which is the biomass that has a surplus production of MSY:

$$H_{MSY} = qE_{MSY} = q\frac{r}{q+qp} = \frac{r}{1+p} \tag{7.16}$$

It is not uncommon to see **Equ(7.16)** depicted as $F_{MSY} = qE_{MSY}$, but this can be misleading as F_{MSY} would often be interpreted as an instantaneous fishing mortality rate, whereas, in this case, it is actually a proportional harvest rate. For this reason I have explicitly used H_{MSY}.

7.2.5 The Trouble with Equilibria

The real-world interpretation of management targets is not always straightforward. An equilibrium is now assumed to be unlikely in most fished populations, so the interpretation of MSY is more like an average, long-term expected potential yield if the stock is fished optimally; a dynamic equilibrium might be a better description. The E_{MSY} is the effort that, if consistently applied, should give rise to the MSY, but only if the stock biomass is at B_{MSY}, the biomass needed to generate the maximum surplus production. Each of these management statistics is derived from equilibrium ideas. Clearly, a fishery could be managed by limiting effort to E_{MSY}, but if the stock biomass starts off badly depleted, then the average long-term yield will not result. In fact, the E_{MSY} effort level may be too high to permit stock rebuilding in this non-equilibrium world. Similarly, H_{MSY}, would operate as expected but only when the stock biomass was at B_{MSY}. It would be possible to estimate the catch or effort level that should lead the stock to recover to B_{MSY}, and this could be termed F_{MSY}, but this would require conducting stock projections and searching for the catch levels that would eventually lead to the desired result. We will examine making population projections in later sections.

It is important to emphasize that the idea of MSY and its related statistics are based around the idea of equilibrium, which is a rarity in the real world. At best, a dynamic equilibrium may be achievable but, whatever the case, there are risks associated with the use of such equilibrium statistics. When it was first developed the concept of MSY was deemed a suitable target towards which to manage a fishery. Now, despite being incorporated into a number of national fisheries acts and laws as the overall objective of fisheries management, it is safer to treat MSY as an upper limit on fishing mortality (catch); a limit reference point rather than a target reference point.

Ideally, the outcomes from an assessment need to be passed through a harvest control rule (HCR) that provides for formal management advice, with regard to future catches or effort, in response to the estimated stock status (fishing mortality and stock depletion levels). However, few of these potential management outputs are of value without some idea of the uncertainty around their values. As we have indicated, it would also be very useful to be able to project the models into the future to provide a risk assessment of alternative management strategies. But first we need to fit the models to data.

7.3 Model Fitting

Details of the model parameters and other aspects relating to the model can also be found in the help files for each of the functions (try ?spm or ?simpspm). In brief, the model parameters are r the net population rate of increase (combined individual growth in weight, recruitment, and natural mortality), K the population carrying capacity or unfished biomass, and B_{init} is the biomass in the first year. This parameter is only required if the index of relative abundance data (usually cpue) only becomes available after the fishery has been running for a few years implying that the stock has been depleted to some extent. If no initial depletion is assumed then B_{init} is not needed in the parameter list and is set equal to K inside the function. The final parameter is sigma, the standard deviation of the Log-Normal distribution used to describe the residuals. simpspm() and spm() are designed for use with maximum likelihood methods so even if you use ssq() as the criterion of optimum fit, a *sigma* value is necessary in the parameter vector.

In Australia the index of relative stock abundance is most often catch-per-unit-effort (cpue), but it could be some fishery-independent abundance index (e.g., from trawl surveys, acoustic surveys), or both could be used in one analysis (see simpspmM()). The analysis will permit the production of ongoing management advice as well as a determination of stock status.

In this section we will describe the details of how to conduct a surplus production analysis, how to extract the results from that analysis as well as plot out illustrations of those results.

7.3.1   A Possible Workflow for Stock Assessment

When conducting a stock assessment based upon a surplus production model one possible workflow might include:

1. reading in time-series of catch and relative abundance data. It can help to have functions that check for data completeness, missing values, and other potential issues, but it is even better to know your own data and its limitations.

2. using a ccf() analysis to determine whether the cpue data relative to the catch data may be informative. If a significant negative relation was found this would strengthen any defense of the analysis.

3. defining/guessing a set of initial parameters containing r and K, and optionally B_{init} = initial biomass, which is used if it is suspected that the fishery data starts after the stock has been somewhat depleted.

4. using the function plotspmmod() to plot up the implications of the assumed initial parameter set for the dynamics. This is useful when searching for plausible initial parameters sets.

5. using nlm() or fitSPM() to search for the optimum parameters once a potentially viable initial parameter set are input. See discussion.

6. using plotspmmod() using the optimum parameters to illustrate the implications of the optimum model and its relative fit (especially using the residual plot).

7. ideally examining the robustness of the model fit by using multiple different initial parameter sets as starting points for the model fitting procedure, see robustSPM().

8. once satisfied with the robustness of the model fit, using spmphaseplot() to plot out the phase diagram of biomass against harvest rate so as to determine and illustrate the stock status visually.

9. using `spmboot()`, asymptotic standard errors, or Bayesian methods from *On Uncertainty*, to characterize uncertainty in the model fit and outputs. Tabulating and plotting such outputs.

10. Documenting and defending any conclusions reached.

Two common versions of the dynamics are currently available inside **MQMF**: the classical Schaefer model (Schaefer, 1954) and an approximation to the Fox model (Fox, 1970; Polacheck *et al*, 1993); both are described in Haddon (2011). Prager (1994) provides many additional forms of analysis that are possible using surplus production models and practical implementations are also provided in Haddon (2011).

```
#Initial model 'fit' to the initial parameter guess  Fig. 7.5
data(schaef); schaef <- as.matrix(schaef)
param <- log(c(r=0.1,K=2250000,Binit=2250000,sigma=0.5))
negatL <- negLL(param,simpspm,schaef,logobs=log(schaef[,"cpue"]))
ans <- plotspmmod(inp=param,indat=schaef,schaefer=TRUE,
                  addrmse=TRUE,plotprod=FALSE)
```

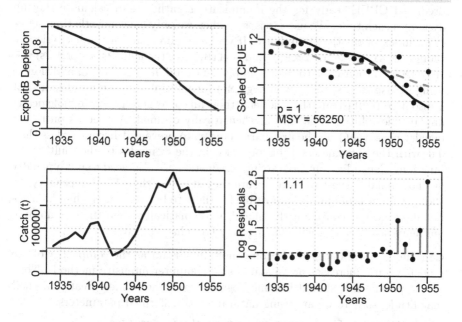

FIGURE 7.5 The tentative fit of a surplus production model to the *schaef* data-set using the initial parameter values. The dashed green line in the CPUE plot is a simple loess fit, while the solid line is that implied by the guessed input parameters. The horizontal red line in the catch plot is the predicted MSY. The number in the residual plot is the root mean square error of the Log-Normal residuals.

The r value of 0.1 leads to a *negatL* = 8.2877, and a strong residual pattern of all residuals below 1.0 up to about 1950 and four large and positive residuals afterwards. The K value was set at about 10 times the maximum catch and something of that order (10x to 20x) will often lead to sufficient biomass being available that the stock biomass and CPUE trajectories get off the x-axis ready for entry to a minimizer/optimizer. We used the plotprod = FALSE option (the default), as before fitting the model to data there is little point in seeing the predicted productivity curves.

With numerical methods of fitting models to data it is often necessary to take measures to ensure that one obtains a robust, as well as a biologically plausible model fit. One option for robustness is to fit the model twice, with the input parameters for the second fit coming from the first fit. We will use a combination of optim() and nlm() along with negLL1() to estimate the negative log-likelihood during each iteration (this is how fitSPM() is implemented). Within **MQMF** we have a function spm(), which calculates the full dynamics in terms of predicted changes in biomass, CPUE, depletion, and harvest rate. While this is relatively quick, to speed the iterative model fitting process, rather than use spm(), we use simpspm() to output only the log of the predicted CPUE ready for the minimization, rather than calculate the full dynamics each time. We use simpspm() when we only have a single time-series of an index of relative abundance. If we have more than one index series we should use simpspmM(), spmCE(), and negLLM(); see the help, ?simpspmM, ?spmCE, ?negLLM, and their code to see the implementation in each case. In addition to using simpspmM(), spmCE(), and negLLM() for multiple time-series of indices it is also used to illustrate that model fitting can sometimes generate biologically implausible solutions that are mathematically optimal. As well as putting a penalty penalty0() on the first parameter, r, to prevent it becoming less than 0.0, with the extreme catch history used in the examples to the multi-index functions, depending on the starting parameters, we also need to put a penalty on the annual harvest rates to ensure they stay less than 1.0 (see penalty1(). Biologically it is obviously impossible for there to be more catch than biomass but if we do not constrain the model mathematically, then there is nothing mathematically wrong with having very large harvest rates.

Considerations about speed become more important as the complexity of the models we use increases or we start using computer-intensive methods. None of our parameters should become negative, and they differ greatly in their magnitude, so here we are using natural log-transformed parameters.

```
#Fit the model first using optim then nlm in sequence
param <- log(c(0.1,2250000,2250000,0.5))
pnams <- c("r","K","Binit","sigma")
best <- optim(par=param,fn=negLL,funk=simpspm,indat=schaef,
              logobs=log(schaef[,"cpue"]),method="BFGS")
outfit(best,digits=4,title="Optim",parnames = pnams)
```

```
cat("\n")
best2 <- nlm(negLL,best$par,funk=simpspm,indat=schaef,
            logobs=log(schaef[,"cpue"]))
outfit(best2,digits=4,title="nlm",parnames = pnams)
```

```
# optim solution:  Optim
# minimum      :  -7.934055
# iterations   :  41 19  iterations, gradient
# code         :  0
#                par        transpar
# r       -1.448503        0.2349
# K       14.560701 2106842.7734
# Binit 14.629939 2257885.3255
# sigma -1.779578        0.1687
# message      :
#
# nlm solution:  nlm
# minimum      :  -7.934055
# iterations   :  2
# code         :  2 >1 iterates in tolerance, probably solution
#                par        gradient      transpar
# r       -1.448508  6.030001e-04        0.2349
# K       14.560692 -2.007053e-04 2106824.2701
# Binit 14.629939  2.545064e-04 2257884.5480
# sigma -1.779578 -3.688185e-05        0.1687
```

The output from the two-fold application of the numerical optimizers suggests that we did not need to conduct the process twice, but it is precautionary not to be too trustful of numerical methods. By all means do single model fits but do so at your own peril (or perhaps I have had to work with more poor to average quality data than many people!).

We can now take the optimum parameters from the *best2* fit and put them into the plotspmmod() function to visualize the model fit. We now have the optimum parameters so we can include the productivity curves by setting the *plotprod* argument to *TRUE*. plotspmmod() does more than just plot the results, it also returns a large list object invisibly so if we want this we need to assign it to a variable or object (in this case *ans*) in order to use it.

```
#optimum fit. Defaults used in plotprod and schaefer Fig. 7.6
ans <- plotspmmod(inp=best2$estimate,indat=schaef,addrmse=TRUE,
                  plotprod=TRUE)
```

The object returned by plotspmmod() is a list of objects containing a collection of results, including the optimum parameters, a matrix (*ans$Dynamics$outmat*) containing the predicted optimal dynamics, the production curve, and numerous summary results. Once assigned to a particular

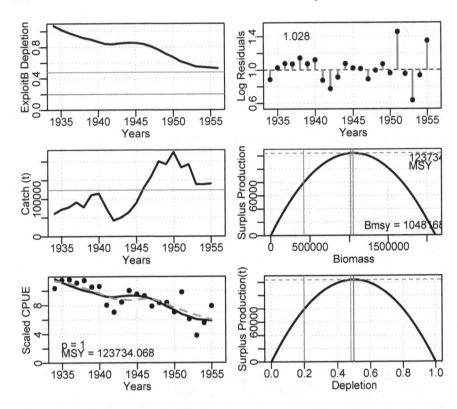

FIGURE 7.6 A summary of the fit of a surplus production model to the *schaef* data-set given the optimum parameters from the final nlm() fit. In the CPUE plot, the dashed green line is the simple loess curve fit while the solid red line is the optimal model fit.

object in the working environment these can be quickly extracted for use in other functions. Try running str() without the *max.level=1* argument or set it = *2*, to see more details. Lots of functions generate large, informative objects, you should become familiar with exploring them to make sure you understand what is being produced within different analyses.

```
#the high-level structure of ans; try str(ans$Dynamics)
str(ans, width=65, strict.width="cut",max.level=1)
```

```
# List of 12
# $ Dynamics :List of 5
# $ BiomProd : num [1:200, 1:2] 100 10687 21273 31860 42446 ...
# ..- attr(*, "dimnames")=List of 2
# $ rmseresid: num 1.03
# $ MSY      : num 123731
```

```
#  $ Bmsy    : num 1048169
#  $ Dmsy    : num 0.498
#  $ Blim    : num 423562
#  $ Btarg   : num 1016409
#  $ Ctarg   : num 123581
#  $ Dcurr   : Named num 0.528
#   ..- attr(*, "names")= chr "1956"
#  $ rmse    :List of 1
#  $ sigma   : num 0.169
```

There are also a few **MQMF** functions to assist with pulling out such results or that use the results from `plotspmmod()` (see `summspm()` and `spmphaseplot()`), which is why the function includes the argument *plotout = TRUE*, so that a plot need not be produced. However, in many cases it can be simpler just to point to the desired object within the high-level object (in this case *ans*). Notice that the *MSY* obtained from the generated productivity curve differs by a small amount from that calculated from the optimal parameters. This is because the productivity curve is obtained numerically by calculating the productivity for a vector of different biomass levels. Its resolution is thus limited by the steps used to generate the biomass vector. Its estimate will invariably be slightly smaller than the parameter-derived value.

```
#compare the parameteric MSY with the numerical MSY
round(ans$Dynamics$sumout,3)
cat("\n Productivity Statistics \n")
summspm(ans) # the q parameter needs more significantr digits
```

```
#           msy        p  FinalDepl   InitDepl       FinalB
#   123734.068    1.000     0.528      1.072 1113328.480
#
#  Productivity Statistics
#        Index   Statistic
# q          1      0.0000
# MSY        2 123731.0026
# Bmsy       3 1048168.8580
# Dmsy       4      0.4975
# Blim       5  423562.1648
# Btarg      6 1016409.1956
# Ctarg      7  123581.3940
# Dcurr      8      0.5284
```

Finally, to simplify the future use of this double model fitting process there is an **MQMF** function, `fitSPM()` that implements the procedure. You can either use that (check out its code, etc), or repeat the contents of the raw code, whichever you find most convenient.

7.3.2 Is the Analysis Robust?

Despite my dire warnings you may be wondering why we bothered to fit the model twice, with the starting point for the second fit being the estimated optimum from the first. One should always remember that we are using numerical methods when we fit these models. Such methods are not foolproof and can discover false minima. If there are any interactions or correlations between the model parameters then slightly different combinations can lead to very similar values of negative log-likelihood. The optimum model fit still exhibits three relative large residuals towards the end of the time-series of cpue, **Figure** 7.6. They do not exhibit any particular pattern so we assume they only represent uncertainty, which should make one question how good a model fit one has and how reliable the output statistics are from the analysis. One way we can examine the robustness of the model fit is by examining the influence of the initial model parameters on that model fit.

One implementation of a robustness test uses the robustSPM() **MQMF** function. This generates N random starting values by using the optimum log-scale parameter values as the respective mean values of some Normal random variables with their respective standard deviation values obtained by dividing those mean values by the *scaler* argument value (see the code and help of robustSPM() for the full details). The object output from robustSPM() includes the N vectors of randomly varying initial parameter values, which permits their variation to be illustrated and characterized. Of course, as a divisor, the smaller the *scaler* value the more variable the initial parameter vectors are likely to be and the more often one might expect the model fitting to fail to find the minimum.

```
#conduct a robustness test on the Schaefer model fit
data(schaef); schaef <- as.matrix(schaef); reps <- 12
param <- log(c(r=0.15,K=2250000,Binit=2250000,sigma=0.5))
ansS <- fitSPM(pars=param,fish=schaef,schaefer=TRUE,    #use
            maxiter=1000,funk=simpspm,funkone=FALSE) #fitSPM
#getseed() #generates random seed for repeatable results
set.seed(777852) #sets random number generator with a known seed
robout <- robustSPM(inpar=ansS$estimate,fish=schaef,N=reps,
                scaler=40,verbose=FALSE,schaefer=TRUE,
                funk=simpspm,funkone=FALSE)
#use str(robout) to see the components included in the output
```

By using the *set.seed* function the outcome of the pseudo-random numbers used to generate the scattered initial parameter vectors are repeatable. In **Table** 7.2 we can see that out of 12 trials we obtained 12 with the same final negative log-likelihood to five decimal places, although there was some slight variation apparent in the actual r, K, and B_{init} values and that led to minor variation in the estimated MSY values. If we increase the number of trials we finally see some that differ slightly from the optimum.

TABLE 7.2: A robustness test of the fit to the *schaef* data-set. By examining the results object we can see the individual variation. The top columns relate to the initial parameters and the bottom columns, perhaps of more interest, to the model fits.

	ir	iK	iBinit	isigma	iLike
6	0.232	2521208	2394188	0.1727	-5.765
10	0.242	2564306	1386181	0.1659	14.306
11	0.237	2189281	2032237	0.1811	-7.025
1	0.239	2351319	3401753	0.1692	-6.351
8	0.244	2201215	2934055	0.1795	-7.078
3	0.233	3164529	1632687	0.1702	22.093
4	0.233	3482370	1584633	0.1683	34.534
12	0.237	3492106	1895315	0.1653	23.789
2	0.247	2359029	2137751	0.1787	-5.575
5	0.234	3057512	1502916	0.1713	23.720
7	0.242	1671149	2512111	0.1687	4.228
9	0.230	1391893	1753155	0.1754	138.808

	r	K	Binit	sigma	-veLL	MSY
6	0.235	2107069	2258144	0.1687	-7.93406	123725
10	0.235	2107034	2258103	0.1687	-7.93406	123726
11	0.235	2107243	2258322	0.1687	-7.93406	123717
1	0.235	2107178	2258293	0.1687	-7.93406	123722
8	0.235	2107119	2258218	0.1687	-7.93406	123720
3	0.235	2107386	2258484	0.1687	-7.93406	123713
4	0.235	2107405	2258514	0.1687	-7.93406	123713
12	0.235	2107417	2258533	0.1687	-7.93406	123713
2	0.235	2106866	2257912	0.1687	-7.93406	123728
5	0.235	2107294	2258319	0.1687	-7.93406	123713
7	0.235	2107319	2258401	0.1687	-7.93406	123712
9	0.235	2106435	2257279	0.1687	-7.93406	123739

Normally one would try more than 12 trials and would examine the effect of the scaler argument. So we will now repeat that analysis 100 times using the same optimum fit and random seed. The table of *results* output by robustSPM() is sorted by the final -ve log-likelihood but even where this is the same to five decimal places notice there is slight variation in the parameter estimates. This is merely a reflection of using numerical methods.

```
#Repeat robustness test on fit to schaef data 100 times
set.seed(777854)
```

```
robout2 <- robustSPM(inpar=ansS$estimate,fish=schaef,N=100,
                     scaler=25,verbose=FALSE,schaefer=TRUE,
                     funk=simpspm,funkone=TRUE,steptol=1e-06)
lastbits <- tail(robout2$results[,6:11],10)
```

TABLE 7.3: The last 10 trials from the 100 illustrating that the last 3 trials deviated a little from the optimum negative log-likelihood of -7.93406.

	r	K	Binit	sigma	-veLL	MSY
12	0.23513	2105553	2256358	0.1687	-7.93405	123770
65	0.23513	2105553	2256358	0.1687	-7.93405	123770
47	0.23513	2105553	2256358	0.1687	-7.93405	123770
11	0.23513	2105553	2256358	0.1687	-7.93405	123770
76	0.23513	2105553	2256358	0.1687	-7.93405	123770
57	0.23513	2105553	2256358	0.1687	-7.93405	123770
23	0.23513	2105553	2256358	0.1687	-7.93405	123770
9	0.23513	2105553	2256358	0.1687	-7.93405	123770
93	0.23514	2105527	2256328	0.1687	-7.93405	123771
55	0.23514	2105510	2256327	0.1687	-7.93405	123772

Table 7.3 is only the bottom 10 records of the sorted 100 replicates, and this indicates that all replicates had the same negative log-likelihood (to 5 decimal places). Again, if you look closely at the values given for r, K, *Binit*, and *MSY* you will notice differences. If we plot up the final fitted parameter values as distributions, the scale of the variation becomes clear, **Figure** 7.7.

```
# replicates from the robustness test          Fig 7.7
result <- robout2$results
parset(plots=c(2,2),margin=c(0.35,0.45,0.05,0.05))
hist(result[,"r"],breaks=15,col=2,main="",xlab="r")
hist(result[,"K"],breaks=15,col=2,main="",xlab="K")
hist(result[,"Binit"],breaks=15,col=2,main="",xlab="Binit")
hist(result[,"MSY"],breaks=15,col=2,main="",xlab="MSY")
```

Even with negative log-likelihood values that are very close in value (to five decimal places), **Figure** 7.7, there are slight deviations possible from the optimum values that occur most often. This emphasizes the need to examine the uncertainty in the analysis closely. Given that most of the trials lead to the same optimum values the median values across all trials can identify optimum values.

An alternative way of visualizing the final variation in parameter estimates from the robustness test is to plot each parameter and model output against

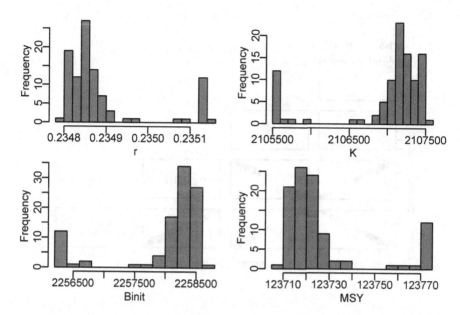

FIGURE 7.7 Histograms of the main parameters and MSY from the 100 trials in a robustness test of the model fit to the *schaef* data-set. The parameter estimates are all close, but still there is variation, which is a reflection of estimation uncertainty. To improve on this, one might try a smaller *steptol*, which defaults to 1e-06, but stable solutions might not always be possible. If you use *steptol = 1e-07* the range of values across the variation becomes much tighter, but some slight variation remains, as expected when using numerical methods. This is another reason why the particular values for the parameter estimates are most meaningful when we also have an estimate of variation or of uncertainty.

each other using the R function `pairs()`, **Figure** 7.8, which illustrates the strong correlations between parameters.

```
#robustSPM parameters against each other  Fig. 7.8
pairs(result[,c("r","K","Binit","MSY")],upper.panel=NULL,pch=1)
```

7.3.3 Using Different Data

The *schaef* data-set leads to a relatively robust outcome. Before moving on it would be enlightening to repeat the analysis using the *dataspm* data-set, which leads to more variable outcomes. Hopefully, such findings should encourage any future modellers reading this not to trust the first solution that a numerical optimizer presents.

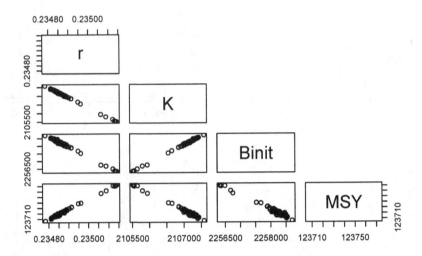

FIGURE 7.8 Plots of the relationships between parameters in the 100 optimum solutions stemming from fitting a surplus production model to the *schaef* data-set. The correlation between parameters is clear, although it needs emphasis that the proportional difference between estimates is very small being of the order of 0.2–0.3%.

```
#Now use the dataspm data-set, which is noisier
set.seed(777854) #other random seeds give different results
data(dataspm);   fish <- dataspm #to generalize the code
param <- log(c(r=0.24,K=5174,Binit=2846,sigma=0.164))
ans <- fitSPM(pars=param,fish=fish,schaefer=TRUE,maxiter=1000,
              funkone=TRUE)
out <- robustSPM(ans$estimate,fish,N=100,scaler=15, #making
                 verbose=FALSE,funkone=TRUE) #scaler=10 gives
result <- tail(out$results[,6:11],10) #16 sub-optimal results
```

TABLE 7.4: The last 10 trials from 100 used with *dataspm*. The last 6 trials deviate markedly from the optimum negative log-likelihood of -12.1288, and 5 gave consistent sub-optimal optima. Variation across parameter estimates with the optimum log-likelihood remained minor, but was large for the false optima.

	r	K	Binit	sigma	-veLL	MSY
46	0.2425	5171.27	2844.29	0.1636	-12.1288	313.537
77	0.2425	5171.51	2843.70	0.1636	-12.1288	313.528
75	0.2425	5171.81	2846.73	0.1636	-12.1288	313.545
79	0.2426	5169.36	2842.83	0.1636	-12.1288	313.555
31	3.5610	149.62	50.74	0.2230	-2.5244	133.201
65	0.0321	36059.56	49.72	0.2329	-1.1783	289.163
60	40.3102	0.26	49.72	0.2329	-1.1783	2.592
57	22.1938	0.00	49.72	0.2329	-1.1783	0.016
38	1.1856	6062.64	49.72	0.2329	-1.1783	1797.041
11	0.5954	4058.97	49.72	0.2329	-1.1783	604.180

In those bottom six model fits with *dataspm* (Table 7.4), we can see cases of very large r values teamed with very small K values, very large K values teamed with small r values, and, in addition, in the last two rows, almost reasonable values for r and K but very small *Binit* values.

7.4 Uncertainty

When we tested some model fits for how robust they were to initial conditions we found that when there were multiple parameters being fitted it was possible to obtain essentially the same numerical fit (to a given degree of precision) from slightly different parameter values. While the values did not tend to differ by much, this observation still confirms that when using numerical methods to estimate a set of parameters, the particular parameter values are not the only important outcome. We also need to know how precise those estimates are, we need to know about any uncertainty associated with their estimation. There are a number of approaches one can use to explore the uncertainty within a model fit. Here, using R, we will examine the implementation of four: 1) likelihood profiles, 2) bootstrap resampling, 3) asymptotic errors, and 4) Bayesian posterior distributions.

7.4.1   Likelihood Profiles

Likelihood profiles do what the name implies and provide insight into how the quality of a model fit might change if the parameters used were slightly different. A model is optimally fitted using maximum likelihood methods (minimization of the -ve log-likelihood), then, while fixing (keeping constant) one or more parameters to pre-determined values, one only fits the remaining unfixed parameters. In this way an optimum fit can be obtained while a given parameter or parameters have been given fixed values. Thus, we can determine how the total likelihood of the model fit will change when selected parameters remain fixed over an array of different values. A worked example should make the process clearer. We can use the *abdat* data-set, which provides for a reasonable fit to the observed data although leaving a moderate pattern in the residuals for the optimum fit and with relatively large final gradients on the optimum solution (try `outfit(ans)` to see the results).

```
# Fig. 7.9 Fit of optimum to the abdat data-set
data(abdat);      fish <- as.matrix(abdat)
colnames(fish) <- tolower(colnames(fish))  # just in case
pars <- log(c(r=0.4,K=9400,Binit=3400,sigma=0.05))
ans <- fitSPM(pars,fish,schaefer=TRUE) #Schaefer
answer <- plotspmmod(ans$estimate,abdat,schaefer=TRUE,addrmse=TRUE)
```

In the chapter *On Uncertainty* we examined a likelihood profile around a single parameter, here we will explore a little deeper to see some of the issues surrounding the use of likelihood profiles. We already have the optimal fit to the *abdat* data-set and we can use that as a starting point. If we consider the r and the K parameters in turn it becomes more efficient to write a simple function to conduct the profiles in each case to avoid duplicating code. As before we use the `negLLP()` function to enable some parameters to be fixed while others vary. As seen in the *Uncertainty* chapter, with one parameter the 95% confidence bounds are approximated by searching for the parameter range that encompasses the minimum log-likelihood plus half the chi-squared value for one degree of freedom ($=1.92$):

$$min(-LL) + \frac{\chi^2_{1,1-\alpha}}{2} \tag{7.17}$$

When plotting each profile we can include this threshold to see where it intersects the likelihood profile, **Figure** 7.10.

```
# Likelihood profiles for r and K for fit to abdat  Fig. 7.10
#doprofile input terms are vector of values, fixed parameter
#Location, starting parameters, and free parameter Locations.
#all other input are assumed to be in the calling environment
doprofile <- function(val,loc,startest,indat,notfix=c(2:4)) {
  pname <- c("r","K","Binit","sigma","-veLL")
```

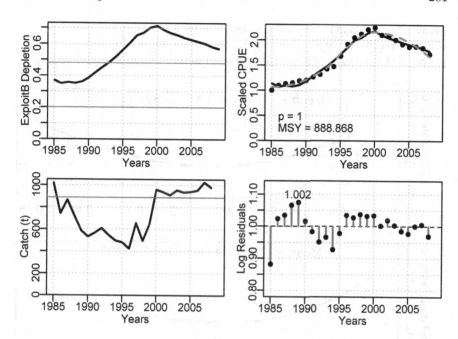

FIGURE 7.9 Summary plot depicting the fit of the optimum parameters to the *abdat* data-set. The remaining pattern in the Log-Normal residuals between the fit and the cpue data are illustrated at the bottom right.

```
  numv <- length(val)
  outpar <- matrix(NA,nrow=numv,ncol=5,dimnames=list(val,pname))
  for (i in 1:numv) { #
    param <- log(startest) # reset the parameters
    param[loc] <- log(val[i]) #insert new fixed value
    parinit <- param   # copy revised parameter vector
    bestmod <- nlm(f=negLLP,p=param,funk=simpspm,initpar=parinit,
                   indat=indat,logobs=log(indat[,"cpue"]),
                   notfixed=notfix)
    outpar[i,] <- c(exp(bestmod$estimate),bestmod$minimum)
  }
  return(outpar)
}
rval <- seq(0.32,0.46,0.001)
outr <- doprofile(rval,loc=1,startest=c(rval[1],11500,5000,0.25),
                  indat=fish,notfix=c(2:4))
Kval <- seq(7200,11500,200)
outk <- doprofile(Kval,loc=2,c(0.4,7200,6500,0.3),indat=fish,
                  notfix=c(1,3,4))
```

```
parset(plots=c(2,1),cex=0.85,outmargin=c(0.5,0.5,0,0))
plotprofile(outr,var="r",defpar=FALSE,lwd=2) #MQMF function
plotprofile(outk,var="K",defpar=FALSE,lwd=2)
```

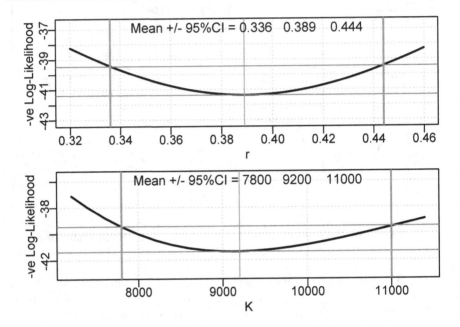

FIGURE 7.10 Likelihood profiles for both the r and K parameters of the Schaefer model fit to the abdat data-set. The horizontal red lines separate the minimum -veLL from the likelihood value bounding the 95% confidence intervals. The vertical green lines intersect the minimum and the 95% CI. The numbers are the 95% CI surrounding the mean optimum value.

An issue with estimating such confidence bounds is that by only considering single parameters one is ignoring the inter-relationships and correlations between parameters, for which the Schaefer model is well known. But the strong correlation that is expected between the r and K parameters means that a square grid search obtained from combining the two separate individual searches along r and K will lead to many combinations that fall outside of even an approximate fit to the model. It is not impossible to create a two-dimensional likelihood profile (in fact a surface), or indeed a profile across even more parameters, but even two parameters usually requires carefully searching small parts of the surface at a time or other ways of dealing with some of the extremely poor model fits that would be obtained by a simplistic grid search.

Likelihood profiles across single parameters remain useful in situations where a stock assessment has one or more fixed value parameters. This will not happen with such simple models as the Schaefer surplus production model but such situations are not uncommon when dealing with more complex stock assessment models for stocks where biological parameters such as the natural mortality, the steepness of the stock-recruitment curve, and even growth parameters may not be known or are assumed to take the same values as related species. Once an optimum model fit is obtained in an assessment, within which some of the parameters take fixed values, it is possible to re-run the model fit while changing the assumed values for one of the fixed parameters to generate a likelihood profile for that parameter. In that way it is possible to see how consistent the model fit is with regard to the assumed values for the fixed parameters. Generating a likelihood profile in this manner is preferable to merely conducting sensitivity analyses in which we might vary such fixed parameters to a level above and a level below the assumed value to see the effect. Likelihood profiles provide a more detailed exploration of the sensitivity of the modelling to the individual parameters.

With regard to simpler models, such as we are dealing with here, there are other ways of examining the uncertainty inherent in the modelling that can attempt to take into account the correlations between parameters.

7.4.2 Bootstrap Confidence Intervals

One way to characterize uncertainty in a model fit is to generate percentile confidence intervals around parameters and model outputs (*MSY*, etc) by taking bootstrap samples of the Log-Normal residuals associated with the cpue and using those to generate new bootstrap cpue samples with which to replace the original cpue time-series (Haddon, 2011). Each time such a bootstrap sample is made, the model is re-fit and the solutions stored for further analysis. To conduct such an analysis on surplus production models one can use the **MQMF** function spmboot(). Once we have found suitable starting parameters, we can use the fitSPM() function to obtain an optimum fit and it is the Log-Normal residuals associated with that optimum fit that are bootstrapped. Here we will use the relatively noisy *dataspm* data-set to illustrate these ideas

```
#find optimum Schaefer model fit to dataspm data-set Fig. 7.11
data(dataspm)
fish <- as.matrix(dataspm)
colnames(fish) <- tolower(colnames(fish))
pars <- log(c(r=0.25,K=5500,Binit=3000,sigma=0.25))
ans <- fitSPM(pars,fish,schaefer=TRUE,maxiter=1000) #Schaefer
answer <- plotspmmod(ans$estimate,fish,schaefer=TRUE,addrmse=TRUE)
```

Once we have an optimum fit we can proceed to conduct a bootstrap analysis. One would usually run at least 1000 replicates, and often more, even though that might take a few minutes to complete. In this case, even in the optimum

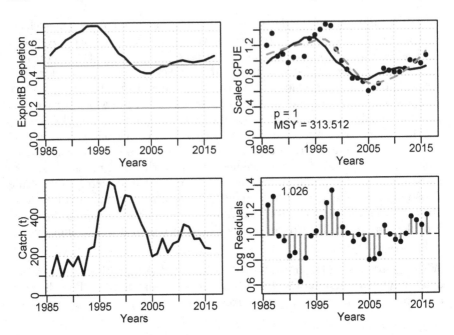

FIGURE 7.11 Summary plot depicting the fit of the optimum parameters to the *dataspm* data-set. The Log-Normal residuals between the fit and the cpue data are illustrated at the bottom right. These are what are bootstrapped and each bootstrap sample is multiplied by the optimum predicted cpue time-series to obtain each bootstrap cpue time-series.

fit there is a pattern in the Log-Normal residuals, suggesting that the model structure is missing some approximately cyclic event affecting the fishery.

```
#bootstrap the Log-Normal residuals from optimum model fit
set.seed(210368)
reps <- 1000 # can take 10 sec on a large Desktop. Be patient
#startime <- Sys.time()  # schaefer=TRUE is the default
boots <- spmboot(ans$estimate,fishery=fish,iter=reps)
#print(Sys.time() - startime) # how long did it take?
str(boots,max.level=1)
```

```
# List of 2
#  $ dynam  : num [1:1000, 1:31, 1:5] 2846 3555 2459 3020 1865 ...
#   ..- attr(*, "dimnames")=List of 3
#  $ bootpar: num [1:1000, 1:8] 0.242 0.236 0.192 0.23 0.361 ...
#   ..- attr(*, "dimnames")=List of 2
```

The output contains the dynamics of each run with the predicted model biomass, each bootstrap cpue sample, the predicted cpue for each bootstrap sample, the depletion time-series, and the annual harvest rate time-series (*reps*=1000 runs for 31 years with 5 variables stored). Each of these can be used to illustrate and summarize the outcomes and uncertainty within the analysis. Given the relatively large residuals in **Figure** 7.11 one might expect a relatively high degree of uncertainty, **Table** 7.5.

```
#Summarize bootstrapped parameter estimates as quantiles  Table 7.6
bootpar <- boots$bootpar
rows <- colnames(bootpar)
columns <- c(c(0.025,0.05,0.5,0.95,0.975),"Mean")
bootCI <- matrix(NA,nrow=length(rows),ncol=length(columns),
                 dimnames=list(rows,columns))
for (i in 1:length(rows)) {
   tmp <- bootpar[,i]
   qtil <- quantile(tmp,probs=c(0.025,0.05,0.5,0.95,0.975),na.rm=TRUE)
   bootCI[i,] <- c(qtil,mean(tmp,na.rm=TRUE))
}
```

TABLE 7.5: The quantiles for the Schaefer model parameters and some model outputs, plus the arithmetic mean. The 0.5 values are the median values.

	0.025	0.05	0.5	0.95	0.975	Mean
r	0.1321	0.1494	0.2458	0.3540	0.3735	0.2484
K	3676.3569	3840.6961	5184.2237	7965.3318	8997.4945	5481.5140
Binit	1727.1976	1845.8458	2829.0085	4935.7516	5603.2871	3041.6876
sigma	0.1388	0.1423	0.1567	0.1626	0.1630	0.1551
-veLL	-17.2319	-16.4647	-13.4785	-12.3160	-12.2377	-13.8150
MSY	280.3701	289.4673	318.4197	352.7195	366.2422	319.5455
Depl	0.3384	0.3666	0.5286	0.6693	0.6992	0.5240
Harv	0.0508	0.0576	0.0877	0.1161	0.1236	0.0871

Such percentile confidence intervals can be visualized using histograms and including the respective selected percentile CI.

One would expect 1000 replicates would provide for a smooth response and representative confidence bounds but sometimes, especially with noisy data, one needs more replicates to obtain smooth representations of uncertainty. Taking 20 seconds for 2000 replicates might seem like a long time, but considering that such things used to take hours and even days, about 20 seconds is remarkable. Note that the confidence bounds are not necessarily symmetrical around either the mean or the median estimates. Notice also that with the final year depletion estimates the 5th percentile CI is well above $0.2B_0$, implying that even though this analysis is uncertain the current depletion level is above the default limit reference point for biomass depletion used most everywhere with more than a 95% likelihood. We would need the central 80th percentiles to find the lower 10% bound, but it is bound to be higher than the 5th percentile. The medians and means exhibited by the K and the *Binit* values differ more than with the other parameters and model outputs, which suggests some evidence of bias, **Figure** 7.12. Because roughness remains in some plots these would be improved by increasing the number of replicates.

```
#boostrap CI. Note use of uphist to expand scale  Fig. 7.12
colf <- c(1,1,1,4); lwdf <- c(1,3,1,3); ltyf <- c(1,1,1,2)
colsf <- c(2,3,4,6)  ;parset(plots=c(3,2))
hist(bootpar[,"r"],breaks=25,main="",xlab="r")
abline(v=c(bootCI["r",colsf]),col=colf,lwd=lwdf,lty=ltyf)
uphist(bootpar[,"K"],maxval=14000,breaks=25,main="",xlab="K")
abline(v=c(bootCI["K",colsf]),col=colf,lwd=lwdf,lty=ltyf)
hist(bootpar[,"Binit"],breaks=25,main="",xlab="Binit")
abline(v=c(bootCI["Binit",colsf]),col=colf,lwd=lwdf,lty=ltyf)
uphist(bootpar[,"MSY"],breaks=25,main="",xlab="MSY",maxval=450)
abline(v=c(bootCI["MSY",colsf]),col=colf,lwd=lwdf,lty=ltyf)
hist(bootpar[,"Depl"],breaks=25,main="",xlab="Final Depletion")
abline(v=c(bootCI["Depl",colsf]),col=colf,lwd=lwdf,lty=ltyf)
hist(bootpar[,"Harv"],breaks=25,main="",xlab="End Harvest Rate")
abline(v=c(bootCI["Harv",colsf]),col=colf,lwd=lwdf,lty=ltyf)
```

The fitted trajectories stored in *boots$dynam* can also provide a visual indication of the uncertainty surrounding the analysis.

```
#Fig 7.13 1000 bootstrap trajectories for dataspm model fit
dynam <- boots$dynam
years <- fish[,"year"]
nyrs <- length(years)
parset()
ymax <- getmax(c(dynam[,,"predCE"],fish[,"cpue"]))
plot(fish[,"year"],fish[,"cpue"],type="n",ylim=c(0,ymax),
    xlab="Year",ylab="CPUE",yaxs="i",panel.first = grid())
for (i in 1:reps) lines(years,dynam[i,,"predCE"],lwd=1,col=8)
lines(years,answer$Dynamics$outmat[1:nyrs,"predCE"],lwd=2,col=0)
points(years,fish[,"cpue"],cex=1.2,pch=16,col=1)
percs <- apply(dynam[,,"predCE"],2,quants)
arrows(x0=years,y0=percs["5%",],y1=percs["95%",],length=0.03,
      angle=90,code=3,col=0)
```

There are clearly some deviations between the predicted and the observed CPUE values, **Figure** 7.13, but the median estimates and the confidence bounds around them remain well-defined.

Remember that whenever using a bootstrap on time-series data, where the values at time $t + 1$ are related to the values at time t, it is necessary to

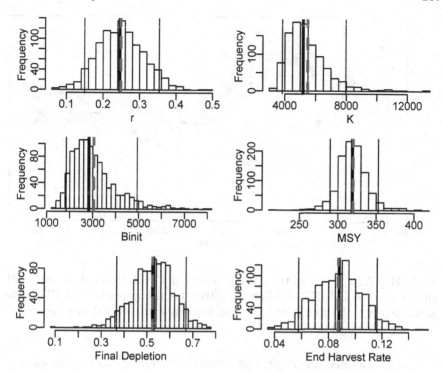

FIGURE 7.12 The 1000 bootstrap replicates from the optimum spm fit to the *dataspm* data-set. The vertical lines, in each case, are the median and 90th percentile confidence intervals and the dashed thick vertical blue lines are the mean values. The function uphist() is used to expand the x-axis in K, *Binit*, and *MSY*.

bootstrap the residual values from any fitted model and relate them back to the optimum fitted values. With cpue data we are usually using Log-Normal residual errors so once the optimal solution is found those residuals are defined as:

$$\hat{I}_{t,resid} = \frac{I_t}{\hat{I}_t} \tag{7.18}$$

where I_t is the observed cpue in year t, $I_t/\hat{I}_t$ is the observed divided by the predicted cpue in year t (the Log-Normal residual $\hat{I}_{t,resid}$). There will be a time-series of such residuals and the bootstrap generation consists of randomly selecting values from the time-series, with replacement, so that a bootstrap sample of Log-Normal residuals is prepared. These are then multiplied by

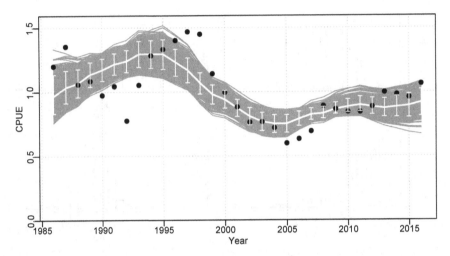

FIGURE 7.13 A plot of the original observed CPUE (black dots), the optimum predicted CPUE (solid line), the 1000 bootstrap predicted CPUE (the grey lines), and the 90th percentile confidence intervals around those predicted values (the vertical bars).

the original optimal predicted cpue values to generate different time-series of bootstrapped cpue:

$$I_t^* = \hat{I}_t * \left[\frac{I}{\hat{I}}\right]^* \tag{7.19}$$

where the superscript $*$ denotes a bootstrap sample, with I_t^* denoting the bootstrap cpue sample for year t, the $\left[\frac{I}{\hat{I}}\right]^*$ denotes a single random sample from the Log-Normal residuals, which is then multiplied by the year's predicted cpue. These equations reflect particular lines of code within the **MQMF** function spmboot().

A worthwhile exercise would be to repeat this analysis but everywhere *schaefer = TRUE* replace that with *FALSE* so as to fit the model using the Fox surplus production model. Then it would be possible to compare the uncertainty of the two models.

```
#Fit the Fox model to dataspm; note different parameters
pars <- log(c(r=0.15,K=6500,Binit=3000,sigma=0.20))
ansF <- fitSPM(pars,fish,schaefer=FALSE,maxiter=1000) #Fox version
bootsF <- spmboot(ansF$estimate,fishery=fish,iter=reps,schaefer=FALSE)
dynamF <- bootsF$dynam
```

```
# bootstrap trajectories from both model fits   Fig. 7.14
parset()
ymax <- getmax(c(dynam[,,"predCE"],fish[,"cpue"]))
```

```
plot(fish[,"year"],fish[,"cpue"],type="n",ylim=c(0,ymax),
     xlab="Year",ylab="CPUE",yaxs="i",panel.first = grid())
for (i in 1:reps) lines(years,dynamF[i,,"predCE"],lwd=1,col=1,lty=1)
for (i in 1:reps) lines(years,dynam[i,,"predCE"],lwd=1,col=8)
lines(years,answer$Dynamics$outmat[1:nyrs,"predCE"],lwd=2,col=0)
points(years,fish[,"cpue"],cex=1.1,pch=16,col=1)
percs <- apply(dynam[,,"predCE"],2,quants)
arrows(x0=years,y0=percs["5%",],y1=percs["95%",],length=0.03,
       angle=90,code=3,col=0)
legend(1985,0.35,c("Schaefer","Fox"),col=c(8,1),bty="n",lwd=3)
```

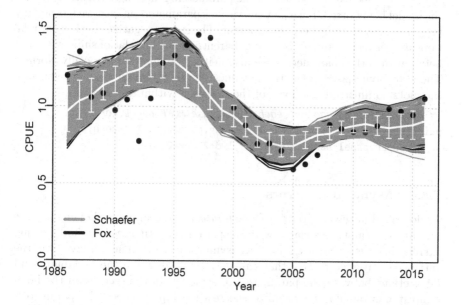

FIGURE 7.14 A plot of the original observed CPUE (dots), the optimum predicted CPUE (solid white line) with the 90th percentile confidence intervals (the white bars). The black lines are the Fox model bootstrap replicates while the grey lines over the black are those from the Schaefer model.

It could be argued that the Fox model is more successful at capturing the variability in this data as the spread of the black lines is slightly greater than that of the grey, **Figure** 7.14. Alternatively, it could be argued that the Fox model is less certain. Overall there is not a lot of difference between the outputs of the Schaefer and Fox models as even their predicted MSY values are very similar (313.512t vs 311.661t). However, in the end it appears the non-linearity in the density-dependence within the Fox model gives it more flexibility hence it is able to capture the variability of the original data slightly better than the more rigid Schaefer model (hence its slighter smaller -ve log-likelihood, see `outfit(ansF)`. But neither model can capture the cyclic property exhibited in the residuals, which implies there is some process not being included in the modelled dynamics, a model misspecification. Neither model is completely adequate although either may provide a sufficient approximation to the dynamics that they could be used to generate management advice (with caveats about the cyclic process remaining the same through time, etc.).

7.4.3 Parameter Correlations

The combined bootstrap samples and associated estimates provide a characterization of the variability across the parameters reflecting both the data and the model being fitted. If we plot the various parameters against each other any parameter correlations become apparent. The strong negative curvi-linear relationship between r and K is very apparent, while the relationships with and between the other parameters are also neither random nor smoothly Normal. There are some points at extreme values but they remain rare; nevertheless, the plots do illustrate the form of the variation within this analysis.

```
# plot variables against each other, use MQMF panel.cor  Fig. 7.15
pairs(boots$bootpar[,c(1:4,6,7)],lower.panel=panel.smooth,
    upper.panel=panel.cor,gap=0,lwd=2,cex=0.5)
```

7.4.4 Asymptotic Errors

As described in the chapter *On Uncertainty*, a classical method of characterizing the uncertainty associated with the parameter estimates in a model fitting exercise is to use what are known as asymptotic errors. These derive from the variance-covariance matrix that can be used to describe the variability and interactions between the parameters of a model. In the section on the bootstrap it was possible to visualize the relationships between the parameters using the `pairs()` function, and they were clearly not nicely multivariate Normal. Nevertheless, it remains possible to use the multivariate Normal derived from the variance-covariance matrix (*vcov*) to characterize a model's uncertainty. We can estimate the *vcov* as an option when fitting a model using either `optim()` or `nlm()`.

```
#Start the SPM analysis using asymptotic errors.
data(dataspm)    # Note the use of hess=TRUE in call to fitSPM
fish <- as.matrix(dataspm)    # using as.matrix for more speed
colnames(fish) <- tolower(colnames(fish))  # just in case
pars <- log(c(r=0.25,K=5200,Binit=2900,sigma=0.20))
ans <- fitSPM(pars,fish,schaefer=TRUE,maxiter=1000,hess=TRUE)
```

We can see the outcome from fitting the Schaefer surplus production model to the *dataspm* data-set with the *hess* argument set to TRUE by using the `outfit()` function.

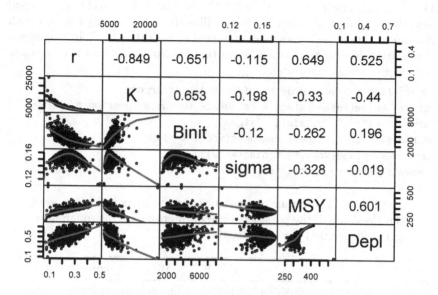

FIGURE 7.15 The relationships between the model parameters and some outputs for the Schaefer model (use `bootsF$bootpar` for the Fox model). The lower panels have a red smoother line through the data illustrating any trends, while the upper panels have the linear correlation coefficient. The few extreme values distort the plots.

```
#The hessian matrix from the Schaefer fit to the dataspm data
outfit(ans)
```

```
# nlm solution:
# minimum     :   -12.12879
# iterations  :   2
# code        :   2 >1 iterates in tolerance, probably solution
#         par        gradient      transpar
# 1 -1.417080  0.0031126661     0.24242
# 2  8.551232 -0.0017992364  5173.12308
# 3  7.953564 -0.0009892147  2845.69834
# 4 -1.810225 -0.0021756288     0.16362
# hessian     :
#              [,1]          [,2]           [,3]         [,4]
# [1,] 1338.3568627 1648.147068   -74.39814471 -0.14039276
# [2,] 1648.1470677 2076.777078  -115.32342460 -1.80063349
# [3,]  -74.3981447 -115.323425    25.48912486 -0.01822396
# [4,]   -0.1403928   -1.800633    -0.01822396 61.99195077
```

The final minimization within `fitSPM()` uses maximum likelihood methods (actually the minimum negative log-likelihood) and so we need to invert the Hessian to obtain the variance-covariance matrix (Table 7.6). The square-root of the diagonal also gives estimates of the standard error for each parameter (see chapter *On Uncertainty*).

```
#calculate the var-covar matrix and the st errors
vcov <- solve(ans$hessian) # calculate variance-covariance matrix
label <- c("r","K", "Binit","sigma")
colnames(vcov) <- label; rownames(vcov) <- label
outvcov <- rbind(vcov,sqrt(diag(vcov)))
rownames(outvcov) <- c(label,"StErr")
```

TABLE 7.6: The variance-covariance (vcov) matrix is the inverse of the Hessian and the parameter standard errors are the square-root of the diagonal of the vcov matrix.

	r	K	Binit	sigma
r	0.06676	-0.05631	-0.05991	-0.00150
K	-0.05631	0.04814	0.05344	0.00129
Binit	-0.05991	0.05344	0.10615	0.00145
sigma	-0.00150	0.00129	0.00145	0.01617
StErr	0.25838	0.21940	0.32581	0.12714

Now that we have the optimum solution and the variance-covariance matrix we can use a multivariate Normal distribution to provide the basis for obtaining multiple plausible combinations of parameters, which can be used to calculate outputs such as the *MSY*, as well as describe the expected dynamics. Base R does not include methods for sampling from a multivariate Normal distribution but there are freely available packages that do. We will use the **mvtnorm** package downloadable from CRAN. When using such a package one can determine who wrote it and other important information by using the `packageDescription()` function. Alternatively, by examining one of the help files for a function within the package, if you scroll to the bottom of the page and click the *Index* hyperlink, this enables you to read the DESCRIPTION file directly.

```
#generate 1000 parameter vectors from multi-variate Normal
library(mvtnorm)   # use RStudio, or install.packages("mvtnorm")
N <- 1000 # number of parameter vectors, use vcov from above
mvn <- length(fish[,"year"]) #matrix to store cpue trajectories
mvncpue <- matrix(0,nrow=N,ncol=mvn,dimnames=list(1:N,fish[,"year"]))
columns <- c("r","K","Binit","sigma")
optpar <- ans$estimate # Fill matrix with mvn parameter vectors
mvnpar <- matrix(exp(rmvnorm(N,mean=optpar,sigma=vcov)),nrow=N,
```

```
                  ncol=4,dimnames=list(1:N,columns))
msy <- mvnpar[,"r"]*mvnpar[,"K"]/4
nyr <- length(fish[,"year"])
depletion <- numeric(N) #now calculate N cpue series in linear space
for (i in 1:N) { # calculate dynamics for each parameter set
  dynamA <- spm(log(mvnpar[i,1:4]),fish)
  mvncpue[i,] <- dynamA$outmat[1:nyr,"predCE"]
  depletion[i] <- dynamA$outmat["2016","Depletion"]
}
mvnpar <- cbind(mvnpar,msy,depletion) # try head(mvnpar,10)
```

With the bootstraps, when the implied CPUE trajectories are plotted, **Figures** 7.13 and 7.14, the outcome appears plausible. On the other hand, with the asymptotic errors, when we plot the implied dynamics, **Figure** 7.16, a proportion, admittedly falling outside the 90th percentile confidence intervals, which are themselves wildly asymmetric, generate wildly fluctuating dynamics, which can even imply the stock goes extinct.

```
#data and trajectories from 1000 MVN parameter vectors    Fig. 7.16
plot1(fish[,"year"],fish[,"cpue"],type="p",xlab="Year",ylab="CPUE",
     maxy=2.0)
for (i in 1:N) lines(fish[,"year"],mvncpue[i,],col="grey",lwd=1)
points(fish[,"year"],fish[,"cpue"],pch=1,cex=1.3,col=1,lwd=2) # data
lines(fish[,"year"],exp(simpspm(optpar,fish)),lwd=2,col=1)# pred
percs <- apply(mvncpue,2,quants)  # obtain the quantiles
arrows(x0=fish[,"year"],y0=percs["5%",],y1=percs["95%",],length=0.03,
       angle=90,code=3,col=1) #add 90% quantiles
msy <- mvnpar[,"r"]*mvnpar[,"K"]/4  # 1000 MSY estimates
text(2010,1.75,paste0("MSY ",round(mean(msy),3)),cex=1.25,font=7)
```

The mean *MSY* estimate from the use of asymptotic errors is very similar to that produced by the bootstrapped estimate (319.546, **Table** 7.5), and the 90th quantile confidence bounds appear meaningful, although far more skewed than in the bootstrap analysis. However, it is clear that the use of the multivariate Normal distribution can lead to some rather implausible parameter combinations that, in turn, lead to implausible cpue trajectories, which deviate very far from the observed cpue. This does not mean that asymptotic errors should not be used, but rather if one does use them then their implications should be examined for plausibility.

In this case we can search for the parameter combinations that led to the extreme results by finding those records where the final cpue level was < 0.4.

```
#Isolate errant cpue trajectories Fig. 7.17
pickd <- which(mvncpue[,"2016"] < 0.40)
plot1(fish[,"year"],fish[,"cpue"],type="n",xlab="Year",ylab="CPUE",
     maxy=6.25)
```

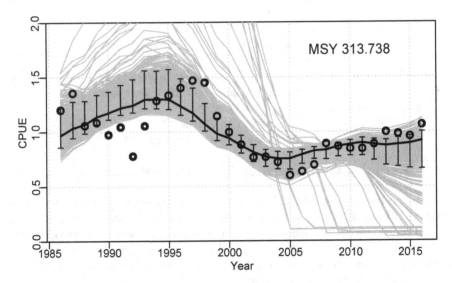

FIGURE 7.16 The 1000 predicted cpue trajectories derived from random parameter vectors sampled from the multivariate Normal distribution defined by the optimum parameters and their related variance-covariance matrix.

```
for (i in 1:length(pickd))
  lines(fish[,"year"],mvncpue[pickd[i],],col=1,lwd=1)
points(fish[,"year"],fish[,"cpue"],pch=16,cex=1.25,col=4)
lines(fish[,"year"],exp(simpspm(optpar,fish)),lwd=3,col=2,lty=2)
```

Now we have identified the majority of the errant trajectories, with their respective parameter vectors we can compare what we take to be the non-errant trajectories by plotting them in a manner so we can identify who is who, **Figure** 7.18.

```
#Use adhoc function to plot errant parameters Fig. 7.18
parset(plots=c(2,2),cex=0.85)
outplot <- function(var1,var2,pickdev) {
  plot1(mvnpar[,var1],mvnpar[,var2],type="p",pch=16,cex=1.0,
        defpar=FALSE,xlab=var1,ylab=var2,col=8)
  points(mvnpar[pickdev,var1],mvnpar[pickdev,var2],pch=16,cex=1.0)
}
outplot("r","K",pickd) # assumes mvnpar in working environment
outplot("sigma","Binit",pickd)
outplot("r","Binit",pickd)
outplot("K","Binit",pickd)
```

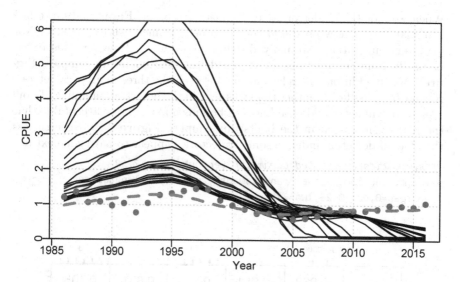

FIGURE 7.17 The 34 asymptotic error cpue trajectories that were predicted to have a cpue < 0.4 in 2016. The dots are the original data and the dashed line the optimum model fit.

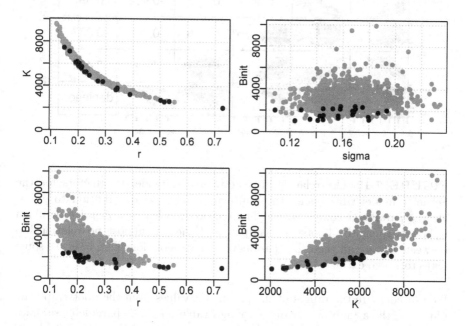

FIGURE 7.18 The spread of parameter values from the asymptotic error samples with the values that predicted final cpue < 0.4 highlighted in black. It appears that low values of *Binit* are mostly behind the implausible trajectories.

When we plot the model variables against each other, **Figure** 7.19, the lack of a relationship between *sigma* and the other parameters is expected for Normally or multivariate Normally distributed variables. However, this differs markedly from the relationship obtained from the bootstrap samples, **Figure** 7.15. In addition, the relationships between the three main parameters r, K, and B_{init} are far smoother than was seen in the bootstrap sampling. To our eyes such symmetry and cleanliness of bounds appears more acceptable than the relationships seen in the bootstrap samples, **Figure** 7.19. Nevertheless, the plots of depletion indicate some trajectories appear to have gone extinct.

```
#asymptotically sampled parameter vectors  Fig. 7.19
pairs(mvnpar,lower.panel=panel.smooth, upper.panel=panel.cor,
      gap=0,cex=0.25,lwd=2)
```

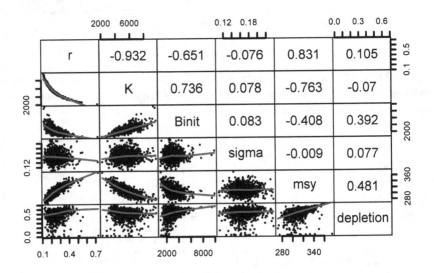

FIGURE 7.19 The relationships between the model parameters for the Schaefer model when using the multivariate Normal distribution to generate the parameter combinations. The relationship between r and K is much tighter than in the bootstrap samples and there is almost no relationship between *sigma* and the other parameters. The depletion plots indicate some trajectories go extinct.

We can compare the ranges of the parameter values from the bootstrap sampling and the asymptotic error sampling (Table 7.7). The parameter samples from the asymptotic errors distribution are less skewed than from the bootstrap, but the bootstrap does not have such low values for B_{init} and K. It needs to be remembered that the use of the multivariate Normal distribution to describe the shape of the likelihood surface around the optimum parameter set remains only an approximation.

```
# Get the ranges of parameters from bootstrap and asymptotic
bt <- apply(bootpar,2,range)[,c(1:4,6,7)]
ay <- apply(mvnpar,2,range)
out <- rbind(bt,ay)
rownames(out) <- c("MinBoot","MaxBoot","MinAsym","MaxAsym")
```

TABLE 7.7: The range of parameter values from the bootstrap sampling compared with those from the asymptotic error sampling.

	r	K	Binit	sigma	MSY	Depl
MinBoot	0.0653	3139.827	1357.264	0.1125	217.164	0.0953
MaxBoot	0.4958	25666.568	8000.087	0.1636	530.652	0.7699
MinAsym	0.1185	2055.714	1003.558	0.1069	271.901	0.0054
MaxAsym	0.7287	9581.273	9917.274	0.2344	374.522	0.6820

7.4.5 Sometimes Asymptotic Errors Work

In some cases the asymptotic errors approach generates results remarkably similar to those from the bootstrap. If we had used the *abdat* data instead of *dataspm* we would have obtained results that appear indistinguishable from the same analysis conducted using bootstraps (see the bootstrap section in the chapter *On Uncertainty* for a comparison). The trajectories generated look very similar, **Figure** 7.20, and the pairwise plots are almost indistinguishable. As in the *On Uncertainty* bootstrap example, we have used rgb() colouring to ease the comparison, **Figure** 7.21.

```
#repeat asymptotice errors using abdat data-set Fig. 7.20
data(abdat)
fish <- as.matrix(abdat)
pars <- log(c(r=0.4,K=9400,Binit=3400,sigma=0.05))
ansA <- fitSPM(pars,fish,schaefer=TRUE,maxiter=1000,hess=TRUE)
vcovA <- solve(ansA$hessian) # calculate var-covar matrix
mvn <- length(fish[,"year"])
N <- 1000   # replicates
mvncpueA <- matrix(0,nrow=N,ncol=mvn,dimnames=list(1:N,fish[,"year"]))
columns <- c("r","K","Binit","sigma")
optparA <- ansA$estimate  # Fill matrix of parameter vectors
mvnparA <- matrix(exp(rmvnorm(N,mean=optparA,sigma=vcovA)),
                  nrow=N,ncol=4,dimnames=list(1:N,columns))
msy <- mvnparA[,"r"]*mvnparA[,"K"]/4
for (i in 1:N) mvncpueA[i,]<-exp(simpspm(log(mvnparA[i,]),fish))
```

```
mvnparA <- cbind(mvnparA,msy)
plot1(fish[,"year"],fish[,"cpue"],type="p",xlab="Year",ylab="CPUE",
      maxy=2.5)
for (i in 1:N) lines(fish[,"year"],mvncpueA[i,],col=8,lwd=1)
points(fish[,"year"],fish[,"cpue"],pch=16,cex=1.0) #orig data
lines(fish[,"year"],exp(simpspm(optparA,fish)),lwd=2,col=0)
```

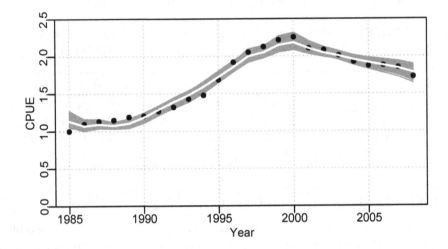

FIGURE 7.20 The use of asymptotic errors to generate plausible parameter
sets and their implied cpue trajectories for the *abdat* data-set. The optimum
model fit is shown as a white line.

```
#plot asymptotically sampled parameter vectors Fig. 7.21
pairs(mvnparA,lower.panel=panel.smooth, upper.panel=panel.cor,
      gap=0,pch=16,col=rgb(red=0,green=0,blue=0,alpha = 1/10))
```

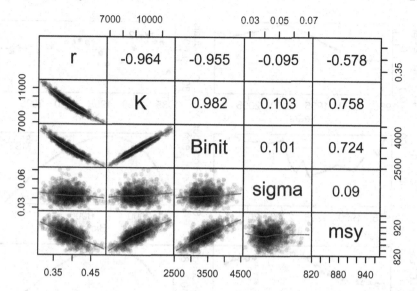

FIGURE 7.21 Model parameter relationships when fitting the Schaefer model to the *abdat* data and using the multivariate Normal distribution to generate subsequent parameter combinations. These are very similar to the bootstrap equivalent in the *On Uncertainty* chapter.

7.4.6 Bayesian Posteriors

We have already seen in the *On Uncertainty* chapter that it is possible to use a Markov Chain Monte Carlo (MCMC) analysis to characterize the uncertainty inherent in a given analysis. Here we will use the *abdat* data-set again as that provides an example of very well-behaved data leading to a relatively tightly fitting model and a well-behaved MCMC analysis. The equations behind the Gibbs-within-Metropolis-Hastings (or single-component Metropolis-Hastings) strategy are given in the *On Uncertainty* chapter. These are all implemented in the do_MCMC() function. To use that it helps to first have an optimally fitting maximum likelihood-based model. This time we will use the Fox model option.

```
#Fit the Fox Model to the abdat data Fig. 7.22
data(abdat); fish <- as.matrix(abdat)
param <- log(c(r=0.3,K=11500,Binit=3300,sigma=0.05))
foxmod <- nlm(f=negLL1,p=param,funk=simpspm,indat=fish,
              logobs=log(fish[,"cpue"]),iterlim=1000,schaefer=FALSE)
optpar <- exp(foxmod$estimate)
ans <- plotspmmod(inp=foxmod$estimate,indat=fish,schaefer=FALSE,
              addrmse=TRUE, plotprod=TRUE)
```

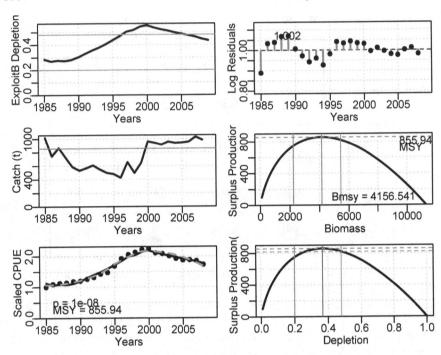

FIGURE 7.22 The optimum model fit for the *abdat* data-set using the Fox model and Log-Normal errors. The green dashed line is a smoother curve while the red line is the optimum predicted model fit. Note the pattern in the Log-Normal residuals indicating that the model has small inadequacies with regard to this data.

With the optimum, which will be somewhere near the posterior mode, we no longer have any need to illustrate the notion of a burn-in period but ideally we do not want to start from exactly the maximum likelihood solution. So we can round off the optimum solution and burn-in the Markov Chain so as to get the sequence of parameter sets to enter the bounds of plausible combinations. We know the r and K parameters are strongly correlated so we might use an initial step size of 128 (4 x 128 = 512) to try to reduce any auto-correlation between any sequentially accepted values, but this is also influenced by the relative *scales* applied to the jumps between parameter iterations. Here we started with values between 1 and 2% but experimented with those values until the acceptance rates lay between 0.2 and 0.4. It is best to do this with reduced N values (using a thinning of 512, even 1000 is half a million iterations). Only when the scales are appropriately set should one expand the replicates, N, out to some larger number to obtain clearer results. We will continue to use the **MQMF** function `calcprior()` to set equal weight on each set of plausible parameters, and to obtain repeatable results you would need to include a set.seed() call on each chain, but generally we would leave this out. The same random number generators are used in R for all operating systems so this should work across computers, but I haven't tried this on all versions. To improve the speed of calculations it would be useful to have something akin to the simpspmC() function using **Rcpp** described in the *On Uncertainty* chapter. Before we run the following MCMC you will need to compile the appendix to this chapter or call simpspm() when using do_MCMC(); note the *schaefer=FALSE* argument is included in order to use the Fox model.

```
# Conduct an MCMC using simpspmC on the abdat Fox SPM
# This means you will need to compile simpspmC from appendix
set.seed(698381) #for repeatability, possibly only on Windows10
begin <- gettime()  # to enable the time taken to be calculated
inscale <- c(0.07,0.05,0.09,0.45) #note large value for sigma
pars <- log(c(r=0.205,K=11300,Binit=3200,sigma=0.044))
result <- do_MCMC(chains=1,burnin=50,N=2000,thinstep=512,
                  inpar=pars,infunk=negLL,calcpred=simpspmC,
                  obsdat=log(fish[,"cpue"]),calcdat=fish,
                  priorcalc=calcprior,scales=inscale,schaefer=FALSE)
 # alternatively, use simpspm, but that will take longer.
cat("acceptance rate = ",result$arate," \n")
cat("time = ",gettime() - begin,"\n")
post1 <- result[[1]][[1]]
p <- 1e-08
msy <- post1[,"r"]*post1[,"K"]/((p + 1)^((p+1)/p))
```

```
# acceptance rate =  0.3136629 0.3337832 0.3789844 0.3660627
# time =  13.30179
```

The Fox model MCMC is currently setup with a thinning rate of 512, with 2000 replicates, and a burn-in of 50, meaning there will be 512 x 2050 = 1049600 iterations used to generate the required parameter traces. On the computer used to write this, even using simpspmC(), this took about 15 seconds; using simpspm() we might expect that to take about 75 seconds. Once this is known for your own system it is obviously possible to plan your analyses and make explicit choices over thinning rates and replicates (do not forget to use the latest version of R for the fastest times).

Once the analysis is complete we can plot each variable against the others using the pairs() function, **Figure** 7.23. But also we can plot the marginal posterior distributions for each main parameter and the derived model output (the MSY). Given we have used 2000 replicates with a chain thinning rate of 512, the posterior distributions are relatively smooth, **Figure** 7.24, although the few bumps present could be smoothed further by more replicates.

```
#pairwise comparison for MCMC of Fox model on abdat  Fig 7.23
pairs(cbind(post1[,1:4],msy),upper.panel = panel.cor,lwd=2,cex=0.2,
      lower.panel=panel.smooth,col=1,gap=0.1)
```

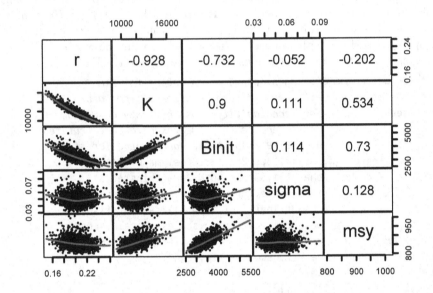

FIGURE 7.23 MCMC output as paired scattergrams. The solid lines are loess smoothers indicating trends and the numbers in the upper half are the correlation coefficients between the pairs. Strong correlations are indicated between *r*, *K*, and *Binit*, and between *K*, *Binit*, and *MSY*, with only minor or no relationships between *sigma* the other parameters or between msy and *r*.

```
# marginal distributions of 3 parameters and msy  Figure 7.24
parset(plots=c(2,2), cex=0.85)
plot(density(post1[,"r"]),lwd=2,main="",xlab="r") #plot has a method
plot(density(post1[,"K"]),lwd=2,main="",xlab="K")   #for output from
plot(density(post1[,"Binit"]),lwd=2,main="",xlab="Binit")  # density
plot(density(msy),lwd=2,main="",xlab="MSY")   #try str(density(msy))
```

Note that to get the acceptance rate for *sigma* down to less than 0.4 required a relatively large scaling factor to be imposed. The other parameters required values between 5% and 9%. If you used 500 replicates to search for the appropriate scaling factors and then reset the replicates to 2000, then the process will take four times as long. If you then increased the step size to say 1024 it would take double as long again. The search for appropriate scaling factors is needed to ensure the posterior space is well-explored by the Markov Chain in a reasonable time. If the scaling were too small the acceptance rate would

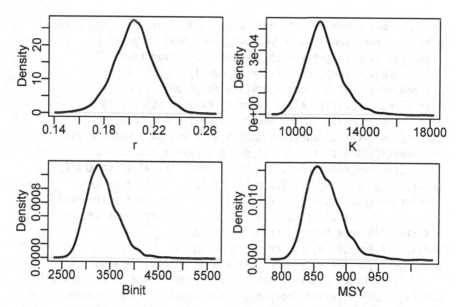

FIGURE 7.24 The marginal distributions for three parameters and the implied MSY from 2000 MCMC replicates for the Fox model applied to the *abdat* data. The lumpiness of the curves suggests more than 2000 iterations are needed.

increase because each trial would effectively be very close to the original so only small steps could ever be taken. The stationary distribution will still be discovered eventually but it could take a very large number of replicates. With the thinning rate of 512, if you were to use the acf() function to plot the auto-correlation of any of the traces, as in acf(post1[,"r"]), you would find significant correlations still occur at steps of 1 and 2. To reduce those would require an increase to at least 1024.

As the number of replicates is increased the observed spread of potential parameter combinations increases. However, if we examine the bounds of the 90th percentile contours, these stay relatively stable. We can do this in two dimensions using the **MQMF** function addcontours(), which can generate contours for any arbitrary cloud of x-y data points (arbitrary but ideally smoothly distributed). The 50th and 90th percentile contours with 2000 observations is not particularly smooth, but even that has bounds for K approximately between about 9500–14000, and r between about 0.17–0.24, **Figure** 7.25. As the numbers increase the contours become smoother but their bounds remain approximately the same even though the x- and y-axes extend in both cases.

```
#MCMC r and K parameters, approx 50 + 90% contours. Fig. 7.25
puttxt <- function(xs,xvar,ys,yvar,lvar,lab="",sigd=0) {
  text(xs*xvar[2],ys*yvar[2],makelabel(lab,lvar,sep="  ",
      sigdig=sigd),cex=1.2,font=7,pos=4)
} # end of puttxt - a quick utility function
kran <- range(post1[,"K"]);  rran <- range(post1[,"r"])
mran <- range(msy)            #ranges used in the plots
parset(plots=c(1,2),margin=c(0.35,0.35,0.05,0.1)) #plot r vs K
plot(post1[,"K"],post1[,"r"],type="p",cex=0.5,xlim=kran,
     ylim=rran,col="grey",xlab="K",ylab="r",panel.first=grid())
points(optpar[2],optpar[1],pch=16,col=1,cex=1.75) # center
addcontours(post1[,"K"],post1[,"r"],kran,rran,  #if fails make
             contval=c(0.5,0.9),lwd=2,col=1)   #contval smaller
puttxt(0.7,kran,0.97,rran,kran,"K= ",sigd=0)
puttxt(0.7,kran,0.94,rran,rran,"r= ",sigd=4)
plot(post1[,"K"],msy,type="p",cex=0.5,xlim=kran,  # K vs msy
     ylim=mran,col="grey",xlab="K",ylab="MSY",panel.first=grid())
points(optpar[2],getMSY(optpar,p),pch=16,col=1,cex=1.75)#center
addcontours(post1[,"K"],msy,kran,mran,contval=c(0.5,0.9),lwd=2,col=1)
puttxt(0.6,kran,0.99,mran,kran,"K= ",sigd=0)
puttxt(0.6,kran,0.97,mran,mran,"MSY= ",sigd=3)
```

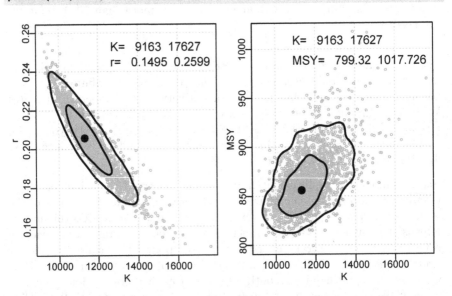

FIGURE 7.25 MCMC marginal distributions output as a scattergram of the r and K parameters, and the K and MSY values. The grey dots are from successful candidate parameter vectors, while the contours are approximate 50th and 90th percentiles. The text gives the full range of the accepted parameter traces.

Finally, we can plot the individual traces for each of the 2000 replicates. This illustrates that even with the smooth marginal distributions occasionally there are spikes of parameter values that serve to illustrate the strong inverse correlation between the main parameters.

```
#Traces for the Fox model parameters from the MCMC   Fig. 7.26
parset(plots=c(4,1),margin=c(0.3,0.45,0.05,0.05),
       outmargin = c(1,0,0,0),cex=0.85)
label <- colnames(post1)
N <- dim(post1)[1]
for (i in 1:3) {
  plot(1:N,post1[,i],type="l",lwd=1,ylab=label[i],xlab="")
  abline(h=median(post1[,i]),col=2)
}
msy <- post1[,1]*post1[,2]/4
plot(1:N,msy,type="l",lwd=1,ylab="MSY",xlab="")
abline(h=median(msy),col=2)
mtext("Step",side=1,outer=T,line=0.0,font=7,cex=1.1)
```

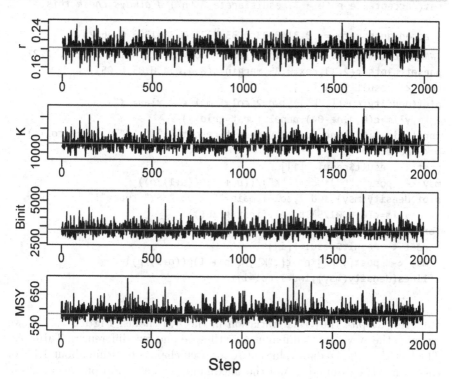

FIGURE 7.26 The traces for the three main Schaefer model parameters and the MSY estimates. The remaining auto-correlation within traces should be improved if the thinning step were increased to 1024 or longer.

Ideally, of course one would conduct such an analysis with multiple chains to ensure that they each converge on the same posterior distribution. In addition there are numerous diagnostic statistics to examine the rate of degree of convergence as the MCMC progresses. Equally ideally it would be best to start each chain from a different location, but the random number sequences eventually lead the chains in very different directions even if they start from the same place. One could use the same method of selecting different random starting points as was used in the robustSPM() function.

```
#Do five chains of the same length for the Fox model
set.seed(6396679)  # Note all chains start from same place, which is
inscale <- c(0.07,0.05,0.09,0.45)  # suboptimal, but still the chains
pars <- log(c(r=0.205,K=11300,Binit=3220,sigma=0.044))  # differ
result <- do_MCMC(chains=5,burnin=50,N=2000,thinstep=512,
                  inpar=pars,infunk=negLL1,calcpred=simpspmC,
                  obsdat=log(fish[,"cpue"]),calcdat=fish,
                  priorcalc=calcprior,scales=inscale,
                  schaefer=FALSE)
cat("acceptance rate = ",result$arate," \n") # always check this
```

```
# acceptance rate =   0.3140023 0.3327271 0.3801893 0.36673
```

```
#Now plot marginal posteriors from 5 Fox model chains     Fig. 7.27
parset(plots=c(2,1),cex=0.85,margin=c(0.4,0.4,0.05,0.05))
post <- result[[1]][[1]]
plot(density(post[,"K"]),lwd=2,col=1,main="",xlab="K",
     ylim=c(0,4.4e-04),panel.first=grid())
for (i in 2:5) lines(density(result$result[[i]][,"K"]),lwd=2,col=i)
p <- 1e-08
post <- result$result[[1]]
msy <-  post[,"r"]*post[,"K"]/((p + 1)^((p+1)/p))
plot(density(msy),lwd=2,col=1,main="",xlab="MSY",type="l",
     ylim=c(0,0.0175),panel.first=grid())
for (i in 2:5) {
  post <- result$result[[i]]
  msy <-  post[,"r"]*post[,"K"]/((p + 1)^((p+1)/p))
  lines(density(msy),lwd=2,col=i)
}
```

However, despite the visual variation in the five chains, **Figure** 7.27, if we examine the K values at different quantiles we find few differences, **Table** 7.8. The fact that the median values of K for each chain are within about 1.1% of each other is encouraging and the maximum percent variation was on 2.7%. As an experiment, using the same random.seed, but running the chains each for 4000 steps, (a total of 5 x 512 x 4050 = 10.368 million iterations, but still less than 5 minutes), the maximum variation dropped to 1.48%, again for the 0.975 quantile, with the others all below 1%. In this case, with such a

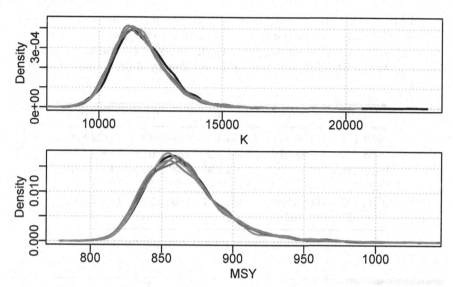

FIGURE 7.27 the marginal posterior for the K parameter and the implied MSY from five chains of 2000 replicates (512 x 2050 = 1049600 iterations). Some variation remains between the distributions, especially at the mode, suggesting that more replicates and potentially a higher thinning rate would improve the outcome.

simple model and each chain only taking a short time, extending the number of steps is worthwhile. For more complex models with many more parameters and more complex calculation of the likelihood, the timing of these analyses can become critical when working to an assessment group deadline.

```
# get qunatiles of each chain
probs <- c(0.025,0.05,0.5,0.95,0.975)
storeQ <- matrix(0,nrow=6,ncol=5,dimnames=list(1:6,probs))
for (i in 1:5) storeQ[i,] <- quants(result$result[[i]][,"K"])
x <- apply(storeQ[1:5,],2,range)
storeQ[6,] <- 100*(x[2,] - x[1,])/x[2,]
```

TABLE 7.8: Five quantiles on the K parameter from the five MCMC chains run on the Fox model applied to the *abdat* data. The last row is the percent difference in the range of the values across the chains, which shows their medians differ by slightly more than 1%.

0.025	0.05	0.5	0.95	0.975
9859.157	10160.471	11633.376	13740.430	14124.828
9893.256	10162.570	11541.118	13689.079	14302.523
9922.313	10157.503	11564.236	13620.369	14150.819
9875.521	10107.843	11541.843	13533.356	13908.780
9835.652	10088.899	11504.845	13640.376	14087.693
0.873	0.725	1.105	1.507	2.753

7.5 Management Advice

7.5.1 Two Views of Risk

A formal stock assessment, even a simple one using a surplus production model, gives one an indication of the stock status for the fishery being assessed, but the question remains how is this used to generate fisheries management advice. Of course, such advice would depend upon what management objectives are in place for the fishery concerned. But even in the absence of a formal fishery policy it should be possible to provide advice on the implications of applying different levels of catch into the future. We can use the optimal model fit to project the model dynamics into the future and such projections are the foundation of the management advice that derives from most stock assessments that are not purely empirical. Once the objective for a fishery is known (or assumed) then, using model projections, it is possible to generate estimate of future effort or catch levels that would be expected to guide the stock to a selected objective.

A common objective is to try to maintain a fishery at a biomass level that can, on average, generate the maximum sustainable yield, this would be termed B_{MSY}. Such an objective would be referred to as a target reference point as it derives from the literature that discussed biological reference points (Garcia, 1994; FAO, 1995, 1997). B_{MSY} may be considered a target but the associated catch (MSY) should really be coinsidered as an upper limit to catches. In addition to a target reference point there is usually a limit reference point, which defines a stock state to be avoided. This is often discussed in terms of stock levels perceived to pose a risk to subsequent recruitment, although this is generally only a guideline. Commonly, a limit reference point of $B_{MSY}/2$

is used, or a common proxy for this is set at $0.2B_0$. Such limit and target reference points are generally set in the context of a formal harvest strategy.

7.5.2 Harvest Strategies

Within a jurisdiction, a harvest strategy defines the decision framework to be used to achieve defined biological, and sometimes economic and social, objectives for different fish stocks. Generally a harvest strategy will have three components (FAO, 1995, 1997; Haddon, 2007; Smith *et al*, 2008):

1) a means of monitoring and collecting data with regard to each fishery of interest.

2) a defined manner in which each fishery is to be assessed, usually relative to pre-selected biological (or other) reference points such as fishing mortality rates, biomass levels, or their proxies.

3) pre-defined harvest control rules, or decision rules, that are used to translate a stock assessment or stock status into management advice relating to future effort or catch levels.

Ideally, such harvest strategies will have been simulation tested to determine the conditions under which they will be effective, and to reject options that fail to achieve the desired objectives (Smith, 1993; Punt *et al*, 2016)

There are a number of well-known examples where management objectives for a jurisdiction's fisheries have been stated explicitly (DAFF, 2007; Deroba and Bence, 2008; Magnuson-Stevens, 2007). In the Australian Commonwealth marine jurisdiction, for example, the target selected is to manage the primary economic stocks towards the biomass that generate the maximum economic yield, B_{MEY} (DAFF, 2007; DAWR, 2018); in fact a proxy of $0.48B_0$ is used for most species as insufficient information is available to reliably estimate B_{MEY}. Similarly a proxy of $0.2B_0$ is defined as a limit reference point for the majority of species "Where information to support selection of a stock-specific limit reference point [$B_{MSY}/2$] is not available..." (DAWR, 2018, p 10). If a stock is estimated to be below the limit reference point then targeted fishing is stopped, although in mixed fisheries a bycatch can still occur. If a stock is above the limit reference point then projections are run to determine what future catches should encourage the stock to increase smoothly to target biomass levels.

7.6 Risk Assessment Projections

Of course, there are a number of important assumptions behind the idea of projecting a stock assessment model forwards. The first is that the model is successful in capturing the important parts of the dynamics controlling the stock biomass. With surplus production models this equates to assuming that the estimate of stock productivity will remain the same into the future. Remember that when using the data-set *dataspm* the residuals retained a relatively large oscillatory pattern indicating that the model was missing something important in the dynamics. Despite such omissions, the model might still retain a sufficient estimate of the approximate average dynamics to allow for useful projections, but that then assumes whatever other factors were influencing the model fit will continue to operate in the manner they did in the past. If an assessment is highly uncertain then future projections will also be highly uncertain, which lowers their value in terms of how confident one can be in providing advice.

The simplest projections are made using the optimal parameter estimates and imposing constant catches or effort. We would require the current stock biomass and the catchability to use the catch equation to convert a specified effort level into a catch level:

$$C_t = qE_tB_t \qquad (7.20)$$

Then the standard dynamics equation with the optimum parameters can be used to project the biomass levels forward under the predicted catch level. If using a specified catch level then we only need use the equation for the dynamics (here we use the Polacheck *et al* (1993) version):

$$B_{t+1} = B_t + \frac{r}{p}B_t\left(1 - \left(\frac{B_t}{K}\right)^p\right) - C_t \qquad (7.21)$$

7.6.1 Deterministic Projections

If we were to use the optimum model parameters then for each of a range of different forward-projected catches we would expect to obtain a different biomass and cpue trajectory. To exemplify this we can once again use the *abdat* data-set (note that we have set the *hessian* option to TRUE as we will use this later on).

```
#Prepare Fox model on abdat data for future projections Fig. 7.28
data(abdat); fish <- as.matrix(abdat)
param <- log(c(r=0.3,K=11500,Binit=3300,sigma=0.05))
```

```
bestmod <- nlm(f=negLL1,p=param,funk=simpspm,schaefer=FALSE,
                logobs=log(fish[,"cpue"]),indat=fish,hessian=TRUE)
optpar <- exp(bestmod$estimate)
ans <- plotspmmod(inp=bestmod$estimate,indat=fish,schaefer=FALSE,
                target=0.4,addrmse=TRUE, plotprod=FALSE)
```

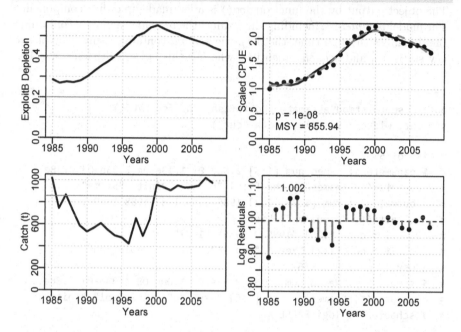

FIGURE 7.28 The optimum model fit for the abdat data-set using the Fox model and Log-Normal errors. The green dashed line is a loess curve while the solid red line is the optimum predicted model fit. Note the pattern in the Log-Normal residuals indicating that the model has some inadequacies with regard to this data.

The MSY estimate is about 854t and the stock appears to be just above a $0.4B_0$ target level that often gets used as a proxy for B_{MSY}. From the plot, and examining the original *abdat* data.frame, we can see that catches from 2000 to 2008 were between 910–1030t each year and that has led to the model predicting that the cpue and biomass would decline. We might thus explore a 10-year projection of catches ranging from 700–1000t, perhaps in steps of 50t. Given the uncertainties in the analysis and that these projections are deterministic, there is little point in examining projections for too many years into the future. Long-term projections can have value for illustrating the implications of different catches, but for practical purposes 10 years would often be more than sufficient depending on the longevity of the species being assessed (longer-lived species can be expected to exhibit slower dynamics than short-lived species). The details of the dynamics that are plotted by the function plotspmmod() are generated by using the function spm() with the optimum parameters (look inside the code of plotspmmod() to see these details). The output of spm(), when run with the optimum parameters, includes the predicted dynamics in the *outmat* table within the *Dynamics* object. Here we will run the model using the Fox model rather than the Schaefer.

The object output by the function spm() is a list made up of five components: The model *parameters*, including the q value; the *outmat* is a matrix containing the dynamics through time, the *msy*; the *sumout* contains a summary of five key statistics; and *schaefer* identifies whether this is for a Schaefer or a Fox model.

```
#
out <- spm(bestmod$estimate,indat=fish,schaefer=FALSE)
str(out, width=65, strict.width="cut")
```

```
# List of 5
#  $ parameters: Named num [1:5] 2.06e-01 1.13e+04 3.23e+03 4.38e..
#   ..- attr(*, "names")= chr [1:5] "r" "K" "Binit" "Sigma" ...
#  $ outmat     : num [1:25, 1:7] 1985 1986 1987 1988 1989 ...
#   ..- attr(*, "dimnames")=List of 2
#   .. ..$ : chr [1:25] "1985" "1986" "1987" "1988" ...
#   .. ..$ : chr [1:7] "Year" "ModelB" "Catch" "Depletion" ...
#  $ msy        : num 856
#  $ sumout     : Named num [1:5] 8.56e+02 1.00e-08 4.34e-01 2.86e..
#   ..- attr(*, "names")= chr [1:5] "msy" "p" "FinalDepl" "InitD"..
#  $ schaefer   : logi FALSE
```

The dynamics in *outmat* include details of the year, the biomass, the cpue, the predicted cpue, and other variables **Table** 7.9.

TABLE 7.9: The first 10 rows of the predicted dynamics of the stock represented by the *abdat* data-set and the optimal Fox model fit.

	Year	ModelB	Catch	Depletion	Harvest	CPUE	predCE
1985	1985	3231.104	1020	0.2856	0.3157	1.0000	1.1259
1986	1986	3043.856	743	0.2691	0.2441	1.0957	1.0607
1987	1987	3122.726	867	0.2761	0.2776	1.1303	1.0882
1988	1988	3082.459	724	0.2725	0.2349	1.1466	1.0741
1989	1989	3182.761	586	0.2814	0.1841	1.1873	1.1091
1990	1990	3426.923	532	0.3030	0.1552	1.2018	1.1942
1991	1991	3736.670	567	0.3303	0.1517	1.2652	1.3021
1992	1992	4020.992	609	0.3555	0.1515	1.3199	1.4012
1993	1993	4267.440	548	0.3773	0.1284	1.4284	1.4871
1994	1994	4575.104	498	0.4045	0.1088	1.4772	1.5943

A projection consists of taking a modelled time-series of results, such as in **Table** 7.9, and continuing the dynamics by sequentially including any new fixed catches and conducting the calculations to fill in the required columns. We can use the **MQMF** function `spmprojDet()`, which takes the list output from the `spm()` function along with some details relating to the deterministic projections, and generates the projected dynamics for us. You should examine the `spmproj()` code to see how the years are set up, the code is surprisingly brief.

```
# Fig. 7.29
catches <- seq(700,1000,50)   # projyr=10 is the default
projans <- spmprojDet(spmobj=out,projcatch=catches,plotout=TRUE)
```

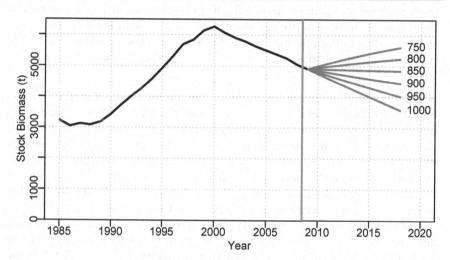

FIGURE 7.29 Deterministic constant catch projections of the optimum Fox model fit to the *abdat* data-set. The vertical green line is the limit of the data available and the red lines to the right of that are the main projections. The numbers are the constant catches imposed.

Using the deterministic constant catch projections, **Figure** 7.29, it can be seen that a constant catch of 850t, which is close to the *MSY* estimate, is the closest catch that predicts relatively stable stock conditions into the future.

7.6.2 Accounting for Uncertainty

An obvious problem with the deterministic projections is exactly the fact that they are deterministic. They fail to take any account of the uncertainty that remains even in the optimal model fit. Ideally we would conduct model projections while taking the estimated uncertainty into account. We can do this in three ways taking as a source of the uncertainty either the asymptotic errors approach, the bootstrap model fits, or the Bayesian approach. In each case, one output from the different analyses are lists of plausible parameter combinations. These can be used to describe the range of plausible dynamics included in the model fit and each of the separate biomass trajectories can be projected forward in the same manner as with the deterministic projections.

We have already calculated the hessian matrix when fitting the optimal model so we will start with an example where we use asymptotic errors to generate a matrix of plausible parameter sets that we can then project forward.

7.6.3   Using Asymptotic Errors

Once we have an optimum model fit, if we have also calculated the hessian matrix, as we have seen, we can use the estimated asymptotic errors to generate a matrix full of plausible parameter vectors. These can then be used to generate replicate biomass trajectories which can then be plotted and summarized.

It is not uncommon to have probabilistic criteria for what constitutes successful fisheries management in terms of avoiding the limit reference point. With large numbers of replicate biomass trajectories, based upon different plausible parameter vectors, it is possible to estimate what proportion of projections will achieve a desired outcome. By tabulating the outcome of such projections managers can select the level of risk they deem appropriate. For example, an acceptable risk in the Australian Commonwealth Fishery Policy is explicitly defined such that "harvest strategies maintain the biomass of all commercial stocks above the Limit Reference Point at least 90% of the time." And this is interpreted such that "the stock should stay above the limit biomass level at least 90% of the time (that is, a 1-in-10-year risk that stocks will fall below B_{LIM})" (DAWR, 2018a, p10).

In the section above we already have the optimum model fit for the *abdat* data-set using the Fox model surplus production (Polacheck *et al*, 1993). Just as we were able to generate the biomass trajectories when characterizing the uncertainty using the asymptotic errors, we can project each of those trajectories forward under a given constant catch and search for the catch level that produces the desired outcome. The first thing to do is to generate multiple plausible parameter vectors from the optimal model fit. We can use the function parasympt() to do this. Once the matrix of plausible parameter vectors is made we can then use spmproj() to project the dynamics forward for a given number of years and a given constant catch (once again check out its working by examining the code). parasympt() is merely a convenient wrapper for calling the rmvnorm() function (part of the package **mvtnorm**) and returning the result as a labelled matrix; spmproj() is slightly more complex. To simplify the projections, the function first extends the input *fish* matrix to include the projection years and their constant catches (filling in the future cpue with

NA). The spmproj() uses the spm() function to calculate the dynamics, while only the modelled biomass is returned in a matrix. Using spm() may appear inefficient, but it means the spmproj() function can be easily modified to return any of the variables estimated in the dynamics. These variables include the model biomass, the depletion levels, the harvest rate, and the predicted cpue (though one can derive the rest from just the biomass, the predicted cpue, and the original data). Run the following code and examine the two output objects, *matpar* contains the parameter vectors, and *projs* contains the biomass trajectories as rows.

```
# generate parameter vectors from a multivariate Normal
# project dynamics under a constant catch of 900t
library(mvtnorm)
matpar <- parasympt(bestmod,N=1000) #generate parameter vectors
projs <- spmproj(matpar,fish,projyr=10,constC=900)#do dynamics
```

Once calculated we can then summarize the results of the projections. First we can plot the 1000 projections using the function plotproj().

```
# Fig. 7.30  1000 replicate projections asymptotic errors
outp <- plotproj(projs,out,qprob=c(0.1,0.5),refpts=c(0.2,0.4))
```

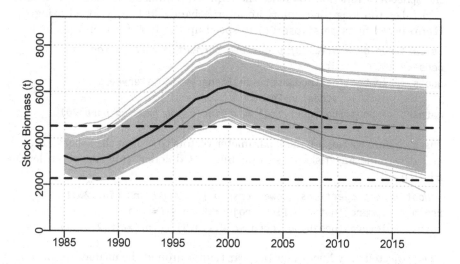

FIGURE 7.30 1000 projections derived from the using the inverse hessian and mean parameter estimates to generate 1000 plausible parameter vectors and projecting each vector forward with the fisheries catches followed by 10 years of a constant catch of 900t. The dashed lines are the limit and target reference points. The blue vertical line is the limit of fisheries data, the thick black line is the optimum fit and the thin red lines parallel to the optimum line are the 10th and 50th quantiles across years.

It is clear that after 10 years, assuming the dynamics remain unchanged, catches of 900t, on average, lead the stock to decline somewhat from the present status but keeps the median outcome close to the target (upper dashed line) and the limit reference point (LRP) is not crossed by more than 10% of the trajectories after 10 years (the lower thin line is above the $0.2B_0$ limit). By exploring different constant catches you will be able to discover that if catches are increased to 1000t then after 10 years the 10th quantile almost breaches the LRP. Tabulating what proportion of trajectories cross the LRP to generate a risk table would clarify the influence of different proposed constant catches and facilitates selection of more defensible management decisions.

7.6.4   Using Bootstrap Parameter Vectors

The essence of the projections is to generate a matrix of plausible parameter vectors from the optimum model fit combined with an estimation of the uncertainty inherent in the analysis. Instead of using the asymptotic errors with their assumed multivariate Normal distribution we can also use a bootstrapping process to generate the needed matrix of parameter vectors. Just as when we were characterizing the uncertainty in an analysis we can use the spmboot() function to create the required parameter vectors. If this function takes too long for comfort an alternative could be developed using the **Rcpp**-based function simpspmC() to speed up the 1000 (or more) model fits.

```
#bootstrap generation of plausible parameter vectors for Fox
reps <- 1000
boots <- spmboot(bestmod$estimate,fishery=fish,iter=reps,
                 schaefer=FALSE)
matparb <- boots$bootpar[,1:4] #examine using head(matparb,20)
```

Just as before we can use these parameter vectors to project the fishery forward and determine any risks to sustainability of different constant catch levels, **Figure** 7.31.

```
#bootstrap projections. Lower case b for boostrap  Fig. 7.31
projb <- spmproj(matparb,fish,projyr=10,constC=900)
outb <- plotproj(projb,out,qprob=c(0.1,0.5),refpts=c(0.2,0.4))
```

The projected grey lines differ (appear tighter around the median) from those generated using the asymptotic errors but the 10th and 50th quantiles look very similar. Certainly the summary outcome remains essentially the same, although, in this case, none of the projections fall below the Limit Reference Point (try *outb$ltLRP* and compare with *outp$ltLRP*).

7.6.5   Using Samples from a Bayesian Posterior

Just as we obtained samples from using asymptotic errors and bootstrap, it is possible to take samples from a Bayesian posterior to generate plausible

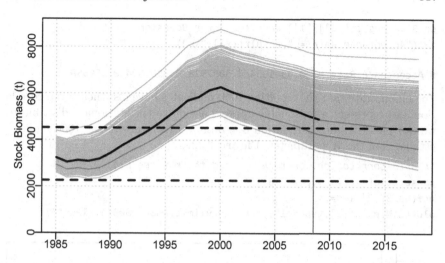

FIGURE 7.31 1000 projections (in grey) derived from the using a bootstrap process to generate 1000 plausible parameter vectors and projecting each vector forward with the fisheries catches followed by 10 years of a constant catch of 900t. The dashed lines are the limit and target reference points. The blue vertical line is the limit of fisheries data, the thick black line is the optimum fit and the red lines are the 10th and 50th quantiles across years.

parameter vectors. In this case we can use the function do_MCMC() to conduct the MCMC. We only need 1000 plausible parameter vectors so we will use a reasonable burn-in from a point close to the maximum likelihood maximum and use a large thinning rate to avoid serial correlation between sequential draws from the posterior distribution. As previously, it is best to use the **Rcpp** derived function simpspmC() to conduct the MCMC as we are still running with 2.145 million iterations. Because we are using the Fox implementation of the surplus production model the scaling factors are rather different from those used by the Schaefer version. If you have not compiled the simpspmC() function (listed in the appendix), then change the following code to use simpspm() and you can leave the as.matrix(fish) as it is to enhance speed.

```
#Generate 1000 parameter vectors from Bayesian posterior
param <- log(c(r=0.3,K=11500,Binit=3300,sigma=0.05))
set.seed(444608)
N <- 1000
result <- do_MCMC(chains=1,burnin=100,N=N,thinstep=2048,
                  inpar=param,infunk=negLL,calcpred=simpspmC,
                  calcdat=fish,obsdat=log(fish[,"cpue"]),
                  priorcalc=calcprior,schaefer=FALSE,
                  scales=c(0.065,0.055,0.1,0.475))
```

```
parB <- result[[1]][[1]] #capital B for Bayesian
cat("Acceptance Rate = ",result[[2]],"\n")
```

```
# Acceptance Rate =  0.3341834 0.3087928 0.3504304 0.3506508
```

To demonstrate that the resulting 1000 replicates have lost their serial correlation and represent a reasonable approximation to the stationary distribution we can plot the auto-correlation diagram and the trace of the 1000 replicate estimates of the K parameter, **Figure** 7.32.

```
# auto-correlation, or lack of, and the K trace Fig. 7.32
parset(plots=c(2,1),cex=0.85)
acf(parB[,2],lwd=2)
plot(1:N,parB[,2],type="l",ylab="K",ylim=c(8000,19000),xlab="")
```

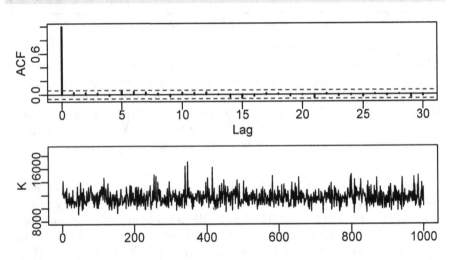

FIGURE 7.32 The lack of autocorrelation within 1000 draws from the posterior distribution for the K parameter for the Fox version of a surplus production model is apparent in the top plot. The trace in the bottom plot shows a typical scattering of values but retaining a few more extreme spikes.

These 1000 plausible parameter vectors can be extracted from the MCMC output and put through the same spmproj() and plotproj() functions as used previously, **Figure** 7.33.

```
#  Fig. 7.33
matparB <- as.matrix(parB[,1:4]) # B for Bayesian
projs <- spmproj(matparB,fish,constC=900,projyr=10) # project them
plotproj(projs,out,qprob=c(0.1,0.5),refpts=c(0.2,0.4)) #projections
```

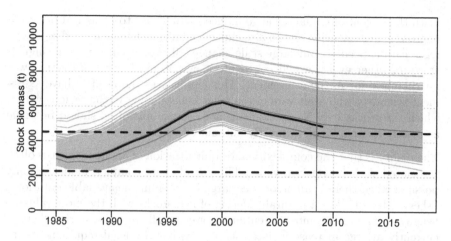

FIGURE 7.33 1000 projections (in grey) of a constant catch of 900t, derived from the using 1000 samples from the Bayesian posterior. The dashed lines are the limit and target reference points. The blue vertical line is the limit of fisheries data, the thick black line is the optimum fit and the red lines are the 10th and 50th quantiles across years.

Note the median thin red line in **Figure** 7.33 (the 50th quantile) deviates slightly from the maximum likelihood optimum model fit line (black). But the 10th quantile remains in approximately the same place relative to the LRP as was observed in the analyses using the asymptotic errors and the bootstrap. Here there is a wider spread of biomass trajectories but the management outcomes would be very similar to the previous two approaches.

7.7 Concluding Remarks

We have examined in some detail how one might use a surplus production model to generate management advice for a fishery. The output might be a tabulation of the expected stock outcome under different catch or effort regimes for the next three to five years. Assuming that some management objective exists for the jurisdiction involved the managers can then make a decision and the fishery assessment scientists would be in a position to defend their results. Of course, given the uncertainty inherent in any fishery's data and the vagaries of recruitment dynamics there can be no firm guarantees, but with the assumptions that the stock dynamics will at least be similar to previous experience then a defense of the results is possible.

With the advent of climate change-induced alterations to biological processes of growth and maturity we can also expect recruitment to be altered so clearly more care is needed. But large-scale changes, if they occur due to a singular storm or other event, would constitute a new form of uncertainty that was unaccounted for in the assessment. This emphasizes the need to have assessment scientists who are aware of the what is happening in the regions within which the stocks live. No assessment, not even simple ones, should be made as merely an analytical, or mostly automated, exercise.

The reality is that the complexity and sophistication of an assessment is often associated with its relative value. The use of more complex models is only possible when there is an associated large increase in the available data for a fishery. There is thus a natural ordering of fish stocks with the most valuable usually receiving the most attention. However, all over the world, there is currently a large increase in interest, often driven by legal requirements, to provide management advice for even data-poor species. Thus, even though we have only reviewed relatively simple assessment methods these should not be spurned or neglected.

7.8 Appendix: The Use of Rcpp to Replace simpspm

In the Bayesian analysis, we wanted to use the Fox model of surplus production. This is certainly doable using the function simpspm() by altering the *schaefer* argument. However,

```
library(Rcpp)

cppFunction('NumericVector simpspmC(NumericVector pars,
            NumericMatrix indat, LogicalVector schaefer) {
   int nyrs = indat.nrow();
   NumericVector predce(nyrs);
   NumericVector biom(nyrs+1);
   double Bt, qval;
   double sumq = 0.0;
   double p = 0.00000001;
   if (schaefer(0) == TRUE) {
     p = 1.0;
   }
   NumericVector ep = exp(pars);
   biom[0] = ep[2];
   for (int i = 0; i < nyrs; i++) {
     Bt = biom[i];
     biom[(i+1)] = Bt + (ep[0]/p)*Bt*(1 - pow((Bt/ep[1]),p)) -
                   indat(i,1);
     if (biom[(i+1)] < 40.0) biom[(i+1)] = 40.0;
     sumq += log(indat(i,2)/biom[i]);
   }
   qval = exp(sumq/nyrs);
   for (int i = 0; i < nyrs; i++) {
     predce[i] = log(biom[i] * qval);
   }
   return predce;
}')
```

References

Akaike, H. (1974) A new look at the statistical model identification. *IEEE Transactions on Autoatic Control*, **19**: 716-723.

Akaike, H. (1979) A Bayesian extension of the minimum AIC procedure of autoregressive model fitting. *Biomatrika* **66(2)**: 237-242.

Andrewartha, H.G. and L.C. Birch (1954) *The Distribution and Abundance of Animals*. The University of Chicago Press, Chicago. 782 p.

Arreguin-Sanchez, F. (1996) Catchability: a key parameter for fish stock assessment. *Reviews in Fish Biology and Fisheries*, **6**: 221-242.

Atkinson, A.C. (1980) A Note on the Generalized Information Criterion for Choice of a Model. *Biometrika*, **67**: 413-418.

Beddington, J.R. and J.G. Cooke (1983) The Potential Yield of Fish Stocks. *FAO Fisheries Technical Paper*, **242**: 1-47.

Begon, M. and M. Mortimer (1986) *Population Ecology*. 2nd edn. Blackwell Scientific Publications, Oxford.

Beverton, R.J.H. and S.J. Holt (1957) *On the dynamics of exploited fish populations*. U.K. Ministry of Agriculture and Fisheries, Fisheries Investigations (Series 2), 19: 1-533.

Beverton, R.J.H. and S.J. Holt (1959) A review of the life spans and mortality rates in nature, and their relation to growth and other physiological characteristics. *Ciba Foundation Colloquia on Ageing* 5: 142 - 180. <https://ia800207.us. archive.org/23/items/cibafoundationco05ciba/ cibafoundationco05ciba.pdf>

Bhansali, R.J. and D.Y. Downham (1977) Some properties of the order of an autoregressive model selected by a generalization of Akaike's EPF criterion. *Biometrika* **64(3)**: 547-551.

Bickel, P. J. and D. A. Freedman (1981) Some asymptotic theory for the bootstrap. *Annals of Statistics*, **9**: 1196-217.

Bickel, P. J. and D. A. Freedman (1984) Asymptotic normality and the bootstrap in stratified sampling. *Annals of Statistics*, **12**: 470-482.

Birkes, D. and Y. Dodge (1993) *Alternative Methods of Regression*. John Wiley & Sons, New York.

Box, G. E. P. and G. C. Tiao (1973) *Bayesian inference in statistical analysis.* Reading, MA: Addison-Wesley.

Bull, B.; Francis, R.I.C.C.; Dunn, A.; McKenzie, A.; Gilbert, D.J.; Smith, M.H.; Bain, R.; Fu, D. (2012) CASAL (C++ algorithmic stock assessment laboratory): CASAL user manual v2.30-2012/03/21 . *NIWA Technical Report 135.* 280 p.

Burnham, K.P. and D.R. Anderson (2002) *Model Selection and Inference. A Practical Information-Theoretic Approach.* 2nd edn. Springer-Verlag, New York. 488 p.

Casella, G. and E.J. George (1992) Explaining the Gibbs sampler. *American Statistician,* **46**: 167-174.

Caughley, G. (1977) *Analysis of vertebrate populations.* London: John Wiley & Sons.

Chambers, J.M. (2008) *Software for Data Analysis. Programming with R.* Springer. 498 p.

Chambers, J.M. (2016) *Extending R* CRC Press. Chapman & Hall, Boca Raton. 364 p.

Chapman, D.G. (1961) Statistical problems in dynamics of exploited fish populations. *Proceedings of the 4th Berkeley Symposium of Mathematics, Statistics and Probability,* **4**: 153-168. University of California Press, Berkeley, CA.

Crawley, M.J. (2007) *The R Book* John Wiley & Sons, Chichester, England. 942p.

Cushing, D.H. (1988) The study of stock and recruitment. pp. 105-128 *In*: Gulland, J.A. (*ed*) *Fish Population Dynamics.* John Wiley & Sons, New York.

DAFF (2007) *Commonwealth Fisheries Harvest Strategy. Policy and Guidelines.* Department of Agriculture, Fisheries and Forestry. 55p.

DAWR (2018) *Commonwealth Fisheries Harvest Strategy Policy,* Department of Agriculture and Water Resources, Canberra, 21p agriculture.gov.au/fisheries/domestic/ harvest_strategy_policy

DAWR (2018a) *Guidelines for the Implementation of the Commonwealth Fisheries Harvest Strategy Policy,* Department of Agriculture and Water Resources Canberra, 76p.

Dennis, B. (1996) Discussion: Should ecologists become Bayesians? *Ecological Applications,* **6**: 1095-103.

Deriso, R. B. (1980) Harvesting strategies and parameter estimation for an age-structured model. *Canadian Journal of Fisheries and Aquatic Sciences* **37**:268-82.

Deroba, J. J., and J. R. Bence. (2008) A review of harvest policies: Understanding relative performance of control rules. *Fisheries Research,* **94**: 210-223.

Dichmont, C. M., R. Deng, A. E. Punt, W. Venables, and M. Haddon. (2006) Management strategies for short-lived species: The case of Australia's Northern Prawn Fishery. 1. Accounting for multiple species, spatial structure and implementation uncertainty when evaluating risk. *Fisheries Research,* **82**: 204–220.

Eddelbuettel, D. and R. Francois (2011). Rcpp: Seamless R and C++ Integration. *Journal of Statistical Software,* **40(8)**: 1-18. URL http://www.jstatsoft.org/v40/i08/.

Eddelbuettal, D. (2013) *Seamless R and C++ Integration with Rcpp.* Springer. New York. 220 p.

Eddelbuettel, D. and J.J. Balamuta (2017) *Extending R with C++: A Brief Introduction to Rcpp.* PeerJ Preprints 5:e3188v1 https://doi.org/10.7287/peerj.preprints.3188v1.

Edwards, A.W.F. (1972) *Likelihood: An Account of the Statistical Concept of Likelihood and its Application to Scientific Inference.* Cambridge University Press, London.

Efron, B. (1979) Bootstrap methods: Another look at the jackknife. *Annals of Statistics,* **7**: 1–26.

Efron, B., and R. J. Tibshirani. (1993) *An introduction to the bootstrap.* London: Chapman & Hall.

Elder, R.D. (1979) Equilibrium yield for the Hauraki Gulf snapper fishery estimated from catch and effort figures, 1960–74, *New Zealand Journal of Marine and Freshwater Research,* **13(1)**: 31-38, DOI: 10.1080/00288330.1979.9515778.

Fabens, A.J. (1965) Properties and fitting of the von Bertalanffy growth curve. *Growth,* **29**: 265-289.

FAO (1995) *Code of Conduct for Responsible Fisheries.* Food and Agricultural Organization of the United Nations, Rome. 41p.

FAO (1996) *Precautionary Approach to Capture Fisheries and Species Introductions. FAO Technical Guidelines for Responsible Fisheries 2.* Food and Agricultural Organization of the United Nations, Rome. 55p.

FAO (1997) *Fisheries Management. FAO Technical Guidelines for Responsible Fisheries 4.* Food and Agricultural Organization of the United Nations, Rome. 84p.

Forbes, C., Evans, M., Hastings, N. and B. Peacock (2011) *Statistical Distributions* 4th edn. John Wiley & Sons, New Jersey. 212 p.

Fournier, D. A., Hampton, J., and Sibert, J. R. (1998). MULTIFAN-CL: a length-based, age-structured model for fisheries stock assessment, with application to south pacific albacore, *Thunnus alalunga*. *Canadian Journal of Fisheries and Aquatic Sciences*, **55**(9):2105-2116

Fournier, D.A., Skaug, H.J., Ancheta, J., Ianelli, J., Magnusson, A., Maunder, M.N., Nielsen, A., and J. Sibert (2012) AD Model Builder: using automatic differentiation for statistical inference of highly parameterized complex nonlinear models. *Optimization Methods and Software*, **27** : 233-249.

Fox, W.W. (1970) An exponential surplus-yield model for optimizing exploited fish populations. *Transactions of the American Fish Society*, **99**: 80-88.

Fox, W.W. (1975) Fitting the generalized stock production model by least-squares and equilibrium approximation. *Fishery Bulletin*, **73**: 23-37.

Francis, R.I.C.C. (1988) Maximum likelihood estimation of growth and growth variability from tagging data. *New Zealand Journal of Marine and Freshwater Research*, **22**: 42-51.

Francis, R.I.C.C. (1992) Use of risk analysis to assess fishery management strategies: A case study using orange roughy (Hoplostethus atlanticus) on the Chatham Rise, New Zealand. *Canadian Journal of Fisheries and Aquatic Science*, **49**: 922-930.

Francis, R.I.C.C. (1995) An alternative mark-recapture analogue of Schnute's growth model. *Fisheries Research*, **23**: 95-111.

Francis, R.I.C.C., and R. Shotton. (1997) "Risk" in fisheries management: A review. *Canadian Journal of Fisheries and Aquatic Sciences*, **54**: 1699-715.

Freedman, D.A. (1981) Bootstrapping regression models. *Annals of Statistics*, **9**:1218–28.

Garcia, S.M. (1994) The precautionary principle: its implications in capture fisheries management. *Ocean & Coastal Management*, **22**: 99-125.

Garstang, W. (1900) The impoverishment of the sea - a critical summary of the experimental and statistical evidence bearing upon the alleged depletion of the trawling grounds. *Journal of Marine Biological Association of the United Kingdom*, **6**: 1-69.

Gause, G.F. (1934) *The Struggle for Existence*. Hafner Publishing Co. reprinted (1964). 163p.

Gelman, A., Carlin, J.B., Stern, H.S., Dunson, D.B., Vehtari, A. and D.B. Rubin (2013) *Bayesian Data Analysis* 3rd edn. CRC Press. Chapman & Hall, Boca Raton 661 p.

Gelman, A. and Rubin, D.B. (1992) Inference from iterative simulation using multiplicative sequences. *Statistical Science*, **7**: 457–472.

Gelman, A. and D.B. Rubin (1992a) A single sequence from a Gibbs sampler gives a false sense of security. Pp. 625-631 In: Bernardo, J.M., Berger, J.O., Dawid, A.P. and A.F.M. Smith (eds) *Bayesian Statistics 4*. Oxford. Oxford Uniersity Press.

Geweke, J. (1989) Bayesian inference in economietric models using Monte Carlo integration. *Econometrica*, **57**: 1317-1339.

Geyer, C.J. (1992) Practical Markov chain Monte Carlo. *Statistical Science*, **7**: 473-483.

Gleick, J. (1988) *Chaos: Making a New Science*. Penguin Books, London.

Gompertz, B. (1825) On the nature of the function expressive of the law of human mortality, and on a new mode of determining the value of life contingencies. *Philosophical Transactions of the Royal Society*, **115**: 515-585.

Greaves, J. (1992) *Population trends of the New Zealand Fur Seal Arctocephalus forsteri, at Taumaka, Open Bay Islands*. B.Sc. Hons. Thesis. Victoria University of Wellington, New Zealand.

Grolemund, G. (2014) *Hands-on Programming with R* O'Reilly Media Inc. Sebastopol. https://rstudio-education.github.io/hopr/

Gulland, J.A. (1983) *Fish Stock Assessment: A Manual of Basic Methods*. John Wiley & Sons, New York.

Haddon, M. (1980) Fisheries science now and in the future: a personal view. *New Zealand Journal of Marine and Freshwater Research* **14**: 447-449.

Haddon, M. (1998) The use of biomass-dynamic models with short-lived species: is B0 = K a sensible assumption? pp 63-78 in: *Risk Assessment*. Otago Conference Series No 3 (eds.) Fletcher, D.J. & B.F.J. Manly. Otago University Press.

Haddon, M. (2001) *Modelling and Quantitative Methods in Fisheries*. Chapman & Hall, CRC Press, Boca Raton. 406 p.

Haddon, M. (2007) Fisheries and their management. Pp 515-532. In *Marine Ecology* (eds) S.D. Connell & B.M. Gillanders. Oxford University Press. 630 p.

Haddon, M. (2011) *Modelling and Quantitative Methods in Fisheries 2nd Ed.* Chapman & Hall/CRC Press, Boca Raton, 449 p.

Haddon, M. (2018) Catch rate standardizations for selected species from the SESSF(data 1986–2012). pp. 43-406 In: Tuck, G.N. (*ed.*) *Stock Assessment for the Southern and Eastern Scalefish and Shark Fishery 2016 and 2017. Part 2, 2017.* Australian Fisheries Management Authority and CSIRO Oceans and Atmosphere Flagship, Hobart. 837p. [Dec 2019] https://afma.govcms.gov.au/sites/default/files/

stock_assessments_for_the_southern_and_eastern_scalefish_and_shark_ fishery_2016_-_2017_part_2.pdf.

Haddon, M. and F. Helidoniotis (2013) Legal minimum lengths and the management of abalone fisheries. *Journal of Shellfish Research* **32**: 197-208.

Hastings, W.K. (1970) Monte Carlo Sampling Methods Using Markov Chains and Their Applications. *Biometrika*, **57**: 97-109.

Helidoniotis, F. and M. Haddon (2012) Growth model selection for juvenile blacklip abalone (*Haliotis rubra*): assessing statistical and biological validity. *Marine and Freshwater Research*, **63**:23-23.

Helidoniotis, F. and M. Haddon (2013) Growth models for fisheries: The effect of unbalanced sampling error on model selection, parameter estimation, and biological predictions. *Journal of Shellfish Research*, **32**: 223-235.

Helidoniotis, F. and M. Haddon (2014) The effectiveness of broad-scale legal minimum lengths for protecting spawning biomass of Haliotis rubra in Tasmania, *New Zealand Journal of Marine and Freshwater Research*, ___48_: 70-85, DOI: 10.1080/00288330.2013.843574

Higgins, K., Hastings, A., Sarvela, J.N., and L.W. Botsford (1997) Stochastic dynamics and deterministic skeletons - population behavior of dungeness crab. *Science*, **276**:1431-1435.

Hilborn, R. (1979) Comparison of fisheries control systems that utilize catch and effort data. *Journal of the Fisheries Research Board of Canada*, **36**: 1477-1489.

Hilborn, R. and C.J. Walters (1992) *Quantitative Fisheries Stock Assessment: Choice, Dynamics, and Uncertainty.* Chapman & Hall, London. 570 p.

Huffaker C. B. (1958) Experimental studies on predation: dispersion factors and predator-prey oscillations. *Hilgardia*, **27(14)**:343-383. DOI: 10.3733/hilg.v27n14p343.

Hurtado-Ferro, F., Punt, A.E., and K.T. Hill (2014) Use of multiple selectivity patterns as a proxy for spatial structure. *Fisheries Research* **158**: 102-115.

Kell, L. T., Pilling, G. M., Kirkwood, G. P., Pastoors, M., Mesnil, B., Korsbrekke, K., Abaunza, P., Aps, R., Biseau, A., Kunznil, P., Needle, C., Roel, B. A., and C. Ulrich-Rescan. (2005) An evaluation of the implicit management procedure used for some ICES roundfish stocks. *ICES Journal of Marine Science*, **62**: 750–759.

Kimura, D.K. (1980) Likelihood methods for the von Bertalanffy growth curve. *Fishery Bulletin*, **77**: 765-776.

Kimura, D.K. (1981) Standardized measures of relative abundance based on modelling log(c.p.u.e.), and their application to pacific ocean perch (*Sebastes alutus*). *Journal du Conseil International pour l'Exploration de la Mer* **39**: 211-218.

Knight, W. (1968) Asymptotic growth: An example of nonsense disguised as mathematics. *Journal of Fisheries Research Board of Canada*, **25**: 1303-1307.

Kristensen, K., Nielsen, A., Berg, C.W., Skaug, H.J., and B. M. Bell (2016) TMB: Automatic Differentiation and Laplace Approximation. *Journal of Statistical Software*, **70(5)** : 1-21.

Lauwerier, H. (1991) *Fractals. Endless Repeated Geometrical Figures.* (translated by Sophia Gill-Hoffstadt. Penguin Books, London.

Legendre, P. and L. Legendre (1998) *Numerical Ecology* 2nd edn. Elsevier Science B.V. Amsterdam, 853 p.

Lotka, A.J. (1925) *Elements of Physical Biology.* Baltimore, Williams and Wilkins. Reprinted (1956) as *Elements of Mathematical Biology.* Dover Publications Inc. New York. 465p.

MacCall, A.D. (2009) Depletion-corrected average catch: a simple formula for estimating sustainable yields in data-poor situations. *ICES Journal of Marine Science*, **66**:2267-2271.

Magnuson-Stevens (2007) *Magnuson-Stevens Fishery Conservation and Management Act. As Amended Through January 12 2007.* U.S. Department of Commerce, National Marine Fisheries Service, National Oceanic and Atmospheric Administration. 170p. https://www.fisheries.noaa.gov/resource/document/ magnuson-stevens-fishery-conservation-and-management-act.

Matloff, N. (2011) *The Art of R Programming. A tour of statistical software design* no starch press. San Francisco. 373 p.

Maunder, M.N. and A.E. Punt (2013) A review of integrated analysis in fisheries stock assessment. *Fisheries Research*, **142**: 61-74

May, R.M. (1973) *Stability and Complexity in Model Ecosystems.* Princeton University Press, Princeton, NJ.

May, R.M. (1976) Simple mathematical models with very complicated dynamics. *Nature*, **261**: 459-467.

May, R.M. and A.R McLean (eds)(2007) *Theoretical Ecology. Principles and Applications* 3rd edn. Oxford University Press, Oxford. 257 p.

Maynard Smith, J. and M. Slatkin (1973) The stability of predator-prey systems *Ecology*, **54**: 384-391

Methot, R.D., Wetzell, C.R. (2013) Stock Synthesis: a biological and statistical framework for fish stock assessment and fishery management. *Fisheries Research* 142:86-99.

Metropolis, N., Rosenbluth, A.W., Rosenbluth, M.N., Teller, A.H. and E. Teller (1953) Equation of State Calculations by Fast Computing Machines. *The Journal of Chemical Physics* **21**: 1087-1092.

Myers, R.A. (2001) Stock and recruitment: generalizations about maximum reproductive rate, density dependence, and variability using meta-analytic approaches. *ICES Journal of Marine Science*, **58**: 937-951.

Myers, R.A and N.J. Barrowman (1996) Is fish recruitment related to spawner abundance? *Fishery Bulletin*, **94**: 707-724.

McAllister, M. K., and J. N. Ianelli. (1997) Bayesian stock assessment using catch-age data and the sampling—Importance resampling algorithm. *Canadian Journal of Fisheries and Aquatic Sciences*, **54**: 284-300.

McAllister, M. K., E. K. Pikitch, A. E. Punt, and R. Hilborn. (1994) A Bayesian approach to stock assessment and harvest decisions using the sampling/importance resampling algorithm. *Canadian Journal of Fisheries and Aquatic Sciences*, **51**: 2673-87.

Metropolis, N., Rosenbluth, A.W., Rosenbluth, M.N., Teller, A.H. and E. Teller (1953) Equations of state calculations by fast computing machines. *Journal of Chemical Physics*, **21**: 1087-1092.

Millar, R.B. and R. Meyer (2000) Non-linear state space mode of fisheries biomass dynamics by using Metropolis-Hastings within Gibbs sampling. *Journal of the Royal Statistical Society Series C Applied Statistics*, **49(3)**: 327-342.

Murrell, P. (2011) *R Graphics* 2nd edn. Chapman & Hall / CRC Press 518 p. https://www.stat.auckland.ac.nz/~paul/RGraphics/rgraphics.html.

Neter, J., Kutner, M.H., Nachtsheim, C.J, and W. Wasserman (1996) *Applied Linear Statistical Models*. Richard D. Irwin, Chicago.

Nievergelt, Y. (2000) A tutorial history of least squares with applications to astronomy and geodesy. *Journal of Computational and Applied Mathematics*, **121**: 37-72.

Pella, J.J. and P.K. Tomlinson (1969) A generalized stock-production model. *Bulletin of the Inter-American Tropical Tuna Commission*, **13**: 421-458.

Penn, J.W. and N. Caputi (1986) Spawning stock-recuitment relationships and environmental influences on the tiger prawn (*Penaeus esculentus*) fishery in Exmouth Gulf, Western Australia. *Australian Journal of Marine and Freshwater Research*, **37**: 491-505.

Petersen, C.G.J. (1896) The yearly immigration of young plaice into the Limfjord from the German Sea. *Report of the Danish Biological Station to the Board of Agriculture (Copenhagen)* **6**: 5-30.

Pitcher, T.J., and P.D M. MacDonald. (1973) Two models of seasonal growth. *Journal of Applied Ecology*, **10**:599-606.

Polacheck, T., Hilborn, R., and A.E. Punt (1993) Fitting surplus production models: Comparing methods and measuring uncertainty. *Canadian Journal of Fisheries and Aquatic Sciences*, **50**: 2597-2607.

Prager, M.H. (1994) A suite of extensions to a nonequilibrium surplus-production model. *Fishery Bulletin*, **92**: 374-389.

Punt, A.E. (1994) Assessments of the stocks of Cape hakes *Merluccius* spps off South Africa. *South African Journal of Marine Science*, **14**: 159-186.

Punt, A.E., Butterworth, D.S., de Moor, C.L., De Oliveira, J.A.A. and M. Haddon (2016) Management strategy evaluation: best practices. *Fish and Fisheries*, **17**: 303-334

Punt, A.E. and J.M. Cope (2019) Extending integrated stock assessment models to use non-depensatory three parameter stock-recruitment relationships. *Fisheries Research* **217**:46-57.

Punt, A.E., Day, J., Fay, G., Haddon, M., Klaer, N., Little. L.R., Privitera-Johnson, K., Smith, A.D.M., Smith, D.C., Sporcic, M., Thomson, R., Tuck, G.N., and S. Wayte (2018) Retrospective in-vestigation of assessment uncertainty for fish stocks off southeast Australia. *Fisheries Research*, **198**: 117-128.

Punt, A. E. and R. Hilborn (1997) Fisheries stock assessment and decision analysis: The Bayesian approach. *Reviews in Fish Biology and Fisheries*, **7**: 35-63.

Quinn II, T.J. (2003) Ruminations on the development and future of population dynamics models in fisheries. *Natural Resource Modeling*, **16(4)**: 341–392. doi:10.1111/j.1939-7445.2003.tb00119.

Quinn II, T.J. and R.B. Deriso (1999) *Quantitative Fish Dynamics* Oxford University Press, Oxford. 542 p.

R Core Team (2019) R: A language and environment for statistical computing. R Foundation for Statistical Computing, Vienna, Austria. URL https://www. R-project.org/.

Racine-Poon, A. (1992) [Practical Markov CHain Monte Carlo]: Comment *Statistical Science*, **7**: 492-493.

Rayns, N. (2007) The Australian government's harvest strategy policy. *ICES Journal of Marine Science* **64**: 596-598.

Restrepo, V.R. and J.E. Powers (1999) Precautionary control rules in US fisheries management: specification and performance. *ICES Journal of Marine Science* **56**: 846-852.

Restrepo, V. R., G. G. Thompson, P. M. Mace, W. L. Gabriel, L. L. Low, A. D. MacCall, R. D. Methot, J. E. Powers, B. L. Taylor, P. R. Wade, and J. F. Witzig (1998) *Technical guidance on the use of precautionary approaches to implementing National Standard 1 of the Magnuson- Stevens Fishery Conservation and Management Act.* NOAA Tech. Memo. NMFS-F/SPO-31. U.S. Dept. Commerce, Washington, D.C. 54 p.

Richards, F.J. (1959) A flexible growth function for empirical use. *Journal of Experimental Botany,* **10**: 290-300.

Ricker, W.E. (1954) Stock and recruitment. *Journal of the Fisheries Research Board of Canada* **11**: 559-623.

Ricker, W.E. (1958) *Handbook of Computations for Biological Statistics of Fish Populations.* Fisheries Research Board of Canada, Bulletin No. 119.

Ricker, W.E. (1973) Linear regressions in fishery research. *Journal of the Fishery Research Board of Canada,* **30**: 409-434.

Ricker, W.E. (1975) Computation and Interpretation of Biological Statistics of Fish Populations. *Fisheries Research Board of Canada, Bulletin* No. **191**.

Robins, C. and I. Somers (1994) Appendix A. Fishery Statistics pp 141–164 in Pownall, P.C. (*ed.*) *Australia's Northern Prawn Fishery: The first 25 years.* NPF25. Cleveland, Austrlaia. 179p.

Rosenberg, A.A., and V.R. Restrepo. (1994) Uncertainty and risk evaluation in stock assessment advice for U.S. marine fisheries. *Canadian Journal of Fisheries and Aquatic Science,* **51**: 2715-2720.

RStudio (2019) https://www.rstudio.com/.

Russell, E.S. (1942) *The Overfishing Problem.* Cambridge University Press, London.

Saila, S. B., J. H. Annala, J. L. McKoy, and J. D. Booth. (1979) Application of yield models to the New Zealand rock lobster fishery. *New Zealand Journal of Marine and Freshwater Research,* **13**: 1-11.

Schaefer, M.B. (1954) Some aspects of the dynamics of populations important to the management of the commercial marine fisheries. *Bulletin, Inter-American Tropical Tuna Commission,* **1**: 25-56.

Schaefer, M.B. (1957) A study of the dynamics of the fishery for yellowfin tuna in the Eastern Tropical Pacific Ocean. *Bulletin, Inter-American Tropical Tuna Commission,* **2**: 247-285

Schnute, J. (1981) A versatile growth model with statistically stable parameters. *Canadian Journal of Fisheries and Aquatic Science,* **38**: 1128-1140.

Schnute, J. (1985) A general theory for analysis of catch and effort data. *Journal du Conseil International pour l'Exploration de la Mer* **42**:414-29.

Schnute, J.T. and L.J. Richards (1990) A unified approach to the analysis of fish growth, maturity, and survivorship data. *Canadian Journal of Fisheries and Aquatic Science*, **47**: 24-40.

Schwartz, G. (1978) Estimating the dimension of a model. *The Annals of Statistics*, **6**: 461-464. https://www.jstor.org/stable/2958889.

Scudo, F.M. and J.H. Ziegler (1978) *Lecture Notes in Biomathematics. The Golden Age of Theoretical Ecology: 1923-1940* Springer-Verlag, Berlin. 490p.

Snedecor, G. W. and W. G. Cochran (1967) *Statistical methods.* 6th edn. Iowa State University Press, Ames.

Snedecor, G. W. and W. G. Cochran (1989) *Statistical methods.* 9th edn. Iowa State University Press, Ames.

Seber, G. A. F. (1982) *The estimation of animal abundance and related parameters.* 2nd edn. Macmillan, New York.

Smith, A.D.M. (1993) Risks of over- and under-fishing new resources. In: Smith, S.J., Hunt, J.J., Rivard, D. (Eds.), Risk Evaluation and Biological Reference Points for Fisheries Management. *Canadian Special Publication of Fisheries and Aquatic Sciences* **120**: 261-267.

Smith, A.D.M., Smith, D.C., Tuck, G.N., Klaer, N., Punt, A.E., Knuckey, I., Prince, J., Morison, A., Kloser, R., Haddon, M., Wayte, S., Day, J., Fay, G., Pribac, F., Fuller, M., Taylor, B., Little., R (2008) Experience in implementing harvest strategies in Australia's south-eastern fisheries. *Fisheries Research* **94**: 373-379.

Smith, T.D. (1994) *Scaling Fisheries: The science of measuring the effects of fishing, 1855-1955.* Cambridge University Press, New York.

Stearns, S.C. (1977) The Evolution of Life History Traits: A Critique of the Theory and a Review of the Data. *Annual Review of Ecology and Systematics*, **8**: 145-171.

Stearns, S.C. (1992) *The Evolution of Life Histories.* Oxford University Press, Oxford.

Summerfelt, R. C., and G. E. Hall (eds.) (1987). *Age and growth of fish.* Ames: Iowa State University Press.

ter Braak, C.J.F. (1992) Permutation versus bootstrap significance tests in multiple regression and ANOVA pp 79-85 In *Bootstrapping and related techniques*, ed. K.-H. Jöckel, G. Rothe, and W. Sendler, Springer-Verlag, Berlin.

Venables, W.N. and B.D. Ripley (2000) *S Programming* Springer, New York. 264 p.

Venables, W.N. and B.D. Ripley (2002) *Modern Applied Statistics with S* 4th edn. Springer, New York. 495 p.

Venzon, D.J. and S.H. Moolgavkar (1988) A method for computing profile-likelihood-based confidene intervals. *Applied Statistics*, **37**: 87-94.

Volterra, V. (1927) *Variazioni e Fluttuazioni del Numero d'individui in Specie Animali Conviventi*. Reprinted and translated by Scudo and Ziegler (1978).

von Bertalanffy, L. (1938) A quantitative theory of organic growth. *Human Biology*, **10**: 181-213.

Walters, C. J., and D. Ludwig. (1994) Calculation of Bayes posterior probability distributions for key population parameters. *Canadian Journal of Fisheries and Aquatic Sciences*, **51**: 713-722.

Watson, J.D. and F.H.C. Crick (1953) Molecular structure of nucleic acids. *Nature* **171**: 737-738.

Wayte, S.E. (2013) Management implications of including a climate-induced recruitment shift in the stock assessment for jackass morwong (*Nemadactylus macropterus*) in south-eastern Australia. *Fisheries Research* **142**: 47-55.

Wickham, H. and G. Grolemund (2017) *R for Data Science* O'Reilly Media Inc. Sebastopol, CA. https://r4ds.had.co.nz/.

Wickham, H. (2019) Advanced R. 2nd edn. CRC Press, Chapman & Hall, Boca Raton, FL. 588 p.

Winker, H., Carvalho, F., and M. Kapur (2018) JABBA: Just Another Bayesian Biomass Assessment. *Fisheries Research*, **204**: 275-288.

Xie, Y. (2017) *bookdown: Authoring Books and Technical Documents with R markdown*. The R Series. CRC Press, A Chapman & Hall Book. Boca Raton, FL. 113 p.

York, A. E. and P. Kozloff (1987) On the estimation of numbers of northern fur seal, Callorhinus ursinus, pups born on St. Paul Island, 1980–86. *Fishery Bulletin*, **85**:367-375.

Index

Printed in the United States
by Baker & Taylor Publisher Services